AF891523

Springer-Verlag Italia Srl.

G. di Prisco • E. Pisano • A. Clarke

Fishes of Antarctica

A biological overview

Springer

GUIDO DI PRISCO
Institute of Protein Biochemistry and Enzymology
CNR
Naples, Italy

EVA PISANO
Institute of Comparative Anatomy
University of Genoa
Genoa, Italy

ANDREW CLARKE
British Antarctic Survey
Cambridge, United Kingdom

© Springer-Verlag Italia 1998
Originally published by Springer-Verlag Italia, Milano in 1998
Softcover reprint of the hardcover 1st edition 1998

ISBN 978-88-470-2182-2 ISBN 978-88-470-2157-0 (eBook)
DOI 10.1007/978-88-470-2157-0

Library of Congress Cataloging-in-Publication Data: Fishes of Antarctica: a biological overview / G. di Prisco, E. Pisano, A. Clarke. p. cm. Papers presented at a conference held May 1997 at the Certosa di Pontignano, Siena. Italy. Includes bibliographical references and index. ISBN 8847000289 (hardcover) 1. Fishes--Antarctic Ocean--Congresses. I. di Prisco, Guido. 1937- . II. Pisano. E. (Eva), 1951- . III. Clarke, A. (Andrew), 1949- . QL637.2.F56 1998 567,177'09167--dc21 98-25736 CIP

This work is subject to copyright. All rights are reserved, whether the whole or part of the material is concerned, specifically the rights of translation, reprinting, re-use of illustrations, recitation, broadcasting, reproduction on microfilms or in other ways, and storage in data banks. Duplication of this publication or parts thereof is only permitted under the provisions of the Italian Copyright Law in its current version, and permission for use must always be obtained from Springer-Verlag. Violations are liable for prosecution under the Italian Copyright Law.

The use of registered names, trademarks, etc. in this publication does not imply, even in the absence of a specific statement, that such names are exempt from the relevant protective laws and regulations and therefore free for general use.

Product liability: the publisher cannot guarantee the accuracy of any information about dosage and application contained in this book. In every individual case the user must check such information by consulting the relevant literature.

Typesetting: Camera-ready by E. Terminiello (edieffe, Napoli)

Cover design: Simona Colombo

SPIN: 10648119

Preface

The Antarctic fish fauna has evolved over a long period of geographic and climatic isolation. In the course of this evolution, Antarctic fish have developed specialized adaptations, some of which characterize these organisms as unique. In strong contrast to the continental shelf faunas elsewhere, the Antarctic shelf ichthyofauna is dominated by a single highly endemic group, the Notothenioidei. This group of perciform fish probably first appeared and diversified in the early Tertiary. The development of the Polar Front (referred to as the Antarctic Convergence in the older literature) resulted in a natural oceanographic barrier to migration in either direction, and thus became a key factor in the evolution of Antarctic fish.

The dominance of the Antarctic continental shelf fauna by a single taxonomic group of fish provides a simplified natural laboratory for exploring the wealth of physiological, biochemical and ecological adaptations that characterize the fauna. Understanding of the patterns of adaptation in this highly specialized group of fish can tell us much about the process of evolution.

In the past 20 years Antarctic fish have become the focus of a large and varied research programme by scientists from many nations. Topics have included morphology, ecology, population dynamics, biochemistry, physiology, genetics and molecular phylogeny. Most of this research has been undertaken on an individual basis, but there has been an increasing realisation that the difficulties of working in Antarctica and the isolation of the few laboratories on the continent mean that international collaboration is important. Such cooperation strengthens the efforts of individuals and small research groups with a well coordinated multinational and multidisciplinary scientific framework.

A significant step in this direction was taken in 1993 when the European Science Foundation launched the 3-year Network "Fishes of the Antarctic Ocean." This Network linked all of the European countries engaged in scientific research on Antarctic fish, but contact was also extended to nonEuropean scientists and especially those from the USA, New Zealand and the former Soviet Union. This Network has facilitated exchanges of ideas and also helped the planning of collaborative research. Existing interactions have been strengthened, with new cooperation established and significant exchange of younger scientists between European laboratories. Of particular importance have been the annual workshops on selected scientific themes.

The Final Conference of the Network was held in May 1997 at the Certosa di Pontignano, Siena, Italy. The contributions presented at the Conference have been collected together in this volume after peer-review. They focus on the evolutionary adaptation of the Antarctic ichthyofauna to the Southern Ocean, how this has been achieved and the mechanisms involved. Several papers also make comparison with Arctic fish. The taxonomic composition and evolutionary history of the Arctic fish fauna differ from that of the Southern Ocean, and such comparisons are thus important in distinguishing factors that might be related to polar conditions from those that are simply a general feature of notothenioid fishes.

This volume provides a comprehensive summary of the current state of research into Antarctic fish. It also emphasizes the interdisciplinary nature of recent work in this field, and particularly the important interactions between taxonomists, ecologists, evolutionary biologists and molecular biologists. The resulting synthesis could not currently be achieved for any other group of Antarctic marine organisms. Five keynote presentations, given by Joe Eastman and Andrew Clarke, George Somero, Luca Bargelloni and Guillaume Lecointre, Chris Cheng DeVries, and Charles Gerday and Guido di Prisco, paved the way for wide-ranging and stimulating discussions. These demonstrated the extent to which the subject has benefitted from the interactions between traditional biologists and molecular biologists in elucidating the pattern and timing of the notothenioid radiation in the Southern Ocean. This was particularly pleasing as it was a major reason behind the setting up of the Network.

We would like to express our gratitude to the speakers, authors and referees of the papers collected together in this volume. The financial support of the European Science Foundation and PNRA (Italian National Programme for Antarctic Research) is gratefully acknowledged. The help offered by Silvano Focardi of the University of Siena ensured that the stimulating conclusive Conference of the Network was held in beautiful surroundings.

May 1998

Guido di Prisco
Eva Pisano
Andrew Clarke

Table of Contents

Introductory Overview

Part 1 Ecology

Part 2 Physiology and Biochemistry

List of Contributors

Introductory Overview

A Comparison of Adaptive Radiations of Antarctic Fish with those of NonAntarctic Fish

Joseph T. Eastman[1] and Andrew Clarke[2]

[1]Department of Biological Sciences, Ohio University, Athens, OH 45701, USA
[2]British Antarctic Survey, High Cross, Madingley Road, Cambridge CB3 0ET, UK

Introduction

Antarctic biologists frequently emphasize the differences between the modern Antarctic environment and its fauna, and aquatic habitats and faunas elsewhere in the world. While it is valid to portray Antarctica as remote and its fauna as endemic and cold adapted, this approach tends to obscure broad scale similarities between Antarctic and non-Antarctic faunas. For example, the Antarctic fish fauna shares an evolutionary response to its habitat with fish in some tropical, temperate and boreal lakes. In this review we compare some well studied lacustrine radiations of fish with the two radiations of marine fish in the Antarctic Region of the Southern Ocean, notothenioids and liparids. We shall first make the case that, unlike other marine habitats, the Antarctic Region fulfills most of the essential parameters of lakes containing radiations of fish and that this large component of the world ocean is equivalent to a closed basin. Therefore in spite of its vastness, the Antarctic Region provides a comparable opportunity for studying evolutionary biology within a confined area. It is likely that notothenioids, and possibly liparids, are the first known examples of species flocks or radiations of marine fish. Thus the high Antarctic shelf and upper slope is an insular evolutionary site, with endemic faunas equally as interesting, but less well known, as those in ancient lakes throughout the world.

We use recent information to compare Antarctic and non-Antarctic radiations of fish with respect to age of habitat, species diversity, molecular divergence times, key innovations and degree of morphological and ecological divergence with respect to phyletic divergence. Also of interest are the geological, climatic and ecological factors that created the conditions leading to these radiations. The radiation of notothenioid and liparid fishes has taken place during a period of significant tectonic activity and substantial climatic change [1-6]. Furthermore the evolutionary history of one group, the notothenioids, has been elucidated by cladistic analyses employing both morphological [1,7] and molecular [8,9] techniques. Although the major features of the radiation of notothenioid fishes are now

G. di Prisco, E. Pisano, A. Clarke (Eds)
Fishes of Antarctica. A biological overview
© Springer-Verlag Italia 1998

reasonably well established, there remain critical areas of uncertainty. These include the nature of the parent stock, and the timing of key evolutionary events in relation to tectonic and climatic changes.

Species Flocks or Radiations of Fish

Evolutionary biologists have long been fascinated by the species flocks or adaptive radiations of organisms in isolated aquatic habitats, especially ancient lakes [10-16]. The distinguishing characteristics of ancient lakes are these: considerable age (5-20 Ma), great depth (permanence) and isolation. They also contain speciose and highly endemic faunas, sometimes including species flocks, and have been centers of evolution for a variety of organisms [15,16].

Species flocks usually exhibit monophyly, endemism and speciosity [15]. With respect to fish species flocks, the importance of monophyly has been considered either essential [17] or not a general criterion [18]. We follow Ribbink [18 p.24] in recognizing a species flock as "an assemblage of a disproportionately high number of closely related species which evolved rapidly within a narrowly circumscribed area to which all or almost all the member species are endemic."

The shelf and upper slope waters around Antarctica share many of the features of isolated aquatic habitats. Although large in size, the Antarctic shelf and upper slope is isolated from other shelf areas in the Southern Hemisphere by large areas of deep water, a major oceanographic feature (the Polar Front), and by being at subzero temperatures. In these respects the fish fauna of the Antarctic continental shelf is as isolated as a lacustrine fauna. The Antarctic continental shelf also resembles a lacustrine habitat in supporting a highly endemic fish fauna, although the species richness might be considered to be somewhat small for the size of the area. This fauna contains two groups which match Ribbink's [18] criteria for a species flock, the notothenioids and the liparids. In Table 1 we summarize the features of these groups with flocks of fish from ancient lakes.

The Fish Fauna of the Antarctic Continental Shelf and Upper Slope

The fish fauna of the Southern Ocean is small and limited in taxonomic diversity, containing only 274 species in 49 families [19]. Thus although representing 10% of the world's ocean, the Southern Ocean accommodates only about 1% of the world's fish fauna of 25 000 species. As will be seen below, however, a count of taxa leads to an underappreciation of the morphological and ecological diversity of the fauna. The fauna is highly

Table 1. Comparison of species flocks or radiations of fish from some ancient lakes and the Antarctic Region

Group & Location	Habitat	Mean depth (m)	Age of habitat (my)	No. families/ genera/species	Species endemism	Nature of divergence	Key innovations
Cichlids Lake Tanganyika	African rift lake	572	5-20	1/50/185	99%	Trophic morph. Behavior/ecology	Modifications of phar. jaw appar.
Cichlids Lake Malawi	African rift lake	290-426 1-2		1/49+/500	99%	Trophic morph. Behavior/ecology	Modifications of phar. jaw appar.
Cichlids Lake Victoria	African lake	40	0.0124	1/21/250	99%	Trophic morph. Behavior/ecology	Modifications of phar. jaw appar.
Cottoids Lake Baikal	Siberian rift lake	730	2-25	3/11/29	93%	Morphological Ecological Physiological	
Cyprinodontids Titicaca Basin	Andean plateau lakes	105	0.8-2	1/1/43	>50%	Head shape & mouth position	
Cyprinids (barbs) Lake Tana	Ethiopian lake	8	2	1/1/1?	0%	14 mouth/head shape morphs	
Notothenioids Antarctic Region	Antarctic shelf & upper slope	<500	12-22	6/40/95	97%	Morphological Ecological Physiological	AFGP derived from trypsinogen
Liparids Antarctic Region	Antarctic shelf & upper slope	>500	12-22	1/5/64	?	Phyletic	

endemic, with 88% of the species confined to the Antarctic Region south of the Polar Front. If just notothenioids are considered, species endemism rises to 97% [20]. The major component of the fauna, that of the shelf and upper slope (Table 2), is represented by only 208 species and is unusual among marine habitats in containing radiations of two unrelated lineages, namely perciform notothenioids in shallower habitats and scorpaeniform liparids in deeper waters.

Table 2. Dominant families of benthic fishes inhabiting the continental shelf and upper continental slope of the Antarctic Region

Taxon[a]	No. of species[b]	Percentage
Myxinidae (hagfishes)	1	0.5
Petromyzontidae (lampreys)	1	0.5
Rajidae (skates)	8	3.8
Carapidae (pearlfishes)	1	0.5
Moridae (deepsea cods)	4	1.9
Muraenolepididae (eel cods)	4	1.9
Gadidae (cods)	1	0.5
Congiopodidae (horsefishes)	1	0.5
Liparidae (snailfishes)	64	30.8
Zoarcidae (eelpouts)	23	11.0
Notothenioidei (includes six Antarctic families[c])	95	45.7
Tripterygiidae (triplefins)	1	0.5
Achiropsettidae (southern flounders)	4	1.9
TOTALS	208	100.0%

[a]Phylogenetic sequence according to [21].
[b]Based on [19] with these exceptions: addition of two new nototheniids from Balushkin [22, 23], a number of liparids from Andriashev and Stein [4] and of zoarcids from Anderson [24].
[c]Balushkin [25].

Notothenioids

The perciform suborder Notothenioidei includes eight families and 121 species. All but 26 of these species are confined to the Antarctic Region. In the entire Southern Ocean, about 35% of species are notothenioids; in subzero shelf and slope waters 46% are notothenioids (Table 1). Here they occupy virtually all benthic and water column habitats, accounting for 90-95% of the biomass [26, 27]. In the eastern and southern Weddell Sea, for example, 94% of the specimens captured are members of the family Nototheniidae [28]. This exceptional degree of habitat saturation by a single taxonomic group is attributable to the fact that notothenioids occupy niches filled by taxonomically diverse groups of fishes in temperate and tropical oceans.

A simple count of taxa does not begin to adequately describe the nature of the notothenioid radiation. In addition to phyletic diversification, and unlike most other fish groups including some lacustrine radiations [29,30], notothenioids have also experienced substantial morphological and ecological diversification in that there are a number of distinct life history or ecological types. Notothenioids underwent a depth-related diversification centered on the evolutionary alteration of buoyancy and the morphology associated with swimming and feeding in the water column.

Although lacking swim bladders, some notothenioids diversified away from the ancestral benthic habitat toward pelagic or partially pelagic zooplanktivory and piscivory. Within the family Nototheniidae, for example, about 50% of the Antarctic species are pelagic, semipelagic, cryopelagic or epibenthic rather than benthic, the presumed ancestral type [1]. Referred to as pelagization, this ecomorphological trend toward life in the water column has arisen independently in at least two different nototheniid clades, although phylogenetically related species are not necessarily morphologically and ecologically similar [3].

Liparids

With the recent description of dozens of new species [4], it appears that a radiation of snailfishes of the scorpaeniform family Liparidae inhabits the deep shelf and upper slope of the Antarctic Region. The 64 liparid species from the Antarctic Region, mostly members of the genera *Paraliparis* and *Careproctus* [4], represent 31% of the 208 species of benthic fishes from the continental shelf and upper slope (Table 2). The Liparidae has surpassed the Nototheniidae as the most speciose fish family in the Southern Ocean.

Unlike notothenioids, liparids have a worldwide distribution and are one of the few fish families having both boreal and austral centers of species diversity [31]. The family includes about 200 species living from the intertidal zone to depths of 7000 m [4,21,32]. Liparids probably originated and radiated in the North Pacific Region with subsequent dispersal southward along the western coast of the Americas [33,34]. In the deep waters of the Antarctic Region liparids underwent a secondary radiation complementing the shallow water radiation of perciform notothenioids. All known species are members of deep-water genera [35]; most are epibenthic at depths of 200-2000 m [4]. As the sea below 200 m is the largest living space on earth, it is reasonable to expect that it would be subject, with some constraints, to the same evolutionary processes occurring in other aquatic environments.

Although phyletically diverse, many Antarctic liparids are known from only a few specimens. They are taxonomically dominant in waters over 500-600 m deep, but may not be represented by large populations and biomass [4]. Liparids lack a swim bladder and phyletically primitive species are inshore benthic fish [36], but the liparid radiation in the Antarctic Region has centered on diversification into the water column, especially epibenthic niches. The amount of morphological divergence relative to lineage splitting reflects the degree of macroevolution and, if families are compared, this has been considerably less in liparids than in nototheniids.

Since they are few in number, deep-living and difficult to maintain in aquaria, we know little of the biology of Antarctic liparids. The sole exception is *Paraliparis devriesi* which lives at 500-650 m in McMurdo Sound. Although lacking a swim bladder, this species is neutrally buoyant through the combined effects of reduced skeletal ossification and expansion of a watery gelatinous subdermal extracellular matrix consisting of glycosaminoglycans. This matrix serves as a low density buoyancy agent comprising about one-third of the body weight, the largest known proportion of any adult fish [37]. At nearly 78°S the population of *P. devriesi* in McMurdo Sound has the most southerly distribution of any known liparid, living in water with a temperature of about -1.9 °C and little seasonal variation. However only small amounts of antifreeze compounds have been inferred from the melting point-freezing point behavior of body fluids. Freezing avoidance in *P. devriesi* may therefore be the result of the combined effects of modest concentrations of a noncolligative antifreeze and elevated osmolality of body fluids [38].

The Environmental Background to the Evolution of the Antarctic Continental Shelf Fauna

The evolution of the Antarctic fish fauna can only be understood in relation to the tectonic and climatic history of the area. We will discuss this before moving to a comparison of the radiations of Antarctic and lacustrine species flocks.

Tectonic History

Although Antarctica is presently covered by a single ice sheet, the underlying rocks comprise two crustal blocks of differing tectonic history: East Antarctica and West Antarctica. The latter includes the Antarctic Peninsula, whose position relative to East Antarctica has changed significantly through time. There is also a cluster of very small crustal plates (microplates) whose precise relationships and tectonic history are the subject of much debate.

The main block of what is now Antarctica includes Precambrian basement rocks, and in the Palaeozoic formed part of a single supercontinent, Pangea. As far back as the Cambrian the southern edge of East Antarctica and Australia supported a shallow-water marine fauna, with a defined trilobite faunal realm [39].

Following the inception of the present cycle of seafloor spreading in the Mesozoic, Pangea split into two major blocks, Laurasia and Gondwana, separated by the Tethys Sea. At this point there were two quite distinct areas of shallow water adjacent to the East Antarctic shield. One was the long coastline formed by South America and the southern/western coast of East Antarctica, facing the Pacific Ocean. The other was the northerly/eastern coast of the East Antarctic shelf and Australia, where they formed the southern boundary to Tethys. It is from the latter that the faunal elements defining the Weddellian Province of the later Mesozoic came. Since Gondwana remained as a single intact unit throughout the Palaeozoic and into at least the Jurassic, those proto-Pacific coastal habitats are considerably older than those around present-day East Antarctic (which faced the Tethys Sea).

It seems likely that the initial opening between Antarctica and South America/Africa was at about 150 Ma BP [40]. Initially this would have been a shallow epicontinental sea, evidence for which comes from the disjunct distribution of certain Jurassic endemics in both the southern Andes and East Africa/Madagascar [41-44]. Faunal exchange between the Pacific and Tethys/Weddellian Provinces could thus have started in the Jurassic, although it was not until much later that the Drake Passage opened sufficiently for deep-water circulation to start.

By the early Eocene, fragmentation of Gondwana was well underway, with Africa and India having moved substantially northwards. Australia was still attached to East Antarctica and there was no deep-water separation of West Antarctica (the Antarctic Peninsula) from South America. Although faunal evidence suggests some shallow water faunal exchange, the deep water basins of the Atlantic/Weddellian Province and the Pacific Ocean were still separated.

The area of West Antarctica contains many microplates, the detailed movements of which have yet to be resolved. Nevertheless it is clear that the Drake Passage between South America and the Antarctic Peninsula first opened in the early Oligocene [45]. By this time Australia had already moved north from East Antarctica, and the CircumAntarctic Current had probably started by about 25 Ma BP.

The movement of West Antarctic microplates and the fragmentation of the Scotia arc allowed a free exchange of fauna between the previously isolated Pacific and Weddellian (ex-Tethys) Provinces. Coupled with the oceanographic isolation of Antarctic by the inception of the Circumantarctic Current and the formation of the Polar Front, this also effectively reduced any exchange of shallow water faunal elements between West Antarctica and South America. The scene was therefore set for the continued evolution of the marine fauna of the Antarctic continental shelf in effective isolation from nearby faunas.

Climatic History

It is widely recognized that the waters around Gondwana in the late Cretaceous were mild, with bottom temperatures about 12 °C [46]. This was followed by a distinct warming phase in the early Tertiary, during which bottom temperatures may have reached 16 °C. After this warm period, which lasted through much of the Eocene, there started the sustained cooling which has dominated much (but by no means all) of the remainder of the Tertiary.

The first evidence of widespread sea-ice and continental glaciation appears in the late Eocene/early Oligocene [46]. This corresponds with a sharp drop in bottom temperature; the seawater temperatures are, however, cold rather than truly polar. The Oligocene appears to have been generally cool or cold at high latitudes, and this matches the time when the physical and oceanographic isolation of Antarctica became complete. This was followed, however, by a period of global warming in the middle Miocene, when Southern Ocean bottom temperatures may have reached 10 °C. High latitude seawater temperatures then fell in the mid to late Miocene, perhaps by as much as 4-5 degrees °C. This was accompanied by a switch from predominantly equatorial to strongly meridional circulation patterns in the large ocean basins, with major consequences for global heat transfer.

After brief periods of warming in the late Miocene and early Pliocene, seawater temperatures then cooled to the present fully polar temperatures. There is some evidence for a general warming in the Pliocene, which may have led to some retreat of the ice-cap but this is still debated (see below).

Key Problems in the Climatic History of the Antarctic in Relation to the Evolution of the Marine Fauna

The broad features of the tectonic and climatic history of Antarctica during the Tertiary are now reasonably well established. Nevertheless there remain important unanswered questions in relation to the evolution of the Southern Ocean marine fauna.

When Did Widespread Continental Glaciation Start?

Prior to the establishment of a significant ice sheet in Antarctica, the continental shelves would have been shallower, shallow seas would probably have been more extensive, and there would have been a greater range of shallow-water habitats available including estuaries and mudflats. All of these are likely to have allowed a more diverse fauna than exists today, and fossil evidence indicates that early Tertiary fish fauna in Antarctica was indeed diverse, including many chondrichthyans and other groups not represented today [1,47]. Ancestral notothenioids probably inhabited the shelf of Gondwana, and later Antarctica, during the late Cretaceous or early Tertiary, although their representation in the late Eocene (40 Ma) fossil fish fauna from Seymour Island is debatable [47, 48].

Although there is evidence from glacial deposits of glaciers discharging at sea-level in the Eocene, these do not necessarily indicate the presence of ice sheets [46]. Evidence from sea-level changes suggests the growth and decay of substantial ice sheets in the Eocene, but isotopic data give no support to significant variations in ice volume prior to about 36 Ma BP.

On present evidence, there are thus indications that widespread continental glaciation started around the Eocene/Oligocene border. It is at this time that there may have been significant changes in habitat area, and many habitat types such as rivers and estuaries may have been eradicated, with consequent impacts on the diversity of the fish fauna.

How Far Have Ice-Sheets Extended in the Past?

The present ice-sheet covers over 99% of Antarctica, but does not extend over all of the continental shelf. Geophysical evidence has shown that in the past, ice-sheets have extended fully to the edge of the Antarctic continental shelf [49]. This would obviously eradicate all available habitat for continental shelf organisms which would either become extinct, survive in refugia not covered by ice, or move into deeper water on the continental slope.

The critical piece of information in relation to the evolutionary history of the Southern Ocean marine fauna is whether during one or more glacial maxima in the past, the continental ice-sheet ever extended so far as to cover all of the Antarctic continental shelf at once. We cannot yet answer this question from geophysical or glaciological data. Nevertheless two pieces of ecological evidence suggest that previous glacial maxima may have covered a large proportion of the Antarctic continental shelf. The first is that the peak of species richness for notothenioid fishes in East Antarctica is at about 500-m depth [20]. The second is that the bathymetric ranges of many Antarctic marine organisms extended deeper than elsewhere on the globe [50].

Taken together this evidence suggests (but no more) that glacial maxima, perhaps as recently as the Pliocene or Pleistocene, may have severely reduced or even eradicated available habitat for fish living on the Antarctic continental shelf. If so, the radiation of notothenioid fish may be a much more recent event than traditionally has been thought to be the case.

How Extensive Were Changes in Ice-Sheet Volume and the Extent of Shallow-Water Marine Habitat in the Pliocene?

There has recently been considerable controversy over the extent, or even existence, of deglaciation in Antarctica during the Pliocene. Although the original evidence which came from diatom fossils has been shown to be unreliable [46,51], other evidence has suggested some limited deglaciation in Pliocene times (for example the Pecten conglomerate on Cockburn Island, dated at 3.5-5.3 Ma BP: [52]). Some highly speculative reconstructions of the Antarctic ice-sheet during the Pliocene suggest the presence of large shallow epicontinental seas, which would have increased greatly the habitat area available for shallow-water fish. On present evidence, however, we can only conclude that the evidence for Pliocene warming is suggestive. The possibility exists that shallow water habitats were more extensive in the Pliocene, but we cannot yet be sure.

What is the Role of Milankovitch Cyclicity in the Evolution of the Southern Ocean Fauna?

Gravitational interactions with other bodies in the solar system induce variations in the Earth's orbital parameters. The main cycles have periods of 23 kyr (precession), 41 kyr (obliquity) and ~100 kyr (eccentricity). It is now recognized that the small variations in received solar radiation caused by these Milankovitch cycles are important drivers of climatic variability [53].

The strongest evidence that Milankovitch cycles induce changes in ice volume during glacial periods comes from the analysis of oxygen isotope fractionation in marine sediment cores. These show variability with strong signals at frequencies corresponding to cycles of 23 kyr and 41 kyr; recent studies have also demonstrated lower frequency cycles at ~100 kyr and ~400 kyr [54].

Although ice volume is only one factor influencing the marine sediment oxygen isotope signal, there is general agreement that orbital variations have been responsible for Milankovitch cycles in the volume of polar ice in the Pleistocene glaciation. Although it would seem reasonable to assume that similar Milankovitch cyclicity in ice volume will have occurred throughout the Cenozoic glaciation of Antarctica, evidence is hard to come by. There is, however, some evidence from varving in marine sediment cores for Milankovitch cyclicity in the Mesozoic [55].

We can therefore conclude that superimposed on large-scale changes in ice volume, such as that suggested for the Pliocene, will be shorter-term fluctuations on time-scales in the range 23 to ~400 kyr. If these fluctuations in ice volume were substantial it is possible that they may have driven significant changes in available shallow-water habitat. This in turn could be a mechanism for diversification through allopatric speciation, through a mechanism termed by Valentine [56] the climatic diversity pump [57].

Milankovitch-driven cycles in habitat area may have been a mechanism contributing to the high diversity of some taxa in the Southern Ocean [5,6,58]. At present, however, we do not have the evidence from the marine glacial record to determine the extent of high-frequency variations in the ice volume of Antarctica.

What Has Been the Role of Sudden Climatic Events in the Evolution of the Southern Ocean Fauna?

High resolution analyses of ice-cores from Greenland have shown that at certain periods, the regional terrestrial climate has changed very rapidly [59,60], confirming the rapid changes first demonstrated by climatic reconstructions from subfossil Coleoptera (beetles) [61,62].

What is not at all clear is the extent to which such rapid changes might occur in the sea. Whilst the thermal mass of water means that changes in the temperature of bulk seawater must inevitably be slow, it has become clear that climatic change in the sea may be exhibited through sudden changes in patterns of current flow. Evidence for such changes is accumulating slowly. In the Southern Ocean the only evidence for rapid changes in temperature comes from the late Palaeocene where Kennett and Stott [63] have demonstrated a possible short-term warm event.

Although significant changes in circulation patterns clearly have the potential to influence distribution and evolutionary process through regional or wider-scale extinction, there is as yet no evidence that these processes have been important in the evolution of the Southern Ocean marine fauna.

The Evolution of the Antarctic Fish Fauna

The broad features of the evolution of the Southern Ocean fish fauna are now well established [1] and so will only be summarized here.

The vast ice-sheets of Antarctica limit the exposure of rock, and so our knowledge of the fossil history of the Southern Ocean fish fauna is frustratingly incomplete. Mesozoic fossils confirm the suggestion from tectonic reconstructions that the shallow waters around Gondwana were populated by faunas typical of the period [1]. Early Cenozoic fossils from Seymour Island indicate that prior to the oceanographic isolation of Antarctica, the fish fauna of Gondwana was diverse and contained representatives of many cosmopolitan taxa. The fauna appears to have been typical of the temperate seas of the time.

Unfortunately there are few postEocene fossil fish known from Antarctica and the modern fauna thus has no fossil record. We are therefore left with the intriguing observation that at some time between the Eocene and today, early Tertiary fish fauna of Antarctica largely disappeared. There was then a substantial radiation of notothenioid and liparid fish to produce the fauna we see today.

What Caused the Demise of the Early Tertiary Fish Fauna?

In the absence of a fossil record, we can only speculate on the causes of the extinction of the early Tertiary fish fauna of Antarctica. Nevertheless it is tempting to assume that a major factor must have been the severe reduction in the amount of shallow water and the loss of many habitats traditionally rich in fish consequent upon the major glaciation of Antarctica. This working hypothesis would then put the extinction event (if needed it can be regarded as a single event) around the late Eocene/early Oligocene (40-35 Ma BP). An explanation based on habitat loss also implies that the associated cooling was not necessarily the primary cause of extinction, but in truth there is no evidence to allow us to distinguish the two causes (which may have acted together).

The Origin of the Present Fauna

The modern Antarctic fauna of continental shelf and slope fishes is dominated in terms of species by liparids and notothenioids.

The liparids (snailfishes) appear to have their center of origin in the north Pacific. It is likely that they reached the Southern Ocean by range extension along the Pacific continental shelf and slope of the Americas, probably at a time when glacial maxima had pushed colder waters to lower latitudes. Once in Antarctica the group has radiated to produce over 60 species, though nowhere do liparids appear to be abundant [4]. The timing of the radiation is not known but if it was assumed to have taken place at the last glacial maximum then the radiation of liparids in the Southern Ocean must be a relatively recent affair.

The notothenioids have both radiated and become the dominant taxa of the Southern Ocean continental shelf fauna [1]. The notothenioids appear to be a monophyletic clade, though the ancestral stock cannot be identified with certainty [1]. Neither can we rely on fossil history, for as yet no unequivocal fossil notothenioids have been identified.

We must assume that the notothenioids evolved from a demersal perciform stock, and probably one which was already living in the shallow waters of Gondwana and probably in the Weddellian/Tethyan rather than the Pacific faunal province. Once the early Tertiary fauna became extinct the notothenioids were able to undergo a radiation to produce the fauna we see today. Diversification of the emerging notothenioid fauna was facilitated by the oceanographic and thermal isolation of Antarctica, by the increasing productivity of the Southern Ocean beginning about 22 Ma [64] and by the absence of competition from nonnotothenioids. This ecological space was filled by a diversifying notothenioid fauna tolerant of deeper and colder habitats, which poses the questions as to why it was the notothenioids that radiated, rather than some other group, and when did the radiation take place?

Why Notothenioids?

This is an impossible question to answer at present. It may be, however, that through an accident of history the notothenioids were simply the group that survived whatever caused the demise of much of the early Tertiary fauna. Certainly a number of features of their biology suggest that the notothenioids were in no way "pre-adapted" to radiate in the Southern Ocean. In particular the secondary loss of a functional swimbladder in the ancestral stock has limited their ability to move into the water column and despite their evolutionary pelagization most species remain benthic or demersal [1].

When Did the Radiation Take Place?

Cladistic analysis of the notothenioids suggests that the suborder is monophyletic, though at present the topology of the split between Harpagiferidae and Artedidraconidae is not resolved (Fig. 1). Recent molecular evidence from sequence analysis of 12S and 16S ribosomal RNA genes from 18 species spread across 5 families has broadly supported the analyses based on morphological data. It has, however, suggested that the most speciose family, Nototheniidae, may be paraphyletic [8].

The notothenioid diversification appears to have been protracted throughout much of the Tertiary, with the initial divergence of some of the basal families having taken place during the early fragmentation of Gondwana. At this time stocks appear to have become established in brackish water on isolated continental blocks (*Pseudaphritis* in Australia, and *Eleginops* in South America).

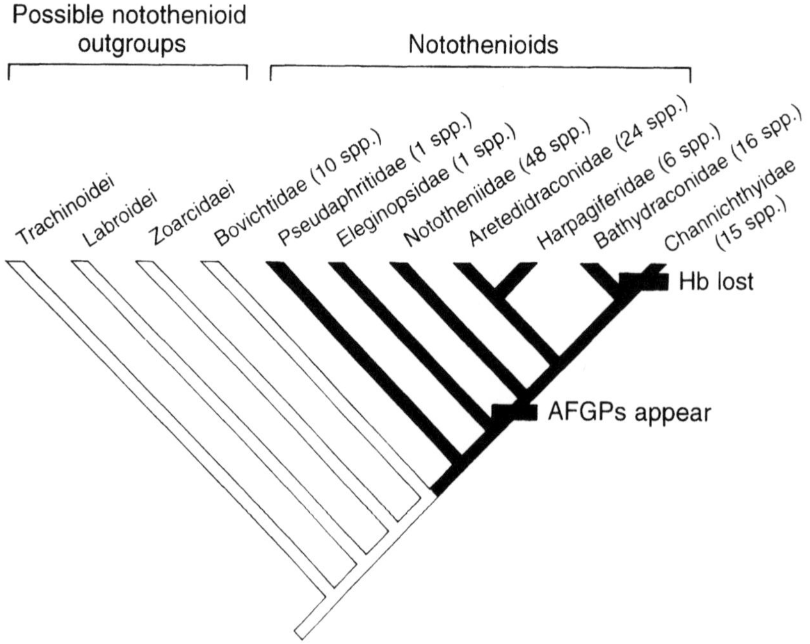

Fig. 1. Notothenioid phylogeny of Lecointre et al. [65] with familial nomenclature of Balushkin [25]. Hb, hemoglobin, AFGPs, antifreeze glycopeptides

McDonald and coworkers [66] used electrophoresis to examine variation at 30 protein-coding loci in six species of notothenioids from McMurdo Sound. Estimated divergence times within the Nototheniidae suggested that *Dissostichus* separated from *Trematomus* at about 28-20 Ma BP, and from *Pagothenia* at about 13 Ma BP. This would be consistent with the major phyletic divergence of notothenioids taking place after the isolation of Antarctica and after the initial period of cooling.

The traditional Bovichtidae is paraphyletic, with *Pseudaphritis* more closely related to nonbovichtid notothenioids than are *Cottoperca* or *Bovichtus*. The position of these latter two genera with respect to various notothenioid outgroups is uncertain [65]. Inference from mtDNA sequences implies that the radiation of the phyletically derived Antarctic notothenioids took place more recently than previously suspected, probably within the last 15 Ma [8]. Recent analysis of partial mitochondrial sequences from 12S and 16S rRNA genes and assuming a conventional rRNA molecular clock puts the average age of the trematomid radiation at only 3.4 Ma (mid-Pliocene). This is relatively recent, and it is intriguing that the age coincides so well with the timing of the controversial Pliocene warming event. If there was a significant warming event at around 2.5-5 Ma BP then this may have resulted in significant changes in available habitat and together with Milankovitch-driven variations in habitat extent may have prompted the radiation of trematomids.

The mapping of the presence of antifreeze glycoprotein onto the cladogram suggests that this feature evolved only once, after the divergence of the Bovichtidae. If the timing of the cooling of the Southern Ocean to subzero temperatures can be fixed precisely, then this in turn could fix the timing of the main diversification.

There are three main possibilities for the timing of Southern Ocean cooling to subzero temperatures. One is the sharp drop in surface temperatures at the start of the Oligocene, 35 Ma BP, the second is the marked cooling event at the middle of the Miocene at about 17 Ma BP, and the third is the final plunge to truly polar temperatures in the late Pliocene.

The third of these possibilities is not tenable on the grounds of biogeography, and all molecular evidence, and hence can probably be dismissed.

More intriguing, however, is the choice between the first two potential dates. The older date would be consistent with the combined evidence from biogeography and the cladistic analysis [1]. The molecular evidence, however, suggests a date for the main notothenioid diversification (after separation of the bovichtid clade) of around 11 Ma BP [8]. If this later date is correct, it implies a much more rapid diversification than has traditionally been believed. It would also imply that either the first cooling event at the start of the Oligocene was not sufficient to require the evolution of antifreeze, or that antifreeze arose independently in several notothenioid clades.

The small degree of sequence divergence between the notothenioid antifreeze glycopeptide and trypsinogen genes suggests that antifreeze evolved somewhere between 5 and 14 Ma BP [67]. This is satisfyingly close to the entirely independent molecular evidence from mtDNA [8] and suggests strongly that the main diversification of the notothenioids has taken place in the past 10-15 Ma. This matches the end of the first phase of continental glaciation [46].

Comparison of the Radiation of Antarctic Fishes with Lacustrine Species Flocks

As a basis for comparison with notothenioids we will mention relevant aspects of the biology of the well studied flocks of fish in the East African Great Lakes and in Lake Baikal in Siberia.

East African Cichlids

Lakes Tanganyika, Malawi and Victoria differ greatly in morphometry, depth and age (Table 1). They have experienced climatic changes that affected both water levels and degree of isolation of their basins. Tanganyika and Malawi are old deep rift lakes that are permanently stratified and anoxic below 250 m. The rocky shorelines with isolated bays and islands provide a complex array of microhabitats. The present Lake Victoria, on the other hand, is a shallow depression variously dated from 750 000 [68] to 12 400 years old [69].

The perciform cichlids in Lakes Tanganyika, Malawi and Victoria (Table 2) are the most speciose and most extensively studied examples of fish species flocks [14,70,71]. The age of the various lakes is reflected in the higher level taxonomic complexity of the fauna. For example, the old flocks of cichlids in Tanganyika show more tribal and generic diversification than those in the other lakes. Furthermore, Tanganyika is home to small flocks of several catfish families and mastacembelids, and to an offshore fauna consisting of endemic clupeids and centropomids. Divergence times and phylogenetic relationships based on mtDNA sequences suggest that the cichlid flock in Tanganyika is polyphyletic, hence there are actually multiple flocks dating from 3.5 to 5 Ma. The

Malawi and Victoria flocks are closely related to one of the Tanganyikan lineages and are monophyletic and younger [71].

Although there has been behavioral and ecological differentiation, the amount of morphological differentiation has been relatively small, with species differences centering on head and tooth shape and body coloration. Modifications of the pharyngeal jaw apparatus, specifically the development of joints and the shift in muscle insertions, may have been a key evolutionary innovation in the radiation of these flocks. The pharyngeal jaws assumed the function of food mastication, freeing the oral jaws for extremely diverse collecting functions which lead to ecological differentiation [72]. Recent evaluation of the trophic diversity of the flocks in Tanganyika lead to the conclusion that this diversity reflects historical (phylogenetic) factors rather than results of competition [73].

Periods of climatic change probably produced fluctuations in water level in all three lakes, and this was important in promoting microallopatric intralacustrine speciation [69,71,74]. This was likely the most important means of speciation in all cichlid flocks [71]. Speciation was also extremely rapid, hundreds to a few thousand years in some small lakes and in isolated parts of the three great lakes. This conclusion is based on the presence of endemic species in areas of some lakes known to be dry a few centuries ago. The youngest major flock, that in Victoria, is dated by molecular methods at 200 000 years [71]; however seismic reflection profiling and piston coring suggest that the lake dried completely 12 400 years ago [75]. This implies that the flock of 250 species evolved in this short period of time or that components of the flock survived in adjacent water and, after levels rose, repopulated Lake Victoria.

Lake Baikal Cottoids

With a maximum depth of 1620 m, Lake Baikal is the world's deepest and possibly oldest lake. The Baikal rift has contained a deep lake since the early Miocene (25 Ma), although the modern configuration was not acquired until 3.2-0.8 Ma [76]. The origins of the modern fauna date to the Miocene or Pliocene. Baikal holds the greatest animal species diversity of any extant lake [16, 76]. Unlike the East African lakes, Baikal is oxygenated to the bottom and therefore contains a significant deep-water fauna including fish that approach marine species in some of their adaptations to depth.

The fish fauna of Lake Baikal includes 56 species from 14 families [77,78]. Of the 50 native species, 29 (58%) are endemic sculpins (Cottoidei) belonging to the families Cottidae, Comephoridae and Abyssocottidae. The latter two are endemic to the lake. Although cottoids are a marine group of North Pacific origin, they have colonized freshwaters in Eurasia and North America. Cottoids from the freshwaters of Eastern Siberia may have been a source of the ancestral stock for the Baikal cottoids [77]. Recent molecular research has provided divergence times for a few species of the family Cottidae. Merged sequences from the cytochrome b and ATPase regions of mtDNA suggest that three species of *Cottus* and *Cottocomephorus* diverged 1-2 Ma [79].

Although lacking swim bladders, some Baikal cottoids diversified away from the ancestral coastal benthic habitat into water column habitats and deeper (abyssal) waters [77,80]. Species diversity peaks at 400-500 m [78]. In an interesting ecological and morphological parallel with the nototheniids *Dissostichus* and *Pleuragramma* [1], the two species of *Comephorus* have become secondarily pelagic. They are close to neutral buoyancy through a combination of reduced skeletal mineralization and increased levels of lipid [78]. Adaptation to deep water habitats in Baikal has involved modification of sensory systems similar to those seen in deep dwelling marine species. For example, in some Baikal cottoids the photoreceptors of the retina consist exclusively of rods [81,82], and canal neuromasts have been replaced by free neuromasts [78].

Final Remarks

Radiations or species flocks of fish are found in isolated aquatic habitats throughout the world, especially in geologically old lakes. The Antarctic shelf and upper slope, an isolated evolutionary site similar to ancient lakes, may also hold flocks of fish. The appearance of some species flocks is coincident with periods of climatic change and habitat instability that may have served as vicariant events. There is good documentation for the rise and fall of water levels in Lakes Malawi and Victoria. On the other hand, periods of habitat instability cannot be associated with the diversification of the deep living Antarctic liparids or Baikal cottoids. While flocks are a repeatable evolutionary response among phyletically diverse groups of fish, the size, age and nature of the diversifications are quite different (Table 1). In the Antarctic, notothenioids dominate the fish biomass of the shelf and their morphological and ecological diversity parallels their phyletic diversity. The deeper living liparids, on the other hand, are speciose but are not present in great numbers and are not morphologically diverse. Diversification in both groups has been keyed to underutilized water column habitats.

A key evolutionary innovation is a novel feature characterizing a clade (synapomorphy) and correlated with the adaptive radiation of the clade [29, p.181]. Innovations are the exception rather than the rule among flocks of fish. It is likely that antifreeze glycopeptides are a key innovation in the Antarctic notothenioids [67,83]. Given their origin from a transformed trypsinogen gene, they are unique and distinct from the antifreeze glycopeptides in Arctic gadids. Their contribution to fitness or survival is obvious since diversification into subzero ice-laden water would have been impossible without them. A novelty or key innovation is responsible for a new evolutionary direction in a lineage. When mapped on the notothenioid cladogram, antifreezes have a single origin at the node for the Antarctic clade (the sister group of *Eleginops*). Based on ages inferred from the divergence of molecular sequences between trypsinogen and antifreeze, the appearance of genes responsible for the synthesis of antifreeze in the basal nototheniid (*Dissostichus*) is coincident with the cooling and appearance of ice in the Southern Ocean 5-14 Ma [67]. This trait allowed considerably more diversification in the Antarctic clade (95 species) than in the nonAntarctic (26 species) clade.

As an independent test of antifreeze glycopeptides as a key innovation in notothenioids, we could ask whether the absence of high levels of antifreeze in Antarctic liparids has hindered their ecological diversification into shallow ice-laden waters. This may not be completely valid since *Paraliparis* and *Careproctus* are deep water genera and may have been unable to occupy niches already filled by the notothenioids which arrived on the Antarctic shelf and diversified before the liparids [4]. It is also interesting that whereas the Antarctic fish fauna is dominated by a single group of fish in which antifreeze capability evolved only once, the Arctic has a range of lineages in which antifreeze has evolved several times. Gadoids, for example possess an antifreeze glycoprotein (unrelated to that in notothenioids) and yet they have not radiated in the Southern Ocean.

In conclusion, the Antarctic shelf and upper slope is an insular evolutionary site, with radiations or species flocks of fish, equally as interesting as but less well known than the radiations in ancient lakes throughout the world. In spite of its vastness, the Antarctic Region provides an excellent opportunity for retrospective study of the evolution of an unusual fish fauna within a confined area during the latter part of the Tertiary.

Acknowledgments

J.T.E. would like to thank the European Science Foundation and the Coordination Committee of the Fishes of the Antarctic Ocean Network, especially Prof. Guido di Prisco and Dr. Eva Pisano, for the invitation to participate and for organizing the final network meeting. J.T.E.'s research is funded by National Science Foundation grant OPP 94-16870. A.C.'s research is funded by the British Antarctic Survey (Natural Environment Research Council). It is part of the Nearshore

Marine Biology research programme, and contributes to the Ecology of the Antarctic Sea-Ice Zone (EASIZ) programme of the Scientific Committee for Antarctic Research (SCAR).

References

1. Eastman JT (1993) Antarctic fish biology: evolution in a unique environment. Academic Press, San Diego
2. Clarke A, Johnston IA (1996) Evolution and adaptive radiation of Antarctic fishes. Trends Ecol Evol 11:212-218
3. Klingenberg CP, Ekau W (1996) A combined morphometric and phylogenetic analysis of an ecomorphological trend: pelagization in Antarctic fishes (Perciformes: Nototheniidae). Biol J Linn Soc 59:143-177
4. Andriashev AP, Stein DL (1998) Review of the snailfish genus *Careproctus* (Liparidae, Scorpaeniformes) in the Antarctic. Contrib Sci Nat Hist Mus Los Angeles Co (in press)
5. Clarke A, Crame JA (1989) The origin of the Southern Ocean marine fauna. In: Crame JA (ed) Origins and evolution of the Antarctic biota. Geological Society Special Publication No. 47. The Geological Society, London, pp 253-268
6. Clarke A, Crame JA (1992) The Southern Ocean benthic fauna and climate change: an historical perspective. Phil Trans Roy Soc Lond B 338: 299-309
7. Iwami T (1985) Osteology and relationship of the family Channichthyidae. Mem Nat Inst Polar Res E(36): 1-69
8. Bargelloni L, Ritchie PA, Patarnello T, Battaglia B, Lambert DM, Meyer A (1994) Molecular evolution at subzero temperatures: mitochondrial and nuclear phylogenies of fishes from Antarctica (suborder Notothenioidei), and the evolution of antifreeze glycopeptides. Mol Biol Evol 11:854-863
9. Ritchie PA, Bargelloni L, Meyer A, Taylor JA, Macdonald JA, Lambert DM (1996) Mitochondrial phylogeny of trematomid fishes (Nototheniidae, Perciformes) and the evolution of Antarctic fish. Mol Phylogen Evol 5:383-390
10. Brooks JL (1950) Speciation in ancient lakes. Quart Rev Biol 25:30-60, 131-176
11. Kozhov M (1963) Lake Baikal and its life. Junk, The Hague
12. Fryer G, Iles TD (1972) The cichlid fishes of the Great Lakes of Africa. TFH Publications, Neptune City, NJ
13. Echelle AA, Kornfield I (eds) (1984) Evolution of fish species flocks. University of Maine at Orono Press, Orono
14. Coulter GW (ed) (1991) Lake Tanganyika and its life. Oxford University Press, Oxford
15. Martens K, Coulter G, Goddeeris B (1994) Speciation in ancient lakes-40 years after Brooks. Arch Hydrobiol Beih Ergebn Limnol 44:75-96
16. Martens K (1997) Speciation in ancient lakes. Trends Ecol Evol 12: 177-182
17. Greenwood PH (1984) African cichlids and evolutionary theories. In: Echelle AA, Kornfield I (eds) Evolution of fish species flocks. University of Maine at Orono Press, Orono, pp 141-154

18. Ribbink AJ (1984) Is the species flock concept tenable? In: Echelle AA, Kornfield I (eds) Evolution of fish species flocks. University of Maine at Orono Press, Orono, pp 21-25
19. Gon O, Heemstra PC (eds) (1990). Fishes of the Southern Ocean. JLB Smith Institute of Ichthyology, Grahamstown, South Africa
20. Andriashev AP (1987) A general review of the Antarctic bottom fish fauna. In: Kullander SO, Fernholm B (eds) Fifth congress of European ichthyologists, proceedings, Stockholm, 1985. Swedish Museum of Natural History, Stockholm, pp 357-372
21. Nelson JS (1994) Fishes of the world (3rd ed). John Wiley & Sons, New York
22. Balushkin AV (1990) Review of blue notothenias of the genus *Paranotothenia* Balushkin (Nototheniidae) with description of a new species. J Ichthyol 30(6):132-147
23. Balushkin AV (1991) Review of green notothenias, *Gobionotothen*, Balushkin (Nototheniidae) of the Antarctic and SubAntarctic. J Ichthyol 31(8):42-55
24. Anderson ME (1994) Systematics and osteology of the Zoarcidae (Teleostei: Perciformes). Ichthyol Bull JLB Smith Inst Ichthyol No. 60:1-120
25. Balushkin AV (1992) Classification, phylogenetic relationships, and origins of the families of the suborder Notothenioidei (Perciformes). J Ichthyol 32(7):90-110
26. DeWitt HH (1971) Coastal and deep-water benthic fishes of the Antarctic. In: Bushnell VC (ed) Antarctic map folio series, folio 15. American Geographical Society, New York, pp 1-10
27. Ekau W (1990) Demersal fish fauna of the Weddell Sea, Antarctica. Antarct Sci 2:129-137
28. Hubold G (1991) Ecology of notothenioid fish in the Weddell Sea. In: di Prisco G, Maresca B, Tota B (eds) Biology of Antarctic fish. Springer-Verlag, Berlin, pp 3-22
29. Brooks DR, McLennan DA (1991) Phylogeny, ecology, and behavior: a research program in comparative biology. University of Chicago Press, Chicago
30. Mayden RL (1992) An emerging revolution in comparative biology and the evolution of North American freshwater fishes. In: Mayden RL (ed) Systematics, historical ecology, and North American freshwater fishes. Stanford University Press, Stanford, California, pp 864-890
31. Stein DL, Meléndez CR, Kong UI (1991) A review of Chilean snailfishes (Liparididae, Scorpaeniformes) with descriptions of a new genus and three new species. Copeia 1991(2):358-373
32. Stein DL, Andriashev AP (1990) Liparididae. In: Gon O, Heemstra PC (eds) Fishes of the Southern Ocean. JLB Smith Institute of Ichthyology, Grahamstown, South Africa, pp 231-255
33. Andriashev AP (1986) Review of the snailfish genus *Paraliparis* (Scorpaeniformes: Liparididae) of the Southern Ocean. Theses zoologicae, vol. 7, Koeltz Scientific Books, Koenigstein

34. Andriashev AP (1991) Possible pathways of *Paraliparis* (Pisces: Liparididae) and some other North Pacific secondarily deep-sea fishes into North Atlantic and Arctic depths. Polar Biol 11:213-218
35. Andriashev AP, Prirodina VP (1990) A review of Antarctic species of the genus *Careproctus* (Liparididae) and notes on the carcinophilic species of this genus. J Ichthyol 30(6):63-76
36. Kido K (1988) Phylogeny of the family Liparididae, with the taxonomy of the species found around Japan. Mem Fac Fish Hokkaido Univ 35:125-256
37. Eastman JT, Hikida RS, DeVries AL (1994) Buoyancy studies and microscopy of skin and subdermal extracellular matrix of the Antarctic snailfish, *Paraliparis devriesi*. J Morphol 220:85-101
38. Jung A, Johnson P, Eastman JT, DeVries AL (1995). Protein content and freezing avoidance properties of the subdermal extracellular matrix and serum of the Antarctic snailfish, *Paraliparis devriesi*. Fish Physiol Biochem 14:71-80
39. Jell PA (1974) Faunal provinces and possible planetary reconstruction of the Middle Cambrian. J Geol 82: 319-350
40. Hallam, A (1994) An outline of Phanerozoic biogeography. Oxford University Press, 246 pp
41. Enay R (1973) Upper Jurassic (Tithonian) ammonites. In: Hallam A (ed) Atlas of palaeobiogeography. Elsevier, Amsterdam, pp 297-307
42. Mutterlose J (1986) Upper Jurassic belemnites from the Orville Coast, Western Antarctica, and their palaeobiological significance. Bull Br Antarct Surv 70:1-22
43. Hallam A, Perez E, Biro L (1986) Facies analysis of the Lo Valdes Formation (Tithonian-Hauterivian) of the High Cordillera of central Chile and the palaeogeographic evolution of the Andean Basin. Geol Mag 123: 425-435
44. Riccardi AC (1991) Jurassic and Cretaceous marine connections between the Southeast Pacific and Tethys. Palaeogeog Palaeoclim Palaeoecol 87:155-189
45. Barker PF, Burrell J (1977) The opening of Drake Passage. Mar Geol 25:15-34
46. Barrett PJ (1997) Antarctic paleoenvironment through Cenozoic times - a review. Proceedings of the VII Antarctic Earth Sciences Symposium, Siena, August 1995 (In press)
47. Eastman JT, Grande L (1991) Late Eocene gadiform (Teleostei) skull from Seymour Island, Antarctic Peninsula. Antarct Sci 3:87-95
48. Balushkin AV (1994) *Proeleginops grandeastmanorum* gen. et sp. nov. (Perciformes, Notothenioidei, Eleginopsidae) from the Late Eocene of Seymour Island (Antarctica) is a fossil notothenioid, not a gadiform. J Ichthyol 34(8):10-23
49. Hambrey MJ, Barrett P (1993) Cenozoic sedimentary and climate record, Ross Sea region, Antarctica. In: Kennett JP, Warnke DA (eds) The Antarctic paleoenvironment: a perspective on global change. Part Two. Antarctic Research Series, 60. American Geophysical Union, Washington, pp 91-124
50. Brey T, Dahm C, Gorny M, Klages M, Stiller M, Arntz WE (1996) Do Antarctic benthic invertebrates show an extended level of eurybathy? Ant Sci 8:3-6

51. Burckle LH, Stroeven AP, Bronge C, Miller U, Wasell A (1996) Deficiencies in the diatom evidence for a Pliocene reduction of the East Antarctic ice sheet. Paleoceanography 11:379-389
52. Dingle RV, McArthur JM, Vroon P (1997) Oligocene and Pliocene interglacial events in the Antarctic Peninsula dated using strontium isotope stratigraphy. J Geol Soc, Lond 154:257-264
53. Berger A (1988) Milankovitch theory and climate. Rev Geophys 26: 624-657
54. Clemens SC, Tiedemann R (1997) Eccentricity forcing of Pliocene - Early Pleistocene climate revealed in a marine oxygen-isotope record. Nature 385:801-804
55. Herbert TD, Fischer AG (1986) Milankovitch climatic origin of mid-Cretaceous black shale rhythms in central Italy. Nature 321:739-743
56. Valentine JW (1968) Climatic regulation of species diversification and extinction. Geol Soc Amer Bull 79:273-276
57. Crame JA (1993) Latitudinal range fluctuations in the marine realm through geological time. Trends Ecol Evol 8:162-166
58. Clarke A, Crame JA (in press) Diversity, latitude and time: patterns in the shallow sea. In: Ormond RFG, Gage J and Angel MV (eds) Marine biodiversity: patterns and processes. Cambridge University Press, pp 122-147
59. Dansgaard W, Johnsen SJ, Clansen HB, Dahl-Jensen D, Gundestrup NS, Hammer CU, Hvidberg CS, Steffensen JP, Sveinbjørnsdottir AE, Jouzel J, Bond G (1993) Evidence for general instability of climate from a 250-kyr ice-core record. Nature 364:218-20
60. Taylor KC, Lamerey GW, Doyle GA, Alley RB, Grootes PM, Mayewski PA, White JWC, Barlow LK (1993) The 'flickering switch' of late Pleistocene climate change. Nature 361:432-436
61. Coope GR (1979) Late Cenozoic fossil Coleoptera: evolution, biogeography and ecology. Ann Rev Ecol Syst 10:247-267
62. Elias SA (1994) Quaternary insects and their environment. Smithsonian Institution Press, Washington
63. Kennett JP, Stott LD (1991) Abrupt deep-sea warming, palaeoceanographic changes and benthic extinction at the end of the Palaeocene. Nature 353:225-229
64. Kennett JP (1982) Marine geology. Prentice-Hall, Englewood Cliffs, NJ
65. Lecointre G, Bonillo C, Ozouf-Costaz C, Hureau J-C (1997) Molecular evidence for the origins of Antarctic fishes: paraphyly of the Bovichtidae and no indication for the monophyly of the Notothenioidei (Teleostei). Polar Biol 18:193-208
66. McDonald MA, Smith MH, Smith MW, Novak JM, Johns PE, DeVries AL (1992) Biochemical systematics of notothenioid fishes from Antarctica. Biochem Syst Ecol 20:233-241
67. Chen L, DeVries AL, Cheng C-HC (1997) Evolution of antifreeze glycoprotein gene from a trypsinogen gene in Antarctic notothenioid fish. Proc Natl Acad Sci USA 94:3811-3816
68. Greenwood PH (1994) Lake Victoria. Arch Hydrobiol Beih Ergebn Limnol 44:19-26

69. Johnson TC, Odada EO (eds) (1996) The limnology, climatology and paleoclimatology of the East African lakes. Overseas Publishers Association, Amsterdam
70. Lowe-McConnell R (1996) Fish communities in the African Great Lakes. Env Biol Fish 45:219-235
71. Meyer A, Montero CM, Spreinat A (1996) Molecular phylogenetic inferences about the evolutionary history of East African cichlid fish radiations. In: Johnson TC, Odada EO (eds) The limnology, climatology and paleoclimatology of the East African lakes. Overseas Publishers Association, Amsterdam, pp 303-323
72. Liem KF (1973) Evolutionary strategies and morphological innovations: cichlid pharyngeal jaws. Syst Zool 22:425-441
73. Liem KF (1991) Functional morphology. In: Keenleyside MHA (ed) Cichlid fishes: behaviour, ecology and evolution. Chapman & Hall, London, pp 129-150
74. Owen RB et al (1990) Major low levels of Lake Malawi and their implications for speciation rates in cichlid fishes. Proc R Soc Lond B 240:519-553
75. Johnson TC et al (1996) Late Pleistocene desiccation of Lake Victoria and rapid evolution of cichlid fishes. Science 273:1091-1093
76. Martin P (1994) Lake Baikal. Arch Hydrobiol Beih Ergebn Limnol 44:3-11
77. Sideleva VG (1994) Speciation of endemic Cottoidei in Lake Baikal. Arch Hydrobiol Beih Ergebn Limnol 44:441-450
78. Sideleva VG (1996) Comparative character of the deep-water and inshore cottoid fishes endemic to Lake Baikal. J Fish Biol 49 (Suppl. A):192-206
79. Grachev MA et al (1992) Comparative study of two protein-coding regions of mitochondrial DNA from three endemic sculpins (Cottoidei) of Lake Baikal. J Mol Evol 34:85-90
80. Smith GR, Todd TN (1984) Evolution of species flocks of fishes in north temperate lakes. In: Echelle AA, Kornfield I (eds) Evolution of fish species flocks. University of Maine at Orono Press, Orono, pp 45-68
81. Pankhurst NW, Sideleva VG, Pankhurst PM, Smirnova O, Janssen J (1994) Ocular morphology of the Baikal sculpin-oilfishes, *Comephorus baicalensis* and *C. dybowskii* (Comephoridae). Env Biol Fish 39:51-58
82. Smirnova OG (1995) Retinal structure of Baikal oilfishes of the endemic family Comephoridae. J Ichthyol 35(1):139-145
83. Chen L, DeVries AL, Cheng C-HC (1997) Convergent evolution of antifreeze glycoproteins in Antarctic notothenioid fish and Arctic cod. Proc Nat Acad Sci USA 94:3817-3822

Part 1

Ecology

Age, Growth and Maximum Size of Antarctic Notothenioid Fish - Revisited

Karl-Hermann Kock[1] and Inigo Everson[2]

[1]Institut für Seefischerei, BFA für Fischerei, Palmaille 9, 22767 Hamburg, Germany
[2]British Antarctic Survey, High Cross, Madingley Road, Cambridge CB3 0ET, UK

Introduction

The temperature of the Southern Ocean has undergone an overall reduction from about 20 °C to -1.8 °C over the last 55-60 million years. To date, the Southern Ocean is characterized by low temperatures ranging from +3 °C close to the South Polar Frontal Zone to -1.86 °C in the vicinity of the Antarctic continent and a strong seasonality and patchiness of primary productivity.

The Antarctic coastal fish fauna consists of about 140 species of 18 families [1]. This fauna is dominated both in terms of species and biomass by a single endemic perciform suborder, the Notothenioidei. Notothenioidei comprise a variety of ecomorphological types ranging from small sculpin-like to large hake-like forms [1-3]. Two coarse groups of coastal species can be distinguished according to their geographical distribution: lesser-Antarctic/subAntarctic species and high-Antarctic species [4]. Although low temperature has most likely played a dominant role in the evolution of the present fish fauna, other factors, such as climatic cycles, isolation, habitat loss, and seasonality in food availability have probably also been important [2].

Antarctic notothenioid fish show a variety of adaptations at the molecular level, such as freezing resistance, tubulins, which polymerize at -2 °C, the conductivity of neurones and high turnover rates of enzymes, in order to maintain function in their extreme thermal environment [2,5]. The rates of complex physiological systems are frequently slower in Antarctic species than in temperate species. Growth is a complex process that is subject to a number of controls and limitations at a variety of levels, which makes it difficult to determine which particular factor is the strongest in controlling growth rate [6]. Temperature and other environmental variables, such as light levels, food availability and quality, reproduction and activity are all likely to affect growth [6-9].

G. di Prisco, E. Pisano, A. Clarke (Eds)
Fishes of Antarctica. A biological overview
© Springer-Verlag Italia 1998

In what follows, we review information on age and growth, reproduction and seasonality in feeding and then consider the results in the light of the hypothesis that Antarctic notothenioid fish grow slowly.

Age and Growth Studies in Antarctic Fish

There are more than 120 publications which contain information on age and growth of Antarctic notothenioid fish. The majority of these were published after 1975 when commercial fishing provided a ready supply of material on a variety of species. Up to that date, there were only 10 papers published on species such as *Notothenia coriiceps*, *N. rossii* or *Trematomus bernacchii* [i.e. 10-13].

Age and growth estimates are limited to a number of nototheniids, a few channichthyids, and *Harpagifer antarcticus*. They are still lacking for members of the other notothenioid families Artedidraconidae and Bathydraconidae. Most age estimates originate from lesser-Antarctic/sub-Antarctic species. Published information on age and growth of high-Antarctic fish is limited to a few species of the nototheniid genus *Trematomus, Aethotaxis mitopteryx, Dissostichus mawsoni, Pleuragramma antarcticum,* and *Chionodraco hamatus*.

Age Determination

Otoliths, scales, fin rays and vertebrae and other skeletal bones have all been shown to exhibit rhythmic patterns that can be related to changes in daily, seasonal or annual growth [1,14,15]. The growth patterns present on these hard structures have been used to estimate age.

Otoliths should be the most suitable structures for ageing as they contain a complete record of growth from hatching to capture [16]. They are not subject to resorption in times of calcium shortage. Furthermore, they can be used to estimate age in fish species which lack scales, such as channichthyids. Ageing using otoliths has been mostly conducted on sectioned otoliths both on seasonal growth increments and on micro-structural growth increments exhibiting daily increments ([8,15] for review).

Scales do not contain the whole growth history of a fish. In nototheniids, they do not form earlier than several months after hatching. Replacement scales are common. The margins of the scales become eroded with age, and annuli may be obscured by the compression or erosion of sclerites in the marginal zone in older fish. Resorption processes may lead to additional checks which may be misinterpreted as representing annuli ('false check'). Scale circuli analysis should thus ideally combine the determination of the onset of scale formation and validation with data from other methods. These prerequisites have rarely been met. However, scale circuli analysis has been the method most commonly applied, in particular to nototheniids [1,15].

Length compositions of a number of channichthyids and nototheniids exhibit distinct peaks at the lower end of their range which show little overlap between modes. The position of these modes is consistent with little variation between years. Shifts in these modes have been followed within a year, clearly indicating that they represent age classes. Despite its potential to estimate growth and its inter-annual variation in a number of species in their first 3-4 years, this method has thus far only been utilized to estimate age in *Champsocephalus gunnari* from South Georgia and Iles Kerguelen [17,18].

Growth has also been independently checked against age estimates from otolith readings by measurement of the growth increments of marked and recaptured fish, such as in *N. coriiceps* [12,19] and *T. bernacchii* [20].

The Reliability of Age Estimates

The correct determination of age is crucial for any growth estimate. However, only few of the age estimates for Antarctic fish are scientifically validated. For example, comparative age determinations by five investigators on the same set of scales of *Gobionotothen gibberifrons* from South Georgia revealed considerable differences between readers ranging from 1-3 years in scales of fish 10 cm long to 6-13 years in fish of 33 cm length ([1]: Fig. 38).

Numerous such examples are available from the scientific literature and from comparative age readings within the BIOMASS and CCAMLR age determination workshops [21-25]. They clearly demonstrate the high degree of subjectivity still inherent in age determinations of Antarctic fish which means that the possibility of bias caused by the expectation that Antarctic fish grow slowly may have influenced the results.

To date, age estimates have been validated for only a few species, with *N. coriiceps*, juvenile *N. rossii*, *T. bernacchii*, and *C. gunnari* among them, all of which, with the exception of *T. bernacchii,* occur in lesser-Antarctic/subAntarctic waters. It is important to bear this in mind for the following discussion on growth of Antarctic fish.

Maximum Ages

Maximum ages reported for small sized (20-23 cm) nototheniids, for example *Lepidonotothen larseni*, *L. nudifrons* or *Patagonotothen guntheri*, in lesser-Antarctic/subAntarctic waters were 8-10 years ([1] for review). Similar-sized (20-30 cm) high-Antarctic nototheniids, such as

Trematomus species were estimated to become 15-21 years ([1] for review) or even 20-30 years old, such as *P. antarcticum* [26]. Some of the channichthyids from lesser-Antarctic/subAntarctic waters, such as *C. gunnari* or *C. aceratus* attain ages of 12-15 years [1]. Ages of 10 years in samples of the high-Antarctic channichthyid *Chionodraco hamatus* from the Ross Sea [27] suggest that they attain similar ages as *C. gunnari* and *C. aceratus*.

Ages of 18-22 years have been estimated for larger-sized (50-90 cm) *N. rossii*, *G. gibberifrons* and *Lepidonotothen. squamifrons* in the vicinity of South Georgia and Iles Kerguelen ([1] for review). Ages of more than 30 years have been determined in *A. mitopteryx* and *D. mawsoni* from high-Antarctic waters and are also very likely for the subAntarctic *Dissostichus eleginoides* [1].

Growth

The growth model most commonly applied to Antarctic notothenioid fish has been the von Bertalanffy growth formula [28]. Growth performance of Antarctic fish has been commonly compared by comparing the parameters K and L_∞ of the von Bertalanffy growth formula. The parameter K is a measure of how fast fish attain their final size. It is usually negatively correlated with size (L_∞, W_∞) and hence is not very suitable for comparing the growth performance of fish of different sizes. This is much better achieved by the index of growth performance P (= log K + log W_∞) which describes the growth rate at the point of inflexion of the size (mass) growth curve [29].

We have used P to compare growth performance among notothenioids and with ecologically similar fish species from boreal and temperate waters (Fig. 1). Populations of some species, such as *L. squamifrons* (points 3A-D), *H. antarcticus* (10A-C), *C. gunnari* (points 11A,B) or *T. bernacchii* (points 23A-D) exhibit considerable differences in their growth parameters K and W_∞; however their indices of growth performance are very similar. The differences in growth parameters may reflect different environmental conditions in which the individual population lives. However, subjectivity in the age estimates could have also substantially contributed to the differences in estimates of growth parameters [1].

Despite these reservations, some trends are apparent from Fig. 1. Growth performance of notothenioids in lesser-Antarctic/subAntarctic waters appears to be higher than in waters in the vicinity of the Antarctic continent. Most species exhibit P values of 2-3. Two of them, *N. rossii* (points 1A-C) and *D. eleginoides* (point 4), even have P values of 3-4. Exceptions are *L. nudifrons* (point 8), *L. larseni* (point 9) and *H. antarcticus* (points 10A-C) which have comparatively high K values

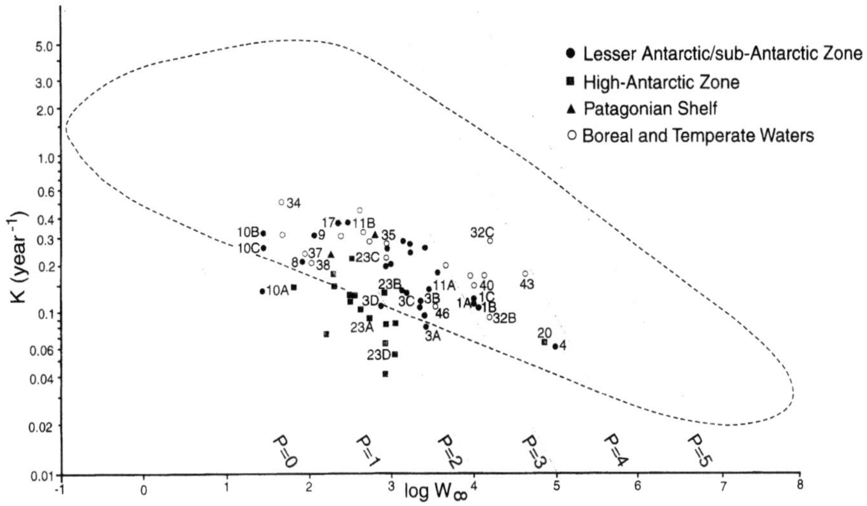

Fig. 1. Auximetric grid showing the growth performance of Antarctic fish and selected fish species from boreal, temperate and freshwater (Lake Baikal) environments. The distance from a point representing a pair of growth parameters K and W_∞, to the baseline P=0 represents P and is therefore a direct indication of growth performance (modified from [1]; for full legend see [1: Fig. 41]). **Lesser-Antarctic/subAntarctic species:** 1. *N. rossii*: A South Georgia, B Antarctic Peninsula, C Iles Kerguelen. 3. *L. squamifrons*: A Iles Kerguelen, B Ob Bank, C Lena Bank, D Iles Crozet. 4. *D. eleginoides*: South Georgia. 8. *L. nudifrons*: Ant. Peninsula. 9. *L. larseni*: South Georgia. 10. *H. antarcticus*: A Arthur Harbour, B Paradise Harbour, C Paradise Harbour, all Ant. Peninsula. 11. *C. gunnari*: A South Georgia, B Iles Kerguelen. 17. *Trematomus newnesi*: Ant. Peninsula. **High-Antarctic species:** 20. *D. mawsoni*: Ross Sea. 23. *T. bernacchii*: A Weddell Sea, B Terre Adelie, C and D Ross Sea. **Boreal and temperate species:** 32. *Gadus morhua*: A East Greenland, B West Greenland, C North Sea, D Labrador. 34. *Callionymus lyra*: North Sea. 35. *Scomber scombrus*: North Sea. 37. *Taurulus bubalis*: Baltic Sea. 38. *Cottus kessleri*: Lake Baikal. 40. *Pollachius virens*: North Sea. 43. *Euthynnus alletteratus*. 46. *Labrus bergylta*: North Sea.

(>0.20), but low values of W_∞, resulting in P values of less than 1.5. Most species from lesser-Antarctic/subAntarctic waters studied to date were water column feeders having krill and fish as their principal diet [1].

High-Antarctic nototheniids typically exhibit P values of 1-2 with the exception of *D. mawsoni* (point 20) which has a growth performance similar to its subAntarctic congener *D. eleginoides*. The bulk of these high-Antarctic nototheniids are bottom or near-bottom feeders with benthos organisms accounting for a considerable part of their diet except in *D. mawsoni* [1].

The growth performance of a number of high-Antarctic nototheniids lies outside the range commonly occupied by marine and brackish-water fish (Fig. 1). This is due to their simultaneously low values estimated for K, L_∞ and W_∞, except in *T. newnesi* (point 17), a high-Antarctic species investigated in the Antarctic Peninsula region for which a relatively high value of K has been described. However, as mentioned above, growth parameters of high-Antarctic species with very few exceptions have been derived from unvalidated age estimates and could be subject to revisions during forthcoming validation studies [1].

Comparing Antarctic species with ecologically similar fish species from temperate and boreal waters, we note that the growth performance of the *Dissostichus* species is comparable to that of North Sea cod (*Gadus morhua*) (point 32C) and little tuna (*Euthynnus alletteratus*) (point 43); that of *N. rossii* compares to Atlantic cod from West Greenland (point 32B) and saithe (*Pollachius virens*) (point 40) in the North Sea. Growth performance of other larger nototheniids and channichthyids in lesser-Antarctic/subAntarctic waters is similar to common North Sea species, such as the ballan wrasse (*Labrus bergylta*) (point 46) or mackerel (*Scomber scombrus*) (point 35). Growth performance in small-sized nototheniids both in lesser-Antarctic/subAntarctic waters and in the vicinity of the Antarctic continent compares well with that of the sea scorpion (*Taurulus bubalis*) (point 37) in the Baltic Sea, of *Cottus kessleri* (point 38) in Lake Baikal, and the dragonet (*Callionymus lyra*) (point 34) in the North Sea.

Feeding Regime

Few studies have followed feeding activity and diet composition year round. Available information indicates a change in food composition of some species in winter ([1] for review). These shifts in prey composition are likely to reflect changes in the availability of prey organisms.

Studies on energy requirements suggested that the daily food demands of *N. rossii* and *G. gibberifrons* in the vicinity of South Georgia were reduced by a factor 2-3 in winter compared with the summer [30,31]. The condition factor in juvenile *N. rossii* around South Georgia exhibited a strong seasonal signal usually being lowest in winter [32]. However, the level of lipid in juvenile *P. antarcticum* in the Weddell Sea was similar in winter and summer [33]. Similarly, the evidence for a reduction in feeding intensity during winter is not unequivocal: a reduction in feeding intensity in winter was observed in some nototheniids [13,34]. It has also been induced in captive *H. antarcticus* [35]. However, observations in other nototheniids provided no evidence for a decline in feeding intensity during winter [7,9,36]. Although observations on a high level of feeding intensity in winter appear to be contradictory to predictions from energy requirements, the change to other prey organisms in winter with a lower energy content and a resulting lower energy intake at the same level of feeding may partly resolve this contradiction [32].

A considerable increase in the proportion of empty stomachs and a reduction in feeding, however, has been observed in the prespawning and spawning period of a number of species, such as *L. larseni*, *C. gunnari*, *N. rossii*, *N. coriiceps*, and *Trematomus eulepidotus* [1,7,18].

Maximum Size

Notothenioids are predominantly small species up to 45 cm in length (Fig. 2). Among the lesser-Antarctic/subAntarctic notothenioids about 45% attain sizes of more than 45 cm, wherereas in the high-Antarctic Zone 90% of the notothenioids are less than 45 cm long and 65% do not grow larger than 30 cm (Fig. 2).

Harpagiferids and artedidraconids (with the exception of some of the *Pogonophryne* species, which grow up to 25-30 cm) are typically less than 15 cm long. Bathydraconids (apart from *Cygnodraco mawsoni* and *Parachaenichthys* species which attain up to 50-60 cm) hardly exceed 30 cm. Most nototheniids do not grow larger than 45 cm. However, the

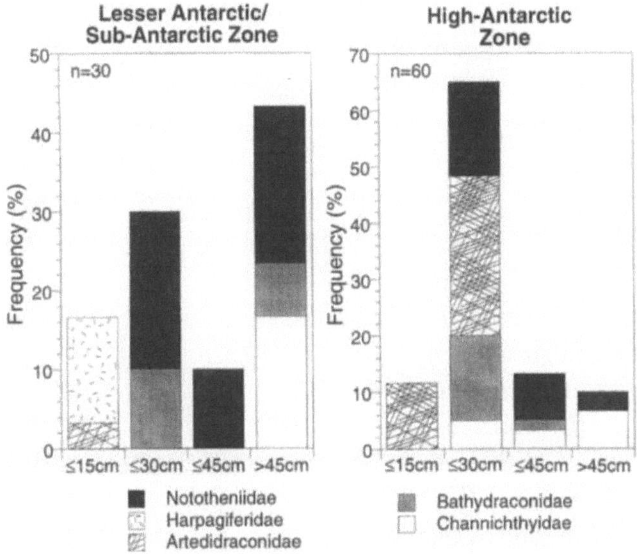

Fig. 2. Maximum lengths observed in notothenioid families from lesser-Antarctic/sub-Antarctic waters and high-Antarctic waters (from [1])

biggest species among Antarctic coastal fish are found in this family: *D. eleginoides* which has a widespread distribution in subAntarctic waters and its high-Antarctic congener *D. mawsoni* attain maximum sizes of more than 200 cm (>100 kg) and 174 cm (ca. 70 kg) respectively. About two thirds of the channichthyids which lack the respiratory pigment hemoglobin grow larger than 45 cm [1].

The Onset of Sexual Maturity

Antarctic fish tend to spawn for the first time when between 55 and 80% of their maximum length. There does not appear to be any difference between notothenioids in the high-Antarctic region in comparison with those of the lesser-Antarctic/subAntarctic waters [1]. A notable exception is the mackerel icefish *C. gunnari* which, at South Georgia, spawns for the first time when only about 40% of its maximum length. Close to spawning, ovaries account for 20-30%(in some species up to 40%) of the body weight [1].

Discussion

Comparatively small maximum sizes, high maximum ages and the extreme thermal environment of Antarctic fish have led to the widespread belief that notothenioids grow slowly. Although our analysis was impaired by the paucity of validated age determinations, the growth performance analysis provides sufficient evidence to suggest that growth performance in a number of Antarctic nototheniids and channichthyids is by no means low as suggested by the low values of K of the von Bertalanffy growth formula, but is in many cases comparable to ecologically similar species from boreal and temperate waters. Growth performance of nototheniids and channichthyids exhibits the large variation one would expect from a group of fish species which occupy a broader range of ecological niches. The results which have been cited in the past as evidence for slow growth fit within this range of values.

Three basic explanations had been offered in support of the hypothesis that Antarctic fish grow slowly:

- A direct limitation by temperature;
- Adaptation to subzero temperatures necessitates an elevation of metabolic rate with a concomitant reduction in the energy available for growth; and
- Seasonality in food availability [6].

After an extensive discussion of the three possible explanations, Clarke and North [6] concluded that the evidence from larval fish (and from polar marine invertebrates) suggests that although low temperature due to its rate-depressing effect on physiological systems is likely to impose some general constraint on growth, it is seasonal food availability and not temperature that normally limits growth rate. The evidence from adult fish appeared to be more equivocal.

Both, otoliths and scales in most nototheniids and channichthyids investigated exhibited a clear seasonal signal [15]. Although our analysis showed that information on seasonality in feeding is not unequivocal, there is sufficient evidence to suggest that seasonal cycles, such as changes in the availability of food and seasonal variation in its energy content [9,37] and the low feeding intensity during the reproductive season [1] may preclude year round growth for many species with the result that under favorable conditions growth can be relatively fast. Hence, ecological constraints are likely to affect growth rate during the year much more than temperature per se and may also largely account for differences in growth performance between species. Our observations are corroborated by findings that differences in activity, metabolic rate, perception of stimuli or behavior among Antarctic notothenioids appear to be much more influenced by the ecological niche a fish species inhabits than by temperature per se [38].

Growth is almost certainly impaired after the fish become sexually mature due to the significant energy costs attendant on reproductive effort. Most Antarctic fish do not start to reproduce before they reach 55-80% of their maximum observed length. Postponing maturity until after achieving a size large enough to exploit abundant food resources, such as amphipods, krill and other euphausids, seems to be one plausible reason for delayed maturity.

We consider that there is an increasing body of evidence now available which does not support the view that Antarctic fish grow slowly. However, age and growth studies should remain an important topic for Antarctic fish research in the future. We suggest that studies on age and growth should focus on

- the validation of ageing, incorporating different techniques, such as the analysis of length compositions of juvenile fish from scientific surveys and the analysis of daily ring structures in otoliths and elemental ratios in microstructure which provide not only powerful tools to validate age but additionally provide information to the individual ecological and physiological past;
- an extension of age and growth estimates to artedidraconids and bathydraconids;

- the comparison of growth estimates in those species whose geographical distribution ranges from the subAntarctic to high-Antarctic waters, such as *N. coriiceps* and *T. hansoni*; and
- controlled laboratory experiments which compare the growth capacities of fish species fed ad libitum in the laboratory [6] with growth estimated in the field.

Acknowledgments

We are grateful to Prof. Dr. G. Duhamel and two anonymous reviewers for their constructive comments on the manuscript. Dr. C. Zimmermann kindly prepared the final layout of the manuscript. We are indebted to Mr. E. Marschmann-Horn for his technical support in the preparation of the figures.

References

1. Kock K-H (1992) Antarctic fish and fisheries. Cambridge University Press, Cambridge, New York
2. Eastman J T (1993) Antarctic fish biology. Evolution in a unique environment. Academic Press, San Diego,
3. Eastman J T (1995) The evolution of Antarctic fishes: questions for consideration and avenues for research. Cybium 19(4):371-389
4. Andriashev, A P (1987) A general review of the Antarctic bottom fish fauna. In: Kullander, S O, Fernholm B (eds), Proceedings of the Fifth Congress of European Ichthyologists, Swedish Museum of Natural History, Stockholm, pp 357-372
5. Clarke A, Johnston, I A (1996) Evolution and adaptive radiation of Antarctic fishes. Trends Ecol Evol 11 (5):212-220
6. Clarke A, North A W (1991) Is the growth of polar fish limited by temperature? In: di Prisco, G, Maresca B, Tota B (eds), Biology of Antarctic fish, Springer, Berlin, Heidelberg, New York, pp 54—69
7. Everson I (1984) Fish biology. In: Laws, R M (ed), Antarctic ecology, Vol. 2, Academic Press, London, pp 491-532
8. Radtke R L, Kellermann, A (1991) Microstructural analysis of growth patterns in early life history stages of Antarctic fishes. In: di Prisco, G, Maresca B, Tota B (eds), Biology of Antarctic fish, Springer, Berlin, Heidelberg, New York, pp 101-115
9. Ashford J R, White M G (1995) An annual profile of growth in the otoliths of immature *Notothenia coriiceps* Richardson in relation to the environment at Signy Island, Antarctica. Antarct Sci 7(4):345-349
10. Olsen S (1954) South Georgia cod, *Notothenia rossi marmorata* Fischer. Norsk Hvalfangst 43 (7):373-382
11. Wohlschlag D E (1962) Antarctic fish growth and metabolic differences related to sex. Ecology 43:589-597

12. Everson I (1970) The population dynamics and energy budget of *Notothenia neglecta* Nybelin at Signy Island, South Orkney Islands. Brit Antarct Surv Bull 23:25-50
13. Hureau J-C (1970) Biologie comparée de quelques poissons antarctiques (Nototheniidae). Bull Inst Océanogr Monaco 68 (1391):1-244
14. Radtke R L, Hourigan, T F (1990) Age and growth of the Antarctic fish *Nototheniops nudifrons*. Fish Bull 88:557-571
15. White M G (1991) Age determination in Antarctic fish. In: di Prisco, G, Maresca B, Tota B (eds), Biology of Antarctic fish, Springer, Berlin, Heidelberg, New York, pp 87-100
16. Radtke R L (1987) Information incorporated in Antarctic fish otoliths. In: Kullander, S O, Fernholm B (eds), Proceedings of the Fifth Congress of European Ichthyologists, Swedish Museum of Natural History, Stockholm, pp 421-425
17. Kock K-H (1980) Graphical analysis of length frequency distributions of *Champsocephalus gunnari* Lönnberg (Channichthyidae) from South Georgia. Cybium 1980 (8):33-42
18. Duhamel G (1987) Ichthyofaune des secteur indien occidental et atlantique oriental de l'océan austral: biogéographie, cycles biologiques et dynamique des populations. Thèse de Doctorat D'Etat. Université Pierre et Marie Curie, Paris, 1-687
19. Barrera-Oro E R, Casaux R J (1996) Validation of age determination in *Notothenia coriiceps*, by means of a tagging experiment at Potter Cove, South Shetland Islands. Arch Fish Mar Res 43 (3):205-216
20. Kawaguchi K, Ishikawa S, Matsuda O, Naito Y (1989) Tagging experiment of nototheniid fish, *Trematomus bernacchii* Boulenger under the coastal fast ice in Lützow-Holm Bay, Antarctica. Proc NIPR Symp Polar Biol 2:111-116
21. Anonymous (1980) Working Party on Antarctic Fish Biology. Report of the Second Meeting, Dammarie-les-Lys, France 27-31 May 1980. BIOMASS Report Series 12, 47 pp
22. Anonymous (1982) Working Party on Antarctic Fish Biology. Report of the Second Workshop on the Ageing of Antarctic Fishes, Orono, Maine, USA, 7-10 June 1982. BIOMASS Report Series 26, 7 pp
23. Everson I (1980) Antarctic fish age determination methods. BIOMASS Handbook 8, 24 pp
24. North A W, White M G, Burchett, M S (1980) Age determination of Antarctic fish. Cybium 8:7-11
25. Kock K-H (1990) Results of the CCAMLR fish otoliths/scales/bones exchange system. SC-CAMLR Select Scient Pap 1989:51-68
26. Radtke R L, Hubold G, Folsom S D, Lenz P H (1993) Otolith structure and chemical analysis: the key to resolving age and growth of the Antarctic silverfish, *Pleuragramma antarcticum*. Antarct Sci 5:51-62
27. Vacchi M, Romanelli M, La Mesa M (1992) Age structure of *Chionodraco hamatus* (Teleostei, Channichthyidae) samples caught in Terra Nova Bay, East Antarctica. Polar Biol 12 (8):735-738

28. Bertalanffy L v (1951) Theoretische Biologie. Vol. 2: Stoffwechsel, Wachstum. A. Francke Verlag, Bern
29. Pauly D (1979) Gill size and temperature as governing factors in fish growth: a generalization of von Bertalanffy's growth formula. Ber Inst Meeresk Kiel 63:1-156
30. Chekunova V I, Naumov A G (1982) Energy metabolism and food requirements of the marbled notothenia, *Notothenia rossi marmorata* (Nototheniidae). Vopr Ikthiol 22 (2): 294-302 (in Russian) (Published in English as J Ichthyol 22(2):112-121
31. Chekunova V I, Naumov A G (1982) Energy metabolism of *Notothenia gibberifrons* Lönnberg off South Georgia. In: Characteristics of the pelagic community from the Scotia Sea and adjacent waters, All-Union Research Institute of Marine Fisheries and Oceanography, Moscow, pp 76-85
32. Burchett M S (1983) Age and growth of the Antarctic fish *Notothenia rossii* from South Georgia. Br Antarct Surv Bull 60:45-61
33. Hubold G, Hagen W (1997) Seasonality of feeding and lipid content in juvenile *Pleuragramma antarcticum* (Pisces: Nototheniidae) in the southern Weddell Sea. In: Battaglia B, Valencia J, Walton, D W H (eds), Antarctic communities - species, structure and survival, Cambridge University Press, Cambridge, pp 277-283
34. Linkowski T B, Presler P, Zukowski C (1983) Food habits of nototheniid fishes (Nototheniidae) in Admiralty Bay (King George Island, South Shetland Islands). Pol Polar Res 4 (1-4):79-95
35. Targett T E, Young K E, Konecki J T, Grecay P A (1987) Research on wintertime feeding in Antarctic fishes. Antarct J U S 22 (1/2):211-213
36. Hubold G (1992) Zur Ökologie der Fische im Weddellmeer. Ber Polarforsch 103:1-157 (in German)
37. Clarke A (1988) Seasonality in the Antarctic environment. In: di Prisco, G, Maresca B, Tota B (eds), Marine biology of Antarctic organisms, Proceedings of the International Conference, Ravello, Italy, Comp Biochem Physiol 90B:461-473
38. Zimmermann C (1997) Zur Ökologie arktischer und antarktischer Fische: Aktivität, Sinnesleistungen und Verhalten. Ber Polarforsch 231:1-137 (in German)

Maternal Output in Polar Fish Reproduction

Jørgen S. Christiansen[1], Svein E. Fevolden[2], Oleg V. Karamushko[3], and Larisa I. Karamushko[3]

[1]Norwegian Institute of Fisheries and Aquaculture, 9005 Tromsø, Norway
[2]The Norwegian College of Fishery Science, University of Tromsø, 9037 Tromsø, Norway
[3]Murmansk Marine Biological Institute, 183010 Murmansk, Russia

Introduction

In oviparous fish, the combination of egg size and fecundity (i.e. the total number of eggs to be spawned) that propagates the largest number of viable offspring is deemed to be favored by evolutionary processes [1,2]. Polar marine fish species tend to produce fewer and larger yolk-rich eggs than their lower latitude counterparts. Rass [3] stated: "The size of eggs of aquatic animals is in accordance with their breeding temperatures" and "... the eggs being greater, the nearer to the pôle lies the habitat of the species."

The hypothesis that the physical environment (i.e. temperature) conditions the increase in egg size towards higher latitudes was, however, opposed by Marshall [4]. He argued that the observed latitudinal cline in egg size and fecundity could be ascribed to sundry behavioral and ecological factors, e.g. feeding conditions, rather than to a single physical variable, such as breeding temperature, per se. The period of egg incubation (i.e. the time between fertilization and hatching) does not seem to correlate with egg size within fish species at any one temperature [5] but large eggs tend to result in large larvae [5,6]. Large larvae subsist longer on endogenous fuel, are less affected by variations in exogenous food supply and this may result in more viable offspring [7].

According to [4], the adaptive value of fish producing relatively large eggs and thus more robust larvae is particularly relevant in habitats which are characterized by relatively sparse food concentrations e.g. the deep-sea and open oceanic waters. Polar fish are also subjected to seasonality extremes in food availability: a short period of high production of planktonic organisms is followed by long periods with little or no food in the water column [8]. Thus, Marshall's hypothesis [4] would also apply to polar fish species.

G. di Prisco, E. Pisano, A. Clarke (Eds)
Fishes of Antarctica. A biological overview
© Springer-Verlag Italia 1998

Polar fish produce relatively few, large eggs although egg size and fecundity in fish are strongly linked to the mode of oviposition [9]. Fish that spawn freely floating eggs (i.e. pelagic oviposition) usually produce smaller eggs and are more fecund than species that deposit their eggs onto/into substrates (i.e. demersal oviposition) [10].

The number of studies dealing with egg size and fecundity in fish has increased considerably in recent years and includes reviews on Antarctic fish reproduction [11,12]. In the light of the new information, this communication outlines how egg size, fecundity, hence reproductive output of marine, teleost species covary with both mode of oviposition (i.e. pelagic vs demersal) and zoogeography (i.e. polar vs boreal).

Methods

The Data Base

A total of 59 marine, teleost fish species, represented by 85 data sets, comprise this study. Data were examined with respect to the mode of oviposition and the zoogeographic distribution of the different species (Table 1). Although the evolutionary history of the polar marine environments (Arctic and Antarctic) is rather different, we have divided the fish into two broad zoogeographic categories viz polar (i.e. Arctic and Antarctic combined) and boreal species. The distinction between boreal and genuinely Arctic species follows the authority of Andriyashev and Chernova [13]. We have also treated Antarctic fish as belonging to the same zoogeographic region (i.e. within the boundary of the Antarctic Convergence) although Kock [6] and Kock and Kellermann [11] have made a further distinction between species living in the High-Antarctic Zone and the Seasonal Pack-ice Zone, respectively. Further, fish were grouped according to their mode of oviposition viz pelagic spawners, in which eggs develop in the water column, and substrate spawners, in which eggs develop on or close to the sea floor.

The data base also includes information about body length (in mm), egg diameter (in mm) and fecundity of sexually mature fish (Table 1). Information has been gathered mainly from Rass [3], Marshall [4], Kock and Kellermann [11], Hubold [12], Rass [14] and Dolgov [15].

Essentially, the reproductive output of female fish depends on two key-parameters, egg size and fecundity. Egg volumes (in cubic millimetres) were calculated from diameters assuming that eggs are spherical. We here define the reproductive output by a female fish, i.e. maternal output (MO), as: MO = EV x FEC, where EV is the average volume of a single egg and

FEC is fecundity (i.e. the total number of eggs to be spawned per individual fish). Thus, MO is the total volume of eggs expressed in mm^3.

Table 1. Reproductive parameters of 59 marine fish species. Species are grouped according to their mode of oviposition, i.e. pelagic and demersal, and zoogeography, i.e. boreal and polar (Arctic and Antarctic combined). BL = body length (mm); ED = egg diameter (mm); EV = egg volume (mm^3); FEC = fecundity (no. ind^{-1}); MO = maternal output (mm^3 ind^{-1}); na = data not available. References to the reproductive parameters are in brackets following the species name

Species [Ref]	BL	ED	EV	FEC	MO
Pelagic spawners					
Boreal					
Trisopterus luscus [3]	150	0.97	0.48	207479	99187
Trisopterus luscus [3]	200	1.23	0.97	835997	814865
Trisopterus esmarki [3]	150	1.00	0.52	61686	32311
Trisopterus esmarki [3]	250	1.13	0.76	384190	290367
Odontogadus merlangus merlangus [3]	300	0.97	0.48	109358	52279
Odontogadus merlangus merlangus [3]	550	1.32	1.20	131712	158677
Merlangius merlangus [24]	300	1.16	0.82	na	na
Pollachius virens [3]	600	1.03	0.57	5000000	2861852
Pollachius virens [3]	900	1.22	0.95	8260000	7856426
Melanogrammus aeglefinus [3]	500	1.19	0.88	170000	150057
Melanogrammus aeglefinus [3]	750	1.67	2.44	1800000	4391242
Gadus morhua morhua [3]	500	1.13	0.76	2700000	2040632
Gadus morhua morhua [3]	800	1.65	2.35	9100000	21412073
Gadus morhua kildinensis [25]	525	1.00	0.52	na	na
Gadus morhua kildinensis [25]	628	1.75	2.81	na	na
Molva molva [9]	1000	1.05	0.61	na	na
Micromesistius poutassou [26]	303	0.72	0.20	122000	23852
Hippoglossoides platessoides[a] [14]	520	2.64	9.64	380000	3662359
Pleuronectes platessa [14]	1000	2.17	5.35	552000	2954498
Limanda limanda [14]	420	1.20	0.91	141000	127623
Platichthys flesus [14]	500	1.13	0.76	2000000	1511579
Polar – Antarctic					
Notothenia rossii [11]	510	4.50	47.73	19271	919829
Notothenia rossii [11]	780	5.00	65.48	99330	6503632
Notothenia coriiceps [11]	340	4.30	41.65	6897	287231
Notothenia coriiceps [11]	550	4.70	54.38	41024	2230987

Table 1. (continued)

Species [Ref]	BL	ED	EV	FEC	MO
Dissostichus eleginoides [11]	930	4.30	41.65	238314	9924769
Dissostichus eleginoides [11]	1310	4.70	54.38	545665	29674620
Champsocephalus gunnari [11]	340	3.50	22.46	3094	69485
Champsocephalus gunnari [11]	500	4.10	36.10	11664	421080
Polar – Arctic					
Boreogadus saida [own data]	237	1.57	2.03	16625	33700
Boreogadus saida [3]	120	1.55	1.95	30000	58517
Boreogadus saida [3]	160	1.84	3.26	67000	218622
Liopsetta glacialis [14]	350	1.40	1.44	125000	179663
Demersal spawners					
Boreal					
Mallotus villosus villosus [27]	153	1.20	0.91	11874	10747
Anarhichas lupus lupus [3]	1250	6.00	113.14	12740	1441414
Anarhichas minor [28]	630	2.70	10.31	4200	43302
Anarhichas minor [28]	1200	4.00	33.52	35200	1180017
Anarhichas denticulatus [3]	1250	6.00	113.14	46500	5261047
Myoxocephalus scorpius scorpius [3]	250	2.51	8.28	2742	22712
Artediellus camchaticus [29]	160	3.40	20.59	370	7617
Artediellus atlanticus [30]	59	na	na	79	na
Artediellus atlanticus [30]	75.5	na	na	77	na
Artediellus atlanticus [30]	97.5	na	na	156	na
Triglops murrayi [15]	122	2.00	4.19	450	1886
Agonus cataphractus [3]	200	2.23	5.81	2400	13941
Liparis liparis [3]	275	1.67	2.44	na	na
Liparis montagui [3]	125	1.19	0.88	793	700
Paraliparis abyssorum [31]	130	3.00	14.14	na	na
Lycodes esmarkii [15]	560	5.50	87.15	1888	164534
Lycodes vahlii gracilis [15]	254	5.00	65.48	100	6548
Cyclopterus lumpus [3]	600	2.60	9.21	194112	1787055
Lumpenus lampretaeformis[b] [15]	258	0.80	0.27	700	188
Leptoclinus maculatus maculatus [15]	128	1.50	1.77	920	1626
Polar – Antarctic					
Pleuragramma antarcticum [32]	154	1.80	3.05	4315	13181
Pleuragramma antarcticum [32]	256	2.00	4.19	17774	74480
Dissostichus mawsoni [11]	1200	4.00	33.52	471900	15819598
Dissostichus mawsoni [11]	1800	4.30	41.65	1343380	55946090
Harpagifer antarcticus [11]	110	2.50	8.18	400	3274
Harpagifer antarcticus [11]	110	2.50	8.18	1000	8184
Nototheniops nudifrons [11]	115	2.00	4.19	544	2280

Table 1. (continued)

Species [Ref]	BL	ED	EV	FEC	MO
Nototheniops nudifrons [11]	191	2.50	8.18	3110	25453
Pagothenia borchgrevinki [11]	165	2.40	7.24	1500	10862
Pagothenia borchgrevinki [11]	235	na	na	3000	na
Trematomus bernacchii [11]	200	3.60	24.44	1200	29326
Trematomus bernacchii [33]	288	3.80	28.74	na	na
Trematomus hansoni [12, 33]	320	3.10	15.60	9600	149803
Trematomus hansoni [12, 33]	440	3.60	24.44	9600	234609
Polar - Arctic					
Eleginus navaga [3]	180	1.50	1.77	20000	35357
Eleginus navaga [3]	200	1.90	3.59	62000	222750
Triglops pingeli [15]	140	2.25	5.97	430	2566
Icelus spatula [15]	90	2.00	4.19	190	796
Gymnocanthus tricuspis [15]	137	1.00	0.52	4710	2467
Leptagonus decagonus [15]	165	1.50	1.77	627	1108
Liparis gibbus [34]	135	1.13	0.76	4000	3023
Liparis gibbus [34]	180	1.70	2.57	12000	30881
Liparis fabricii [15]	90	0.85	0.32	300	97
Careproctus reinhardti [15]	170	4.50	47.73	154	7351
Careproctus micropus [35]	69	3.00	14.14	na	na
Careproctus ranula [35]	54	3.50	22.46	na	na
Rhodichthys regina [35]	310	7.00	179.66	na	na
Lycodes rossii [15]	280	2.50	8.18	387	3167
Lycodes pallidus pallidus [15]	179	4.00	33.52	42	1408
Lycodes reticulatus [15]	260	2.00	4.19	242	1014
Cottunculus microps [15]	175	5.00	65.48	145	9494
Eumicrotremus spinosus [15]	102	3.00	14.14	1187	16787

[a]Subspecies *limandoides*
[b]Subspecies *lampretaeformis*

The dependence of the reproductive parameters (i.e. egg size, fecundity and maternal output) on body length and zoogeography (polar vs boreal) was tested by means of analysis of variance using zoogeography as a fixed class variable and body length as a continuous variable (Table 2). The GLM procedure in SAS was used to estimate F-values and significance levels [16]. To meet the requirements of normality, fecundity and maternal output were log-transformed for the variance analyses. The difference between maternal outputs of pelagic vs demersal spawning species was analyzed using Student's t-test for the total sample set and for fish arbitrarily chosen as larger or smaller than 300 mm in body length.

Results

More than one data set was available for some species (Table 1). For clarity we have not identified the species in Figs. 1 to 4.

Egg Size

Egg size seems to be a characteristic of both zoogeography and mode of oviposition (Fig. 1; Table 2). Within the group of pelagic spawning species, polar species produce significantly larger eggs than boreal species. By contrast, the size of demersal eggs does not differ between zoogeographic regions (Table 2). Further, egg size generally increases with body length irrespective of zoogeography and mode of oviposition. The boreal pelagic spawners form an exception. For this group of species, egg size is consistently small at all body lengths (Fig. 1). Thus, 1) egg size usually increases with increasing body length and 2) egg sizes of pelagic spawning species are largest towards the poles, whereas 3) substrate spawners produce similar sized eggs irrespective of zoogeography.

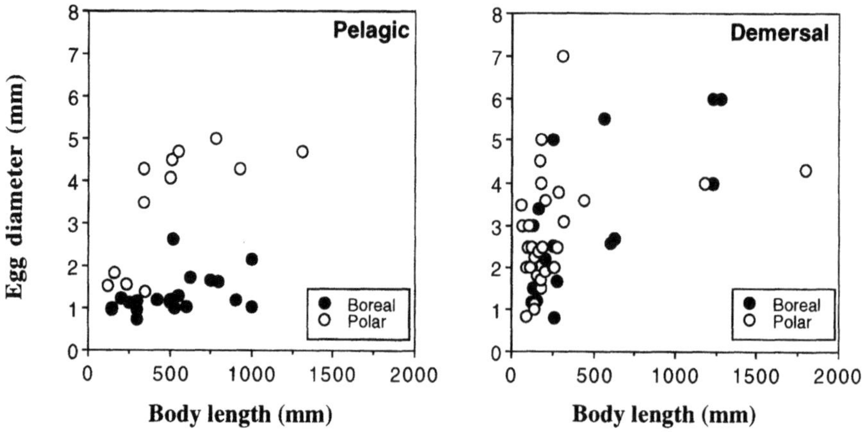

Fig. 1. Scatter plots showing egg diameters and body lengths of boreal and polar fish species with pelagic and demersal spawning modes.

Fecundity

Fecundity increases with body length irrespective of zoogeography and mode of oviposition (Fig. 2; Table 2). Further, among pelagic spawners the polar species are significantly less fecund at a given body length than boreal species. This contrasts with substrate spawners in which fecundity does not differ between polar and boreal species (Table 2).

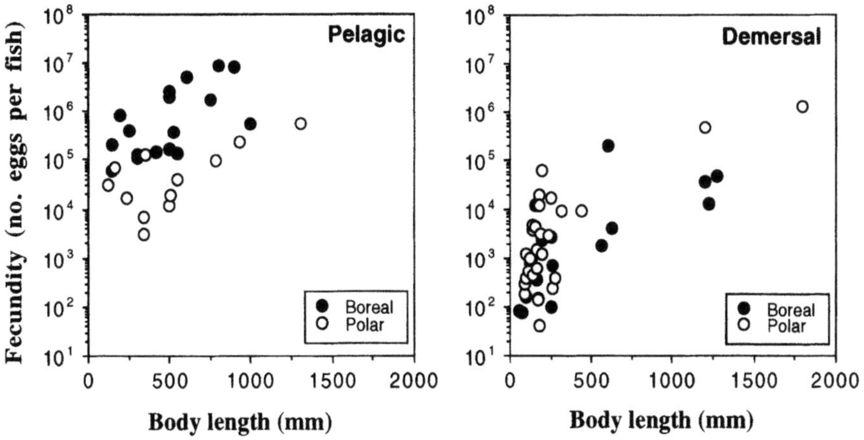

Fig. 2. Scatter plots showing fecundities and body lengths of boreal and polar fish species with pelagic and demersal spawning modes. Note the logarithmic scale for fecundity

Maternal Output

Maternal output, i.e. the product of individual egg volume and fecundity, increases with body length irrespective of zoogeography and mode of oviposition (Fig. 3; Table 2). Moreover, there are no zoogeographic differences in maternal output within modes of oviposition.

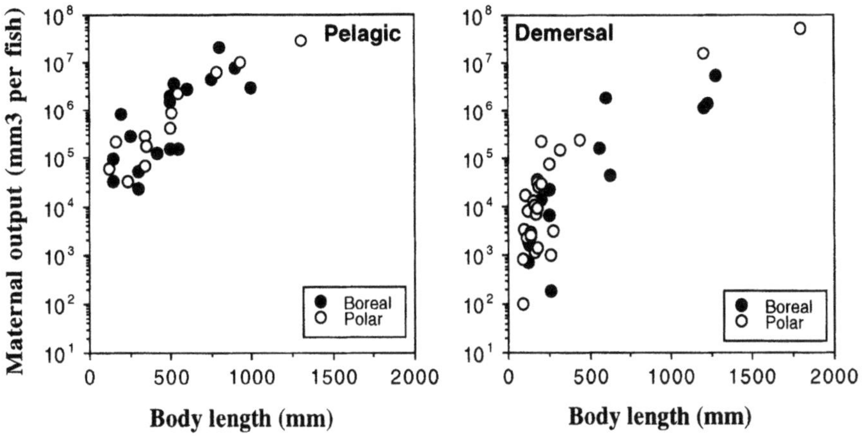

Fig. 3. Scatter plots showing maternal outputs and body lengths of boreal and polar fish species with pelagic and demersal spawning modes. Maternal output is defined as the product of individual egg volume and number of eggs. Note the logarithmic scale for maternal output

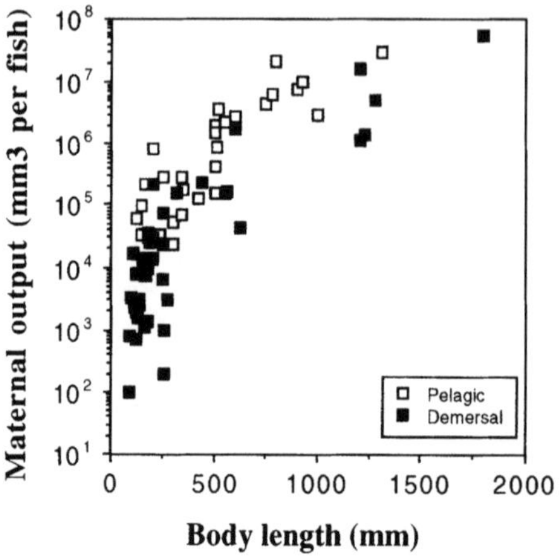

Fig. 4. Scatter plots showing maternal outputs and body lengths of pelagic and demersal spawning fish species (boreal and polar combined). Note the logarithmic scale for maternal output

However, when the maternal output between modes of oviposition, i.e. pelagic vs demersal spawners, is compared (Fig. 4), pelagic spawners have a significantly higher maternal output than substrate spawning species ($p < 0.001$, t-test), albeit only fish smaller than about 300 mm in body length contribute to this difference (for fish larger than 300 mm in body length $p = 0.897$).

Table 2. Analysis of variance for the effects of body length and zoogeography (boreal or polar) on reproductive parameters. F-values and significance levels (p) are given. ns = not significant

Reproductive parameters	Body length *F*	*p*	Boreal/Polar *F*	*p*
Pelagic spawners				
Egg diameter	14.75	0.0006	60.48	0.0001
Fecundity	17.44	0.0003	35.51	0.0001
Maternal output	61.93	0.0001	0.05	ns
Demersal spawners				
Egg diameter	16.09	0.0002	0.11	ns
Fecundity	46.46	0.0001	2.75	ns
Maternal output	91.96	0.0001	1.11	ns

Discussion

Information regarding the modes of oviposition in polar fish is sparse, but most members of the Antarctic Nototheniidae are considered to be substrate spawners. For this family, the development of fully mature ova seems to require two or more successive seasons, and eggs are probably deposited in one batch annually thereafter [11,17].

Morphological characteristics of egg-shells are species-specific and developing pelagic eggs have been identified from a few species i.e. *Notothenia rossii* and *Notothenia coriiceps* [18]. Other species such as *Champsocephalus gunnari* and *Dissostichus eleginoides* may also produce pelagic eggs [18] and they have, therefore, been considered as being pelagic spawners in this study (Table 1). On the other hand, *Pleuragramma antarcticum* (G. Hubold, pers. comm.) and *Dissostichus mawsoni* have been classified as demersal spawners although the latter have many features in common with the pelagic spawning species, i.e. a high fecundity and maternal output (Table 1).

Knowledge about the mode of oviposition among Arctic species is also sparse. Most species are, however, believed to be substrate spawners [3]. Information is available for only two pelagic spawning species, *Boreogadus saida* and *Liopsetta glacialis* [3,14]. The eggs of *Boreogadus saida* are found in the uppermost part of the water column (own unpubl. data) and often develop in close association with the Arctic sea-ice [19]. One could, therefore, argue that *Boreogadus saida* is, in reality, a substrate spawner, with sea-ice providing the substrate. Similar spawning traits may apply for the Arctic gadoid species viz *Arctogadus glacialis* and perhaps also *Arctogadus borizovi* [13].

Egg size and fecundity, and hence maternal output, may show a high degree of intraspecific variation, which is modulated by ecological and environmental conditions [20]. Further, estimations of egg size and fecundity may be biased if samples are taken from ovaries in which the oocytes are not fully developed (K.-H. Kock, pers. comm.). Although the above mentioned uncertainties and limitations should be borne in mind, we would, nevertheless, suggest the following general reproductive traits for polar and boreal marine fish species:

- Polar fish produce relatively few, large eggs when compared with their boreal counterparts, but this trait appears to apply only among pelagic spawning species.
- Substrate spawning species produce relatively few, large eggs irrespective of zoogeographic regions.

- For a given mode of oviposition, maternal output does not differ between polar and boreal species.
- Maternal output is significantly higher in species (body length < 300 mm) with pelagic than in those with demersal oviposition. It should be noted that there need not be an obvious relationship between maternal output and the energetic costs of reproduction [21]. Maternal output, as we define it, merely reflects the volume of eggs spawned without any consideration of the energetic costs involved in e.g. the gonad development (very long among Antarctic species), the energetic content of ova (usually higher in demersal eggs) [22], spawning behavior and, among some species, parental care (see [23] for a comprehensive discussion). However, one might expect that a high maternal output would also require a relatively large amount of energy to be invested.

In conclusion, maternal output is linked to mode of oviposition in phylogenetically distant species and not directly with latitude (i.e. boreal vs polar). However, maternal output of polar fish may be lower than that of boreal species. This is not due to the low temperatures of the habitat per se, but rather because many polar species are small-sized and have demersal oviposition. Consequently, our results corroborate with the suggestion made by Marshall [4], that behavioral and ecological factors seem to be the major determinants governing reproductive traits of marine teleost fish.

Acknowledgments

We wish to thank Prof. A. Clarke, Dr. G. Hubold, Prof. M. Jobling, Dr. K.-H. Kock, Dr. A.W. North and Dr. M.G. White for valuable comments.

References

1. Economou AN (1991) Is dispersal of fish eggs, embryos and larvae an insurance against density dependence? Env Biol Fish 31: 313-321
2. Wootton RJ (1994) Life histories as sampling devices: optimum egg size in pelagic fishes. J Fish Biol 45: 1067-1077
3. Rass TS (1941) Analogous or parallel variations in structure and development of fishes in northern and Arctic Seas. Jubilee Publications of the Moscow Society of Naturalists 1805-1940: 1-60 (in Russian, summary in English)
4. Marshall NB (1953) Egg size in Arctic, Antarctic and deep-sea fishes. Evolution 7: 328-341

5. Blaxter JHS (1988) Pattern and variety in development. In: Hoar WS, Randall DJ (eds) Fish physiology, Vol 11A. Academic Press, New York, pp 1-58
6. Kock K-H (1992) Antarctic fish and fisheries. Cambridge University Press, Cambridge
7. Marteinsdottir G, Able KW (1992) Influence of egg size on embryos and larvae of *Fundulus heteroclitus*. J Fish Biol 41: 883-896
8. Ursin E (1984) The tropical, the temperate and the Arctic Seas as media for fish production. Dana 3: 43-60
9. Blaxter JHS (1969) Development: eggs and larvae. In: Hoar WS, Randall DJ (eds) Fish physiology, Vol 3. Academic Press, New York, pp 177-252
10. Tyler CR, Sumpter JP (1996) Oocyte growth and development in teleosts. Rev Fish Biol Fisheries 6: 287-318
11. Kock K-H, Kellermann A (1991) Reproduction in Antarctic notothenioid fish. Antarct Sci 3: 125-150
12. Hubold G (1992) Zur Ökologie der Fische im Weddellmeer. Ber Polarforsch 103: 1-157
13. Andriyashev AP, Chernova NV (1995) Annotated list of fishlike vertebrates and fish of the Arctic Seas and adjacent waters. J Ichthyol 35: 81-123
14. Rass TS (1992) Metamerism of backbone and size of eggs in specimens of fish from northern seas. Doklady Akademii Nauk 324: 1131-1135 (in Russian)
15. Dolgov AV (1995) Some questions on the biology of non-commercial fish species in the Barents Sea. In: The problems of fishery science in creation of young scientists. PINRO, Murmansk: 69-94 (in Russian)
16. SAS Institute Inc. (1985) SAS user's guide: statistics, version 5 edition. SAS Institute, Cary NC
17. Everson I (1977) The living resources of the Southern Ocean. FAO, GLO/SO/77/1
18. White MG, Veit RR, North AW, Robinson K (1996) Egg-shell morphology of the Antarctic fish, *Notothenia rossii* Richardson, and the distribution and abundance of pelagic eggs at South Georgia. Antarct Sci 8: 267-271
19. Rass TS (1968) Spawning and development of polar cod. ICES Rapp P-v Reun 158: 135-137
20. Daoulas Ch, Economou AN (1986) Seasonal variation of egg size in the sardine, *Sardina pilchardus* Walb., of the Saronikos Gulf: causes and a probable explanation. J Fish Biol 28: 449-457
21. Wootton RJ (1985) Energetics of reproduction. In: Tytler P, Calow P (eds) Fish energetics. New perspectives. Croom Helm Ltd., Great Britain, pp 231-254
22. Kamler E (1992) Early life history of fish. An energetic approach. Chapman and Hall, London

23. Clarke A (1987) Temperature, latitude and reproductive effort. Mar Ecol Prog Ser 38: 89-99
24. Hislop JRG (1984) A comparison of the reproductive tactics and strategies of cod, haddock, whiting and Norway pout in the North Sea. In: Potts GW, Wootton RJ (eds) Fish reproduction: strategies and tactics. Academic Press, London, pp 311-329
25. Tseyeb RY, Pozdnyakov YuF (1975) Ecology of the cod from Kildin Island: reproduction. In: The relict Lake Mogilnoye. Nauka, Leningrad, pp 227-247 (in Russian)
26. Mazhirina GP (1994) Reproductive charateristics of blue whiting, *Micromesistius poutassou*, in the northeastern Atlantic. J Ichthyol 34: 132-139
27. Pozdnyakov YuF (1957) On the fecundity of the Barents Sea capelin. Doklady Akademii Nauk SSSR 112: 777-778 (in Russian)
28. Gusev EV (1995) Individual fecundity of wolffish genus *Anarhichas* in the Barents Sea. In: The problems of fishery science in creation of young scientists. PINRO, Murmansk: 252-270 (in Russian)
29. Tokranov AM (1988) Some aspects of the biology of the Kamchatkan sculpin, *Artediellus camchaticus*, on the eastern coast of Kamchatka. Voprosy Ikhtiologii 3: 415-420 (in Russian)
30. Dorrien CF von (1996) Reproduction and larval ecology of the Arctic fish species *Artediellus atlanticus* (Cottidae). Polar Biol 16: 401-407
31. Andriyashev AP, Chernova NV (1997) Two new species of liparid fishes (*Liparidae, Scorpaeniformes*) from the abyssal part of the eastern North Atlantic. J Ichthyol 37: 479-484
32. Gerasimchuck VV (1987) On the fecundity of Antarctic sidestripe, *Pleurogramma antarcticum*. Voprosy Ikhtiologii 5: 858-860 (in Russian)
33. Vacchi M, Williams R, La Mesa M (1996) Reproduction in three species of fish from the Ross Sea and Mawson Sea. Antarct Sci 8: 185-192
34. Chernova NV (1987) About breeding of hump-backed snailfish *Liparis gibbus* Bean (Scorpaeniformes, Liparidae). In: Ecology-morphological investigations of commercial fishes in the northern basin. Nauka, Leningrad, pp 87-94 (in Russian)
35. Chernova NV (1991) Snailfishes (Liparidae) from the Eurasian Arctic. Apatity (in Russian)

Development, Dispersal and Recruitment: A Paradox for Survival Among Antarctic Fish

Martin G. White

British Antarctic Survey, NERC, High Cross, Madingley Road, Cambridge CB3 0ET, UK

Introduction

At first sight, the life-history strategies adopted by Antarctic fish appear to be a paradox. The majority of fish species are demersal and inhabit the peri-Antarctic islands that are isolated, separated by deep ocean, and subject to the influence of strong currents of the West Wind Drift. In common with Antarctic benthic invertebrates, nearly all of the Antarctic fish have protracted early development but by contrast most Antarctic fish have retained a pelagic larval phase that extends over several months.

The presence of a protracted pelagic phase vulnerable to losses by advection from favourable habitats seems inconsistent with a successful survival strategy.

Antarctic Fish Fauna Composition and Life-History

The Antarctic fish fauna is unusual because it exhibits low species diversity but has high levels of endemism. There are some 274 recognized species of fish that inhabit the Southern Ocean and more are being described, as the less well researched taxa are re-examined. The majority (>71%) are bottom dwelling and this demersal group is dominated in terms of numbers and biomass by the sub-order Notothenioidei [1]. Many notothenioids are endemic and confined to specific island groups or show circum-continental distribution patterns [2].

Notothenioids lack swim bladders, most are demersal and are dependent upon shelf habitats as adults. Only a few species have become adapted to the pelagic system.

The life-history is fully known only for a small number of species but it is generally accepted that most Antarctic notothenioids are characterized by delayed maturation, have low relative fecundity, and produce large yolky eggs [3,4]. The majority of species spawn demersally. The eggs may remain benthic and the adults of some species are known to protect the egg mass while other species have pelagic eggs [5].

G. di Prisco, E. Pisano, A. Clarke (Eds)
Fishes of Antarctica. A biological overview
© Springer-Verlag Italia 1998

Incubation from spawning to hatching is often protracted (>100 days) [6,7]. Once hatched, the larvae are usually large (>12 mm) and develop pelagically over several months before recruiting to the demersal juvenile population [8,9]. Some species have a specialist neustonic juvenile phase or have a juvenile morphology adapted for a pelagic habit, such as the post-larval and juvenile stages of many channichthyids.

Affinities

Despite the apparent potential for extensive dispersal because of the opportunities for distribution of the pelagic life-history stages on ocean currents, endemism is high and each peri-Antarctic island group has a discrete species composition and therefore constraints over dispersal must prevail. Biogeographical studies on Antarctic fish clearly divide the Southern Ocean ichthyofauna into geographically separated regions [1,10].

There is evidence for the influence of the West Wind Drift to disperse species eastwards (cf. [3,10]). Commonly, the closest affinities are between neighbouring island groups.

Physical Environment

The Southern Ocean is well known for low ambient temperatures and seasonal sea-ice development; however, the most striking feature is the oceanographic current system that is dominated by the circum-continental West Wind Drift, that influences most of the peri-Antarctic islands, and its complementary counter current adjacent to the Continent. Major gyres are well known, particularly those in the Weddell and Ross Seas, and these are strongly implicated in the life-history of euphausiids [11] and some high Antarctic fish e.g. *Pleuragramma antarcticum* [12].

The largely circum-polar East Wind Drift and the near continuous shelf habitats around the Antarctic continent offer few barriers to pelagic dispersal and juvenile recruitment among high-Antarctic species. Here, high-Antarctic species exhibit circum-continental distribution patterns with little regionalization. By contrast, the peri-Antarctic island groups are separated by deep ocean and great distances and these result in considerable barriers to colonization. All are strongly influenced by the West Wind Drift.

South Georgia is a typical example. The Discovery Expedition [13] demonstrated that strong currents approach South Georgia from the south east and south west, flow around the shelf and leave the island on the northern side. Studies undertaken more recently [14] support the major features of the large- and meso-scale circulation patterns. The fine-scale circulation patterns have been observed but the temporal and spatial persistence of these has yet to be described in detail.

The flow of a strong current past a sea-mount or island will cause the development of gyres and eddies. An island such as South Georgia with a shelf that is heavily dissected by glacial canyons will also be expected to have eddies associated with these features. Such topographically generated eddies should act as retention zones for the local zooplankton communities.

The Paradox

The fish populations are dependent upon the continental shelf as nursery grounds for juveniles and habitat for adults. These shelf environments particularly those of islands are subject to the strong current systems.

Benthic Antarctic invertebrates have usually been strongly selected for species that have reduced or eliminated the pelagic larval phase [15]. Demersal fish, on the other hand, have maintained an extended pelagic larval phase - even though most species have low relative fecundity and are dependent on shelf habitats as adults. The strong current systems should advect the eggs and larval stages away and cause substantial losses.

Therefore the maintenance of a protracted pelagic early life-history among the dominant groups of Antarctic fish in a strongly advective oceanic system appears to be inconsistent with a successful survival strategy.

Is There Evidence for Dispersal or Retention of Pelagic Stages?

The early juvenile stages of some demersal fish species are found widely dispersed away from the shelf habitats over which they were spawned [6], but these are the exception because the eggs and larvae of the majority of species are usually found concentrated in neritic waters over the continental shelf [5]. Examples of the distribution patterns of the larvae of demersal shelf species at South Georgia, South Shetland Islands, Kerguelen and the Weddell Sea [11,16-19] show that their distribution is largely over the continental shelf and adjacent waters - with a relatively well defined junction between shelf and oceanic species at the shelf/slope junction.

For example, Fig. 1 shows the distribution patterns for eggs and larvae of common demersal nototheniids, a channichthyid, a muraenolepidid as well as an oceanic paralepidid, *Notolepis coatsi* Dollo collected during a

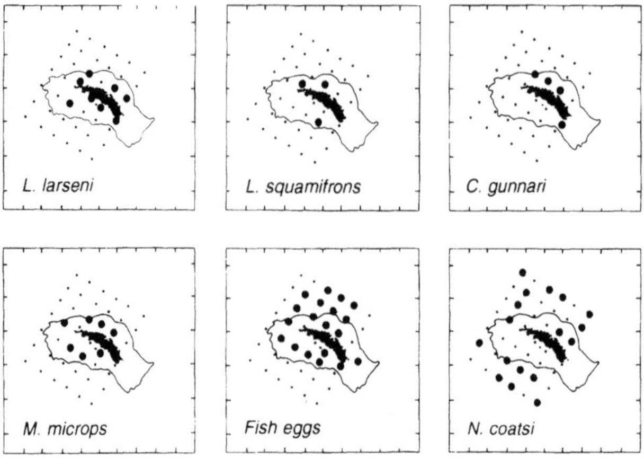

Fig. 1. Examples of ichthyoplankton distribution at South Georgia during BAS Offshore Biological Programme Cruise JB04 using RMT8 nets. Small circles indicate sample sites, large circles indicate samples sites at which each species occurred. Solid line = 500-m isobath

routine zooplankton survey at South Georgia. The distribution patterns of the most passive stages such as pelagic eggs did show some off shelf dispersal conforming with the known local current circulation patterns. However, observations on the distribution patterns of larval stages of demersal species clearly demonstrate that these occurred in neritic water over the continental shelf. By contrast, the larval stages of oceanic fish, such as the myctophids and paralepidids, were found to be mostly distributed beyond the shelf/slope junction.

Experiments to examine the distribution and abundance of larval fish at South Georgia were undertaken during research cruises by the British Antarctic Survey (BAS) Offshore Biological Programme in 1987 and 1991 by sampling along transects across the north shelf (Fig. 2). The results of these experiments supported the conclusions that fish larvae were mostly found over the shelf and their abundance fell with increased distance from the coast (Fig. 3).

Similar observations have been made by French research workers studying ichthyoplankton at Kerguelen. The eggs and larval stages of demersal notothenioids were distributed in neritic water over the continental shelf and nearby sea mounts while the larvae of mesopelagic species were most abundant in oceanic water [16].

Is it clear from observations of the distribution patterns of fish eggs and larvae at the Kerguelen, South Georgia and the South Shetland Islands [11,16-19] that although there is some evidence that pelagic eggs of demersal species may be dispersed into the open ocean [5,20], most larval stages do not appear to be rapidly advected off the shelves of the peri-Antarctic islands but are retained in neritic habitats.

Post-larval and early juvenile fish may well have sufficient locomotory ability to maintain their position in areas of favourable conditions but eggs and early larvae are essentially passive particles for several months after spawning. The mechanisms by which these early stages are distributed must by influenced by the prevailing physical conditions.

The observation that the larval stages of demersal and oceanic fish are separated by a junction coincident with the shelf/slope edge suggests some

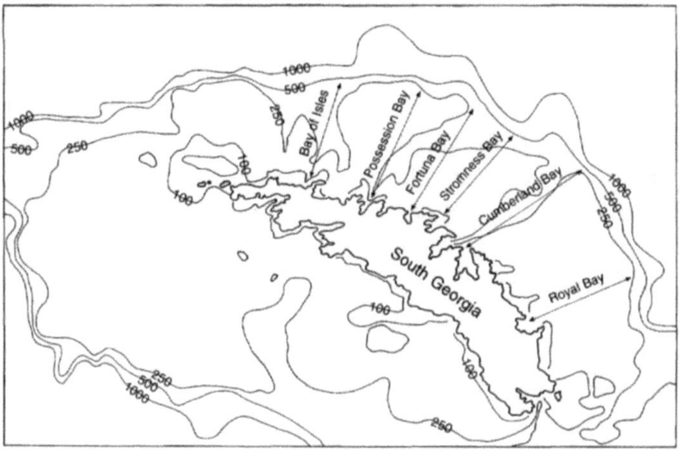

Fig. 2. Positions of transects occupied on the northern shelf of South Georgia during austral summer 1987 and 1991

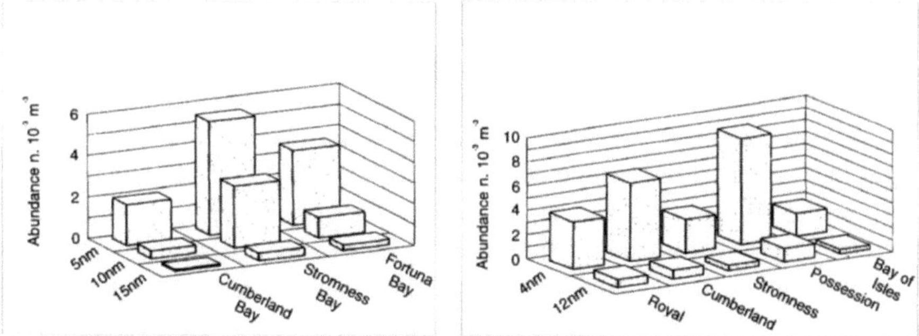

Fig. 3. Histograms of total ichthyoplankton abundance collected after dark on the northern shelf of South Georgia during BAS Offshore Biological Programme during austral summer of 1987 and 1991. Illustration represents view from seaward end of transects

form of physical barrier at the shelf/slope break or mechanisms that retain the demersal species near to the continental shelf.

The development of lee gyres at oceanic islands has been established in the oceanographic literature for some years [21]. More recently the importance of such features to biological processes has been recognized, for example, Gilmartin and Revelante [22] describe the elevated production resulting from eddy formation at Hawaii, and Heywood et al. [23] describe the development, maintenance and the influence on productivity of lee gyres at Aldabra in the Indian Ocean.

Evidence for lee gyres and the development of a near-shelf eddy system at South Georgia is given by Latogursky et al. [24] during their studies about the variations in abundance and distribution of Antarctic Krill, *Euphausia superba* Dana, at South Georgia. They also showed that the development of local eddy systems varied between years and so inferred a mechanism for inter-annual variations in the aggregation of krill at South Georgia. These features, though not stated by the authors [24], will also serve to retain neritic plankton displaced off the shelf.

The presence and development of shelf-break frontal systems in the Antarctic has thus far been inferred from the biological data such as the ichthyoplankton/zooplankton distribution patterns in the Weddell Sea [17], Antarctic Peninsula [18] and at South Georgia ([9,19,25], White and North personal observations).

Brandon and others [26] working with the most recent oceanographic data from British Antarctic Survey research cruise JR11 on "RRS James Clark Ross" have shown strong evidence of a shelf-break front on the northern shelf of South Georgia during January 1996. Data collected from ca. 80 km onshore-offshore transects at right angles to the coast demonstrated that warm low salinity water over the shelf was separated from colder more saline off-shelf water by a well defined hydrographic front more or less coincident with the margin of the continental shelf.

Thus there is evidence of mechanisms that would help to maintain near passive particles over or near the shelf at South Georgia and retain fish eggs and larvae in favourable conditions for their early development.

Indeed it is clear that this *must* be so if the species such as Antarctic notothenioids; with protracted pelagic phases in their early life-history, relative low fecundity and a dependence on shelf habitats, are to be successful. Therefore, although the life-history strategies adopted by Antarctic fish appear to be counter-intuitive with successful survival, we can show that there are environmental mechanisms which permit an extended pelagic component to be retained. The fact that a vulnerable pelagic phase is retained suggests that it is important as a successful adaptation to the polar environment.

Consequences for Recruitment and Fisheries

Notothenia rossii is reported to have been the major component of the commercial-scale catches in the Southern Ocean during the late 1960s and early 1970s [4]. Subsequent to this period little targeted fishing on this species has occurred. Despite this, the species has not returned to its former abundance. The lack of a recovery is a puzzle and implies that factors controlling the population dynamics of Antarctic fish are not fully understood. The larval phase is considered to be the most vulnerable stage in the life-history of any fish species [27].

Observations of the occurrence and relative abundance of larval fish on the shelf at South Georgia demonstrate that species occurrence and abundance are highly variable between years (Table 1). The species composition and species dominance varied markedly between years.

The occurrence, the abundance of young-of-the-year ('O' group) and the subsequent recruitment to the juvenile population will be influenced by many factors such as spawning success, feeding success, larval mortality due to predation, and dispersal from favourable habitats. The relative importance of such factors is not known for any Antarctic fish population but each factor will have some influence on recruitment.

Table 1. Inter-annual variation of larval fish composition and abundance during summer (January) on northern shelf at South Georgia from RMT8 net catches

Species	1988	1991	1993
Lepidonotothen larseni	23033	7	3
Trematomus hansoni	1552		623
Gobionotothen gibberifrons	1275	7	803
Champsocephalus gunnari	1220	24	104
Chaenocephalus aceratus	103	31	
Artedidraco mirus	84		9
Electrona antarctica	74		1
Pseudochaenichthys. georgianus	57		1
Psilodraco breviceps	34		4
Parachaenichthys georgianus	27		
Racovitzia glacialis	25		
Harpagifer georgianus	7		25
Notolepis coatsi	6		1
Lepidonotothen nudifrons		110	7
Notothenia rossii			20
Total larvae	27397	179	1602
Total species	13	5	12
Abundance = n. 10^{-6} m^{-3}			

It is likely that the maintenance of larvae in retention zones is very important to the survival of species with a protracted pelagic phase in their early life-history. It has been shown that wind-sheer has an important effect on fine-scale oceanographic features because at speeds above a critical wind velocity the lee gyre is disrupted and eddies detach, and with them parcels of neritic water and the entrained plankton are advected away from the island shelf [23]. Therefore the maintenance and longevity of the shelf-break front and near shelf gyres will have a major influence on the success or failure of recruitment in any given season and may be a primary forcing factor resulting in the large inter-annual variations in larval abundance. This process will influence the recruitment and demography of fish populations at all the peri-Antarctic islands because of the importance of retaining pelagic larvae in suitable habitats to complete their development and recruitment to the juvenile phase in their early life-history.

The combination of intermittent recruitment due to physical factors as described above, competition with species that have shorter life cycles and faster growth rates, and by-catch effects, are likely to be factors delaying the re-establishment of fish stocks depleted by commercial-scale fishing.

Consequences for Genetic Exchange and Affinities

Many life-history characteristics among Antarctic fish will influence the exchange of genetic information. Gene-flow is likely to be restricted by: geographical isolation of populations, the dependence of shelf habitats, delayed maturation, low fecundity, annual spawning, and the potential losses of pelagic eggs and larvae by advection. On the other hand, gene-flow will be promoted by: the protracted development of eggs, larvae and juveniles, prolonged development in pelagic habitats as well as the longevity and eurybathyal distribution of adults. The balance between the factors restricting and promoting dispersal will be important in determining the levels of genetics exchange between populations of Antarctic fish.

Studies on the genetic affinities among species with wide but disjunct distributions and the distribution of pelagic early life history stages need to be addressed to progress an understanding of the population dynamics and associations of Antarctic fish. The dependence and duration within pelagic habitats during each stage of a species life-history will have fundamental implications for resolving such questions as genetic exchange, stock identity and species affinity among Antarctic fish.

An understanding of the detailed characteristics of the early life-history of Antarctic fish - particularly in relation to temporal and spatial distribution processes - and the relative influence of biological events and physical factors are central to evaluating the survival strategies of the Southern Ocean ichthyofauna.

References

1. Eastman JT (1993) Antarctic fish biology: evolution in a unique environment. Academic Press, San Diego, CA
2. Andriashev AP (1987) A general review of the Antarctic bottom fish fauna. In: Kullander SO, Fernholm B (eds) Proceedings of the Fifth Congress of European Ichthyologists, 1985. Swedish Museum of Natural History, Stockholm, pp 357-372
3. Everson I (1984) Fish biology. In: Laws RM (ed) Antarctic ecology. Academic Press, London, pp 491-532
4. Kock K-H (1992) Antarctic fish and fisheries. Cambridge University Press, Cambridge
5. White MG, Veit RR, Robinson K, North AW (1996) Egg shell morphology of the Antarctic fish, *Notothenia rossii* Richardson, and the distribution and abundance of pelagic eggs at South Georgia. Antarct Sci 8:267-271
6. White MG, North AW, Twelves EL, Jones S (1982) Early development of *Notothenia neglecta* from the Scotia Sea, Antarctica. Cybium 6:43-51
7. White MG, Burren P (1992) Reproduction and larval growth of *Harpagifer antarcticus* Nybelin (Pisces, Notothenioidei). Antarct Sci 4:421-430
8. Kock K-H, Kellermann A (1991) Reproduction in Antarctic notothenioid fish. Antarct Sci 3:125-150
9. Loeb VJ, Kellermann AK, Koubbi P, North AW, White MG (1993) Antarctic larval fish assemblages: a review. Bull Mar Sci 53:416-449
10. DeWitt HH (1971) Coastal and deep-water benthic fishes of the Antarctic. Washington, American Geophysical Union. Ant Map Fol Ser, No 15, pp 10
11. Hempel G (1987) Krill-dominated pelagic system of the Southern Ocean. Env Int 13:33-36
12. Hubold G (1984) Spatial distribution of *Pleuragramma antarcticum* (Pisces: Nototheniidae) near the Filchner and Larsen ice shelves (Weddell Sea/Antarctica). Polar Biol 3:231-236
13. Hardy AC, Gunther ER (1935) The plankton of the South Georgia whaling grounds and adjacent waters, 1926 - 1927. Disc Rep 11:1-456
14. Webb DT, Killworth PD, Coward AC, Thomson SR (1991) The FRAM atlas of the Southern Ocean. NERC, Swindon
15. White MG (1984) Marine benthos. In: Laws RM (ed) Antarctic ecology. Academic Press, London and New York, pp 421- 461
16. Koubbi PG, Duhamel G, Camus P (1990) Early life stages of Notothenioidei from Kerguelen Islands. Cybium, 14:225-250

17. White MG, Piatkowski U (1993) Abundance, horizontal and vertical distribution of fish in the eastern Weddell Sea micronekton. Polar Biol 13:41-53
18. White MG, North AW (1987) Postlarval Nototheniodei and midwater fish collected during the SIBEX cruise by British Antarctic Survey, 1985. In: Kullander SO, Fernholm B (eds), Proceedings of the Fifth Congress of European Ichthyologists, Stockholm 1985, Swedish Museum of Natural History, Stockholm. pp 405-411
19. North AW (1988) Distribution of fish larvae at South Georgia: horizontal, vertical and temporal distribution and early life-history relevant to monitoring year-class strength and recruitment. SC-CCAMLR selected scientific papers 1987 Hobart Tasmania CCAMLR, pp 105-141
20. Kellermann A (1991) Egg and larval drift of the Antarctic fish *Notothenia coriiceps*. Cybium, 30:199-210
21. Mann KH, Lazier JRN (1991) Dynamics of marine ecosystems. Blackwell Scientific Publications, Oxford
22. Gilmartin M, Revelante N (1974) The islands' 'mass' effect on the phytoplankton and primary production of the Hawaiian islands. J Exp Mar Biol Ecol 16:181-204
23. Heywood KJ, Stevens DP, Bigg GR (1996) Eddy formation behind the tropical island of Aldabra. Deep-Sea Res 43:555-578
24. Latogursky VI, Makarov RR, Maklygin LG (1991) Distribution, biomass and abundance of the *Euphausia superba* fishery around South Georgia (sub-area 48.3). SC-CCAMLR Selected scientific papers 1990, Hobart Tasmania CCAMLR 1991, pp 123-127
25. Atkinson A, Ward P, Peck JM, Murray AWA (1990) Mesoscale distribution of zooplankton around South Georgia. Deep-Sea Res 37:1213-1227
26. Brandon MA, Murphy EJ, Whitehouse MT, Heywood KJ, Trathan PN, Bone DG, Priddle J. The shelf break front to the east of South Georgia. Cont Shelf Res (in press)
27. Cushing DH (1975) Marine ecology and fisheries. Cambridge University Press, Cambridge

The Pelagic Fish Community of the Polar Frontal Zone off the Kerguelen Islands

Guy Duhamel

Muséum National d'Histoire Naturelle, Laboratoire d'Ichtyologie générale et appliquée, 43 rue Cuvier, 75231 Paris Cedex 05, France

Introduction

Because of the clearly observed increase of the majority of fur seal and penguin populations in the Southern Ocean new developments in marine studies are attempting to quantify the impact of predators on marine resources. The predation in the subAntarctic zone of this ocean is directed mainly to the mesopelagic fish as noted from studies at the breeding sites of predators. Guinet et al. [1], studying the Indian sector of the Southern Ocean, have estimated that the annual consumption of marine seabirds and fur seals off the Kerguelen Islands was about 1.8×10^6 tonnes during 1985, including 26% (0.46×10^6 tonnes) of Myctophidae. At that time they considered that the king penguin (*Aptenodyptes patagonicus*) population (173 000 pairs) takes about 146 236 tonnes of Myctophidae annually. The nearby Crozet Islands, sheltering 1 771 000 king penguins, consume about 744 700 tonnes of myctophids [1]! The potential impact of these top predators on the marine resources can be estimated when the world population size is considered. These birds are recognized to be the most specialized in the predation of mesopelagic fish, specially the Myctophidae [2-5]. However no real-time study of the pelagic fish community present in the foraging range of a top predator has ever been published. A joint ornitho-ichthyological study in summer, in the offshore Polar Frontal Zone near the eastern coast of the Kerguelen Islands (distance: 20 - 110 miles; bottom depth: 450 - 2300 m), was the first to investigate the pelagic fish community of the upper water column. The aims were to study the distribution and abundance of the prey and to test whether predators could be used as bioindicators of the marine resources.

Only the fish results were presented subsequently.

Material and Methods

The fish sampling strategy was directly dependent on knowledge of the geographic position and foraging range of a satellite tracked predator (king penguin, *A. patagonicus*), breeding on the eastern coast of the Kerguelen

G. di Prisco, E. Pisano, A. Clarke (Eds)
Fishes of Antarctica. A biological overview
© Springer-Verlag Italia 1998

Islands (Ratmanoff colony: 49°23'S - 70°30'E). The period of sampling (first half of March 1995) corresponded to the intense summer feeding period of chicks.

While the birds (n = 9) were being fitted with their instrumentation, the 25 m long scientific vessel "La Curieuse" was positioned in front of the breeding king penguin colony. At the first reception of a satellite signal, the vessel followed the given positions to cross the foraging range of some (n = 4) of the selected birds. The adopted fishing strategy was to conduct simultaneous standardized trawls in the geographic area, limiting the deeper hauls to the known maximum diving depth (300 m) of the king penguin [6]. Four depth horizons were sampled (50, 150, 250, 300 m) day and night to take into account the diel vertical migration of pelagic fish [7]. Hauls were undertaken during the day (09 00 – 16 00 h LMT) and at night (20 00 - 00 30 h LMT) (LMT = GMT + 4 h). The fishing gear was the IYGPT (International Young Gadoid Pelagic Trawl; width:12 m, height: 7 m) with a 10 mm mesh size in the codend. Each trawl, towed at the selected depth, was fished for 30 minutes at 3 knots. Fishing depth was estimated with wire-out tables (R. Williams personnal communication) and regularly tested with a bathykimograph (TSURUMI 0-300 m). Two periods have been studied (1-5 and 10-13 March 1995) which totaled 65 (37 + 28 respectively) midwater trawls including 51 successful standardized hauls. Sorting of the catch was conducted on board and three categories retained: fish, squid and macrozooplankton. The present study includes only the fish. Fish were either deepfrozen or fixed in a 10% sea water buffered formalin. Identification, counting and weighing were conducted in the laboratory, because the sea conditions did not allow the work on board. The fish identification key used was the recent revision by Gon and Heemstra [8] and, when the specimens were in poor condition, of the available otolith keys [9-10]. The analysis was conducted on the 14 232 fish specimens (total weight: 37 483 g) from the standardized 51 hauls. The standard length (SL) (nearest mm below) has been used in the measurements because numerous damaged caudal fins precludes the use of the total length, especially in myctophid specimens. A caliper connected to a computer with approriate software (DataQ) allowed the data to be entered directly into the spreadsheet (Lotus123), which was used to establish the length frequency distributions (LFD). Where sexual dimorphism was visible from external characters (e.g. photophores), the sex was also recorded. To compare the catch data an index of abundance was used. The number of specimens of each species per standard fishing hour of the IYGPT was calculated.

Results

Thirty three fish species (Table 1) were identified in the foraging range of the king penguin; these included nektonic early stages and fingerlings of 8 demersal species. Myctophidae (15 species) dominated the true pelagic species both in number and biomass.

Fish Assemblages

The day composition of species (Table 1) was poor with only five species dominant (>5%) in the total catches by number (*L. squamifrons, K. Anderssoni, M. marmoratus, P. andriashevi* and *P. bolini*). The last four were dominant by biomass. The day distribution showed a low abundance and low biomass in the whole sampled water column. The assemblage was more diverse at night with six myctophid species dominant both in number and in biomass but with an increasing contribution of nonmyctophid species. Abundance and biomass were comparatively higher than in daylight samples.

Table 1. Fish composition (number, mass and standard length, SL, range) from the day and night trawls conducted during March 1995 in the foraging range of the king penguins breeding at the Ratmanoff colony (Kerguelen Islands)

	Day trawl (n = 24)		Night trawl (n = 27)	
Taxa	Total number (No.) and SL range (*mm*)	Total mass (g)	Total number (No.) and SL (*mm*)	Total mass (g)
Bathylagidae				
Bathylagus tenuis			226 (*76-15)*	2829.9
Stomiidae				
Stomias boa			115 (*87-281)*	492.1
Scopelarchidae				
Benthalbella elongata			4 (*182-228)*	168.3
Paralepididae				
Arctozenus risso			2 (*269-272)*	37
Notolepis coatsi	6 (*32-180)*	7.7	46 (*36-150)*	23.2
Myctophidae				
Electrona antarctica	8 (*31-81)*	9.7	3193 (*25-100)*	5523.8
E. carlsbergi	3 (*89-97)*	29.9	152 (*72-101)*	1782.4
E. subaspera			125 (*52-124)*	1373.3
Gymnoscopelus bolini			76 (*169-247)*	6989.2
G. braueri			2178 (*35-135)*	5811.4
G. nicholsi	10 (*36-140)*	35.9	157 (*35-146)*	1124.2
G. fraseri			787 (*33-98)*	3345.7

G. piabilis			70 (*86-152*)	1435.4
Krefftichthys anderssoni	285 (*14-63*)	180	1560 (*32-64*)	969.2
Metelectrona ventralis			2 (*84-90*)	17
Protomyctophum andriashevi	200 (*43-55*)	320	114 (*31-68*)	157.9
P. bolini	115 (*24-59*)	164.2	1774 (*29-61*)	1589.2
P. choriodon	7 (*64-79*)	31.9	10 (*52-82*)	29.7
P. gemmatum	1 (*48*)	1.3		
P. tenisoni			1536 (*26-46*)	747.8
Chiasmodontidae				
Kali macrura			8 (*116-138*)	131.7
Muraenolepididae				
Muraenolepis marmoratus[a]	180 (*33-56*)	96.6	538 (*31-57*)	348.9
Macrouridae				
Macrourus carinatus[b]	2 (*32-33*)	0.5		
Congiopodidae				
Zanclorhynchus spinifer[b]	4 (*18-20*)	0.5	1 (*16*)	0.1
Liparidae				
Paraliparis gracilis			1 (*46*)	0.5
Zoarcidae				
Melanostigma gelatinosum			1 (*130*)	7.5
Nototheniidae				
Dissostichus eleginoides[b]	1 (*48*)	0.8		
Lepidonotothen squamifrons[b]	661 (*17-29*)	36.6	32 (*19-29*)	2.7
Notothenia rossii[a]			2 (*59-60*)	3.8
Paranotothenia magellanica[a]	1 (*52*)	1.3		
Channichthyidae				
Champsocephalus gunnari[b]	1 (*52*)	0.7		
Centrolophidae				
Icichthys australis			2 (*198-316*)	609.8
Gempylidae				
Paradiplospinus gracilis			30 (*172-404*)	1019.7

[a] Fingerlings
[b] Postlarvae

The assemblages showed a nearly monospecific fish composition (*L. squamifrons*) in the day 50 m depth horizon. A slight increase in the number of species and changes in the percentage occurrence were observed in the two following depths. The deepest horizon showed the highest number of species (Table 2). At night the 50 m layer was dominated by five myctophids, of which *E. antarctica* was the most abundant species. *G. braueri* then *K. anderssoni* took the dominant position in the deeper horizons (Table 2).

Table 2. Day and night percentage occurrences (%) and mean relative abundance (A: in number of specimens per fishing hour) of the species within each depth horizon from the trawls conducted in the foraging range of the king penguins breeding at the Ratmanoff colony (Kerguelen Islands) in March 1995

		Day				Night			
	Depth (m)	50	150	250	300	50	150	250	300
	Total No.	600	266	201	423	6067	1891	1923	2861
Myctophidae									
E. antarctica	%				1.9	39.4	17.9	11.7	8.4
	A				*3.1*	*743.0*	*84.5*	*75.0*	*68.6*
E. carlsbergi	%				0.7			0.7	4.8
	A				*11.4*			*4.7*	*39.4*
E. subaspéra	%					1.3	1.2	0.4	0.5
	A					*25.2*	*5.5*	*2.7*	*4.0*
G. bolini	%					0.0	1.6	0.7	1.1
	A					*0.3*	*7.8*	*4.3*	*8.9*
G. braueri	%					18.0	27.6	13.0	11.1
	A					*338.5*	*130.3*	*83.3*	*90.9*
G. nicholsi	%				2.4	2.0	1.0	0.3	0.5
	A				*13.7*	*37.3*	*4.8*	*1.7*	*3.7*
G. fraseri	%					7.2	7.5	4.2	4.5
	A					*135.2*	*35.3*	*27.3*	*36.9*
G. piabilis	%					0.5	1.2	0.3	0.4
	A					*9.9*	*5.5*	*1.7*	*3.1*
K. anderssoni	%	1.8	44	59.2	9.0	2.0	9.5	31.1	23.1
	A	*3.1*	*39.7*	*41.0*	*14.5*	*37.9*	*44.8*	*199.3*	*188.9*
P. andriashevi	%				47.3	0.0	0.4	4.1	1.0
	A				*76.4*	*0.3*	*1.8*	*26.0*	*8.0*
P. bolini	%				27.2	14.2	8.8	14.7	16.3
	A				*43.9*	*267.4*	*41.8*	*94.0*	*132.9*
P. tenisoni	%					14.4	13.3	10.0	7.5
	A					*272.3*	*63.3*	*64.0*	*61.4*
Others	%				1.9	0.1	0.1	0.2	
	A				*0.4*	*1.9*	*0.5*	*1.4*	
NonMyctophidae									

B. tenuis	%						1.0	3.9	4.5
	A						*4.8*	*25.0*	*37.7*
N. coatsi	%			1.5	0.7		0.5	1.0	0.6
	A			*1.0*	*1.1*		*2.3*	*6.3*	*4.9*
S. boa	%					0.4	1.6	1.4	1.2
	A					*8.1*	*7.5*	*8.7*	*9.4*
M. marmoratus	%		48.1	18.9	3.3	0.3	5.8	1.6	13.4
	A		*43.4*	*13.1*	*5.4*	*5.3*	*27.3*	*10.0*	*109.4*
L. squamifrons	%	97.5	7.1	16.9	5.4	0.1	0.5	0.4	0.4
	A	*167.1*	*6.4*	*11.7*	*8.8*	*1.6*	*2.3*	*2.7*	*2.9*
P. gracilis	%					0.1	0.5	0.3	0.4
	A					*1.2*	*2.3*	*1.7*	*3.4*
others	%	0.7	0.8	5.0	1.0	0.0	0.2	0.2	0.4
	A	*1.1*	*0.7*	*1.0*	*0.4*	*0.6*	*1.0*	*1.3*	*3.1*

Circadian Changes

Very large circadian changes could be observed in the distribution of all species, when the sample of each species was large enough to be analyzed in terms of relative abundance (Table 2). First, one group of species (*E. subaspera, G. braueri, G. fraseri, G. piabilis, P. tenisoni*) was totally absent in the whole sampled water column during daylight and was concentrated at night in the shallow layer (50 m). A second group (*E. antarctica, G. nicholsi, P. bolini, P. choriodon*), present in low numbers at the deeper layer (300 m) during the day, exhibited the same circadian change with a 50 m maximum abundance at night. A third group comprised species which invaded only a part of the water column at night and were either absent from (*B. tenuis, B. elongata, G. bolini, K. macrura)* or were present (*E. carlsbergi, N. coatsi, P. andriashevi*) in the deeper layer during the day. The last main category comprised species exhibiting different depth distributions day and night (*K. anderssoni, L. squamifrons, M. marmoratus*). Finally *S. boa* and *P. gracilis* exhibited a nearly uniform abundance (about 8 and 2 specimens/fishing hour respectively) in the water column, but only at night.

Size Structure

The pooled size structure of the sampled species differs between day and night (Fig. 1). Only fish under about 80 mm SL occur in the day with different modal values in each depth stratum (20 mm at 50 m, 40 mm at 150 m, 10 and 40 mm at 250 m and 50 mm at 300 m). In contrast the size distributions are homogeneous at night with the bulk of fish under 130 mm SL, uninterrupted distributions to 230 mm SL and records to 400 mm SL. The modal length is about 40 mm SL in all depths.

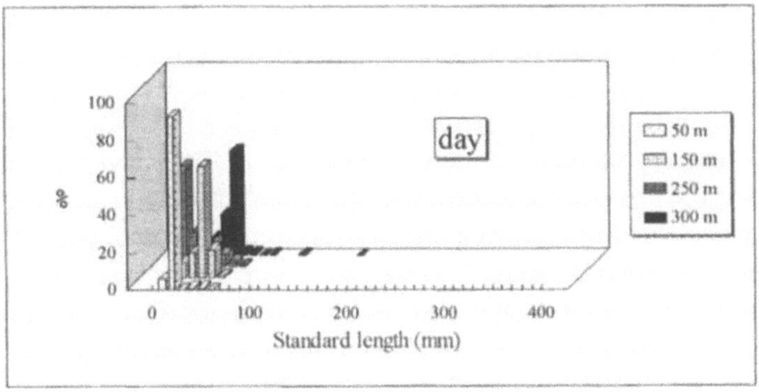

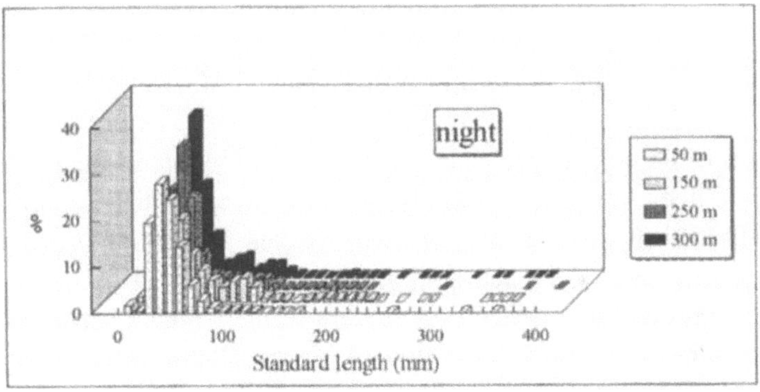

Fig. 1. Pooled size distribution of fish (day and night) at four sampled depths in the foraging range of the king penguins breeding on the Kerguelen Islands during March 1995

When representative samples were available, the general distribution differed from species to species. *L. squamifrons* had the smallest length range distribution followed by *P. tenisoni* and *M. marmoratus* (Table 1). Other species exhibited wider distributions (*E. antarctica, P. andriashevi, G. fraseri, G. nicholsi, N. coatsi*), sometimes associated with clear multimodal peaks (*K. anderssoni* at 16, 42 and 56 mm, *G. braueri* especially at 45 and 65 mm, *P. bolini* at 41 and 52 mm) Some species exhibited only partial LFD with small size fish missing (*E. carlsbergi, G. bolini, B. tenuis, G. piabilis, P. choriodon, P. gracilis, S. boa*) and sometimes also with modal peaks (*E. subaspera* at 64 and 94 mm).

The distributions seemed to be similar in each depth horizon except in some species like *G. braueri* in which specimens up to 90 mm are absent in the 50 m night samples. The mean length in the *L. squamifrons* daylight samples increases markedly with depth (22.1 mm at 50 m, 23.6 mm at 300 m). Finally large differences in the LFD are observed day/night and at each depth for *K. anderssoni*.

Discussion

Some limitations due to the methods used can affect the above results and must be first considered. The most important is the selectivity of the 10 mm mesh size probably affecting the catches of the small sized specimens. Nevertheless the LFD showed that undersampling could only occur for specimens less than 20 mm. Another bias could be introduced by the use of a nonclosing net inducing catches of specimens from the untargeted depths. Such hypothesis does not seem to occur because the quick retrieval of the net automatically closes the mouth when hauling.

Two categories of fish are recognized from the trawl samples in the foraging range of the king penguin off the Kerguelen Islands. The first comprises the early life stages of demersal species occurring on the nearby Plateau and the surrounding banks. *L. squamifrons, D. eleginoides, M. carinatus, C. gunnari, Z. spinifer* postlarvae and *M. marmoratus, N. rossii* and *P. magellanica* fingerlings belong to this group. The presence of early-life stages of demersal species is a special situation due to the proximity of a shelf. It does not seem to apply to other areas of the Antarctic Polar Front. The observed density of two species (*L. squamifrons* and *M. marmoratus*), over deep-sea bottom (> 2000 m, > 100 miles off the shelves) supports the colonization process of the isolated seamounts in the Southern Ocean, via the dispersion of young stages. The second group comprises the major part of the mesopelagic species, mainly undersampled before and often recorded from single records [11-14]. The present cruise provides the first data on dominance and densities for each species in the upper mesopelagic layers. Some species such as *K. macrura* and *B. elongata* have never been recorded in the area and, in the case of the first species, in the Southern Ocean. The majority of these species are Myctophidae (15 species) which allows consideration of the distribution pattern types *sensu* Hulley [15]. All species (86.9% in number and 84.5% in weight of the total collected fish specimens) belong to the cool water group [15] which is divided into the temperate (south) and the polar subgroups. Six species (*P. andriashevi, P. gemmatum, E. carlsbergi, E. subaspera, G. bolini* and *G. fraseri*) have been recognized as holosubAntarctic subpattern and three species (*P. choriodon, M. ventralis* and *G. piabilis*) as semisubAntarctic subpattern in the first subgroup.

These subpatterns include respectively 28.5% and 0.7% of the specimens. The polar subgroup consisted of five Broadly Antarctic Pattern species (*K. anderssoni, P. bolini, P. tenisoni, G. braueri* and *G. nicholsi*) and of only *E. antarctica* in the Antarctic Pattern (respectively 44.8% and 26% of the specimens). Hydrographic data collected recently off the Kerguelen shelf [16], in the exact area on the foraging study, have pointed out the presence of the Antarctic Polar Front running northward off the eastern slope with the typical 2 °C isotherm at 200 m depth. The biogeographic affinities and the dominance of the polar subgroup in terms of number of specimens are in accordance with the conclusion on the presence of the water masses off the islands [17]. In terms of species dominance the results form a continuity with Bekker's study [18] conducted in the area between the subAntarctic and the subtropical fronts, north of the Kerguelen Plateau.

The day distribution showed a low diversity, abundance and biomass in the whole water column sampled, with patchy distributions observed in the main species. *L. squamifrons* early life stages, *M. marmoratus* pelagic juveniles, *K. anderssoni*, *P. bolini* and *P. andriashevi* occupied successively deeper layers. It appears that, within the myctophids, only *K. anderssoni* occurs in the intermediate upper layers (150 - 250 m) of the water column in the day. Other species, partly noted in the day's sampled deepest range, invaded all depths at night with maximum abundance and biomass in the upper surface layer. Some species (*E. antarctica*, *G. braueri*, *G. fraseri*, *P. bolini* and *P. tenisoni*) formed the bulk of the deeper living migratory species. Others (*K. anderssoni*, *E. carlsbergi*, *B. tenuis*) did not reach the surface. In the case of *E. carlsbergi* this situation is probably related to the natural migration of the species from the epipelagic layer in summer to deeper zone (350 – 500 m) in winter [19] The high proportional biomass of *G. bolini* at night, which is due to the large individual size of the sampled specimens of this world-largest myctophid species, was surprising. It contrasted with the numerous but individually light (< 1 g) species such as *K. anderssoni*, *P. bolini* or *P. tenisoni* and the small mass of *L. squamifrons* postlarvae.

Modal values in the pooled size structure of the species in the day corresponded with species dominance at each depth. The 50 m peak corresponded to 20 mm SL *L. squamifrons,* probably undersampled due to the mesh size. The 40 mm SL peak at 150 m was a mixture of two species (unimodal *M. marmoratus* LFD and second mode in the *K. anderssoni* LFD). The two modal values at 250 m comprised the smallest size group (10 mm) of *K. anderssoni*, a group previously recorded in the ichthyoplankton surveys conducted in the same area [20], and the 40 mm peak of *M. marmoratus*. Finally the 300 m peak (50 mm) concerned a mixture of three species *P. andriashevi, P. bolini* and the largest specimens of *K. anderssoni*. The night depth LFD showed a uniformity in the mixture of species (mainly myctophids) occurring at all the depths.

The results show a total or a partial size structure of the fish populations in the investigated area, taking into account the known maximal size of the species [15,21]. All the myctophid species with a complete LFD belong to the polar subgroup, dominant in terms of specimens (70.8% of the total) in the sample. This confirms the prevalence of the pattern and a probable entire life cycle of the species off the islands. Incomplete LFDs are observed for the myctophid species belonging to the temperate subgroup which are probably at the southern limit of their distribution or are living in greater depths in this region. Another explanation would be a functional division of the habitat as suggested for *E. carlsbergi* by Zemsky and Zozulya [22] who consider the SubAntarctic and Subtropical Frontal Zone as reproduction areas and the Polar Frontal Zone as a feeding ground for a part of the populations. Such conclusions could also apply to the non-myctophid species: *N. coatsi*, *P. gracilis* would relate to the polar subgroup, while *B. tenuis* belongs to the temperate subgroup. *S. boa*, *B. elongata*, *A. risso*, *K. macrura*, *I. australis*, *M. gelatinosum* were not adequately sampled to comment further. Finally the multimodal structure, observed in some LFD (*K. anderssoni, G. braueri, G. nicholsi, P. bolini, E. subaspera*), is probably related to the age structure of these fast length growing species. Age determination from otoliths will be necessary to confirm the hypothesis. The increase in the mean size with depth observed in the case of *L. squamifrons* indicates a probable ontogenetic migration from the surface to the deeper layer in postlarvae, before the near-bottom settlement.

The fish results need now to be compared with the stomach contents and the feeding strategy of the top predators. The present real time study will allow us to advance in the knowledge of the relationships between predators and prey organisms in this particular area of the Southern Ocean.

Acknowledgments

I wish to thank the IFRTP for the equipment used at sea and the opportunity to participate in the cruise "IPEKER" of "La Curieuse". I appreciate the help of the master D. Rochard, the crew and J.M. Le Petitcorps from TAAF. The results would never have been obtained without them. I am indebted to C.A. Bost for the collaboration during and after the cruise. I have appreciated the help of C. Ravier and Y. Cherel during the laboratory work and the assistance of P. Pruvost in finalizing the presentation. I am grateful to P.A. Hulley and referees for useful comments to improve the manuscript.

References

1. Guinet C, Cherel Y, Ridoux V, Jouventin P (1996) Consumption of marine resources by seabirds and seals in Crozet and Kerguelen waters: changes in relation to consumer biomass 1962-85. Antarct Sci 8: 22-30
2. Adams NJ, Klages NT (1987) Seasonal variation in the diet of the king penguin *Aptenodytes patagonicus* at sub-Antarctic Marion Island. J Zool Lond 212: 303-324
3. Hindell MA (1988) The diet of the king penguin *Aptenodytes patagonicus* at Macquarie Island. Ibis 130: 193-203
4. Klages NTW, Pemberton D, Gales RP (1990) The diets of the king and gentoo penguins at Heard Island. Aust Wild Res 17: 53 - 60
5. Ridoux V (1994) The diets and dietary segregation of seabirds at the subAntarctic Crozet Islands. Marine Ornithol 22: 1-192
6. Kooyman GL, Cherel Y, Le Maho Y, Croxall JP, Thorson PH, Ridoux V, Kooyman CA (1992) Diving behaviour and energetics during foraging cycles in king penguins. Ecol Monogr 62 (1): 143-163
7. Gjøsaeter J, Kawaguchi K (1980) A review of the world resources of mesopelagic fish. FAO Fisheries Technical Paper 193: 1-151
8. Gon O, Heemstra PC (1990) Fishes of the Southern Ocean. J.L.B. Smith Institute of Ichthyology. Grahamstown
9. Williams R, McEldowney A (1990) A guide to the fish otoliths from waters off the Australian Antarctic Territory, Heard and Macquarie Islands. ANARE Res Notes 75: 1-173
10. Smale MJ, Watson G, Hecht T (1995) Otolith atlas of Southern African marine fishes. Ichthyological Monographs of the J. L. B. Smith Institute of Ichtyology. 1: 1-253
11. Duhamel G, Hureau JC (1982) Données complémentaires sur l'ichtyofaune des îles australes françaises. Cybium 6(1): 65-80
12. Duhamel G (1987) Ichtyofaune des secteurs indien occidental et atlantique oriental de l'océan austral. Biogéographie, cycles biologiques et dynamique des populations. Université Paris 6 thesis. Microfiche SN 87 200 447 institut d'Ethnologie
13. Hulley PA, Camus P, Duhamel G (1989) Ichthyological results of cruise MD-42/SIBEX II. Part 1. Fishes from RMT-8 stations, with additional records of lanternfishes (Myctophidae: Osteichthyes) from the Indian and Atlantic sectors of the Southern Ocean. Cybium 13 (1): 83-99
14. Duhamel G, Hulley PA (1993) Ichtyofaune méso et bathypélagique: In: Campagnes SKALP 1987 et 1988 aux îles Kerguelen à bord des navires "SKIF" et "KALPER". Institut français pour la Recherche et la Technologie Polaires. Les rapports des campagnes à la mer. 93 - 01 (2 vol): 285-295
15. Hulley PA (1981) Results of the research cruises of FRV "Walther Herwig" to South America. LVIII. Family Myctophidae (Osteichthyes. Myctophiformes). Arch FischWiss 31 (1): 1-300
16. Charriaud E (1993) Résultats hydrologiques et météorologie. In: Campagnes SKALP 1987 et 1988 aux îles Kerguelen à bord des navires "SKIF" et

"KALPER". Institut Français pour la recherche et la Technologie Polaires. Les rapports de campagnes à la mer. 93-01. (2 vol.): 53-56

17. Park YH, Gamberoni L, Charriaud E (1993) Frontal structure, water mass and circulation in the Crozet Basin. J Geophys Res 98: 12361-12385
18. Bekker VE (1985) Distribution of myctophid fishes and the position of the biogeographical border between the islands of Saint-Paul and Kerguelen. J Ichthyol 25 (2): 159-162
19. Lubimova TG, Shust KV, Popkov VV (1987) Specific features of the ecology of mesopelagic fish of family Myctophidae of the Southern Ocean. In: Biological resources of hydrosphere and their utilisation. Biological resources of the Arctic and Antarctic. Moscow. pp 320-337
20. Koubbi P (1992) L'ichtyoplancton de la partie indienne de la province kerguelenienne (Bassin de Crozet et plateau de Kerguelen): identification, distribution spatio-temporelle et stratégies de développement larvaire. PhD thesis. University Paris VI [Unpublished]
21. Hulley PA (1990) Family Myctophidae. In: Gon O. Heemstra (ed) Fishes of the Southern Ocean. J.L.B. Smith Institute of Ichthyology. Grahamstown. pp 146-178
22. Zemskiy AV, Zozulya (1991) A functional division of *Electrona carlsbergi* (Taning. 1932) area of habitat in relation to the latitudinal zonation of the Southern Ocean. In: Selected Sci Papers. 1990 (SC-CCAMLR-SSP/7). CCAMLR. Hobart. Australia. pp 369-396

Preliminary Investigations on the Evolution of the Tribe Electronini (Myctophiformes, Myctophidae)

P. Alexander Hulley

South African Museum, Cape Town, Republic of South Africa

Introduction

Lanternfishes are the most speciose family of oceanic teleosts [1]. Stiassny [2] has provided the most recent generic analysis. More than 35 myctophid species are known to occur in the Southern Ocean [3]. They are distributed offshore of the continental shelf break and form the dominant component of mesopelagic fish biomass in the Southern Ocean [4]. A cladogram for Southern Ocean myctophids is lacking, but there is reason to suspect that speciation in this region arose through a series of vicariance events [5-8].

Fossil myctophids, represented mainly by otoliths, are known from the Eocene [9]. They are apparently absent from the Cretaceous [10]. Fossil material further suggests that the family may have approached its present level of generic differentiation by the Late Miocene [10]. Both a cold-water [11] and a warm-water [5,6,12,13] origin for the family have been postulated. It has been proposed that recent Southern Ocean species of *Protomyctophum*, *Electrona* and *Gymnoscopelus* represent derivatives of an Eocene fauna that lived in high latitudes of the Pacific sector; that vicariance events were responsible for isolating populations of this high latitude fauna; and that modern Southern Ocean species appeared during the pronounced cooling periods of the Pliocene and Pleistocene [6].

This paper addresses the hypothesis in greater detail. It attempts to resolve the evolution of one of the main groups of Southern Ocean lanternfishes, the tribe Electronini [*Krefftichthys, Protomyctophum (Protomyctophum), Protomyctophum (Hierops), Electrona, Metelectrona*].

Material and Methods

Biogeographic data were derived from two sources: (1) Cruise WH-1976: FFS "Walther Herwig" [14]: 40 916 specimens (63 species) from 39 stations between 80 and 2 650 m (MATRIX-I); and (2) Cruise ANT XIV/2: RS "Polarstern" [4]: 16 290 specimens (11 species) from 17

G. di Prisco, E. Pisano, A. Clarke (Eds)
Fishes of Antarctica. A biological overview
© Springer-Verlag Italia 1998

stations between 295 and 825 m (MATRIX-II). Data from two "Polarstern" stations, POL 053 (haul aborted) and POL 073 (net snagged on bottom) [4], were not included in the analyses. The data matrices were subjected to cluster analysis to produce dendrograms. Multidimensional scaling ordination (MDS) was undertaken to confirm the groups generated by cluster analysis, and to demonstrate the spatial relationships in two-dimensions. The suite of multivariate statistical programmes (PRIMER Ver. 4.0) was used. A percentage similarity analysis (SIMPER) was undertaken to determine the species responsible for the formation of the station groups in the MATRIX-II series.

Table 1. Character states for species of the tribe Electronini

1. Dn absent (0), present (1)
2. Vn present (0), absent (1)
3. So absent (0), present (1)
4. Body shape fusiform (0), slightly compressed (1), highly compressed (2)
5. PVO2 to PVO1 level (0), elevated (1)
6. PLO to PVO1 level (0), slightly elevated (1), elevated and displaced (2)
7. PO5 level (0), elevated (1)
8. VO2 level (0), elevated (1)
9. VLO nearer Vbase (0), midway ventral base-LL (1), nearer LL (2)
10. Number of SAO2 (0), SAO3 (1)
11. SAO level (0), inclined straight (1), inclined obtuse (2), inclined angulate (3)
12. SAO1 behind VO4 (0), above VO4 (1), above VO3 (2)
13. Prc spacing normal (0), touching (1), widely spaced (2) #
14. Prc1 in front procurrent rays (0), above procurrent rays (1)
15. Interorbital distance wide (0), narrow (1)
16. "Pol" grouping absent (0), present (1)
17. Caudal peduncle glands sexually dimorphic (0), not sexually dimorphic (1)
18. Males with supra caudal gland only (0), with supra + infra gland (1)
19. Females with infra caudal gland only (0), with infra + supra gland (1)
20. Last AO photophores level (0), depressed (1)
21. Op1 well below lower margin of orbit (0), at level of orbit (1)

0 = plesiomorphic; 1 = apomorphic; 2 = advanced apomorphic Stage 2; 3 = advanced apomorphic Stage 3; # = unordered

Hennig86 (Ver. 1.5) [15] was used for the cladistic analysis. Twenty-one characters (mainly photophore groupings and orientation) (Table 1) were identified for all species of Electronini. Plesiomorphic and apomorphic states were allocated by outgroup comparison with *Krefftichthys*. Unpublished investigations of the molecular phylogeny of

Myctophiformes, derived from 1300 bp of the mitochondrial 16SrRNA gene, confirm the ancestral nature of this monotypic genus in the Electronini (Misaki Miya, personal communication). All character states could be ordered except for the Prc spacing (Table 1: Character 13).

Results

The percentage similarity between stations of MATRIX-I is given in Fig. 1A. Three groups of stations were delineated. Group 3 could be further delineated into two subgroups. Group 1 stations were characterized by temperatures at 200 m ranging between 13.6 and 13.8 °C; Group 2 stations by temperatures between 5.9 and 13.2 °C; Group 3A stations by a temperature range 1.0-1.9 °C; and Group 3B by a range of 2.4-4.8 °C (with three exceptions: WH 21-01/76, WH 21-02/76, WH 22/76). The groupings were confirmed by MDS ordination (Fig. 1B).

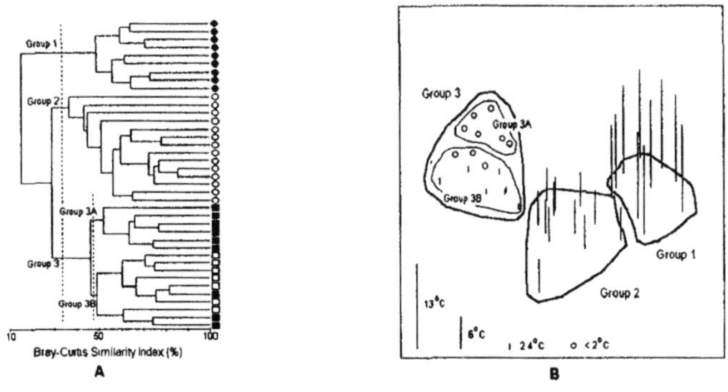

Fig. 1. MATRIX-I (root-root transformed). **A.** Dendrogram of station affinities. 13.6-13.8 °C (solid circle), 5.9-13.2 °C (open circle), 2.4-4.8 °C (open square), 1.0-1.9 °C (solid square). **B.** MDS plot of station affinities with isotherms at 200 m superimposed

The dendrogram of the species distribution relationships in MATRIX-I is given in Fig. 2A. Four major groups of species were delineated. Two of these groups (Groups B, D) could be subdivided. Pattern types (Fig. 2B) and genera and subgenera (Fig. 2C) have been superimposed on the MDS plot. Group A species represent a Tropical Pattern for the pseudoceanic population of *Diaphus dumerilii* found off the slope regions of South America [14].

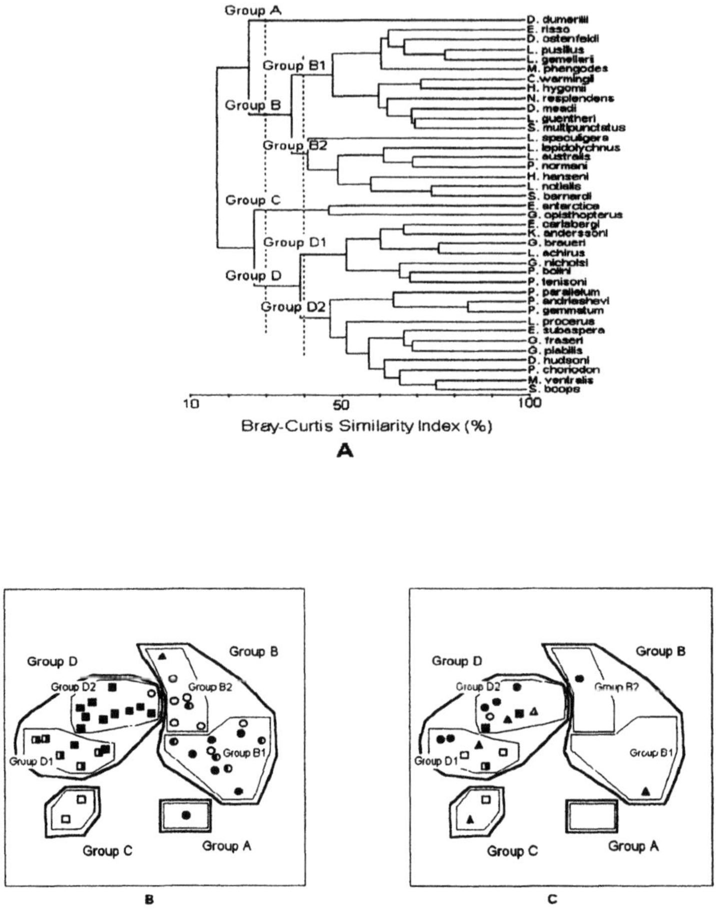

Fig. 2. MATRIX-I (root-root transformed). **A.** Dendrogram of species distribution affinities. **B.** MDS plot of species distribution affinities, pattern types in accordance with Hulley [14]. Widespread + Tropical (solid circle), Subtropical (half-solid circle), Convergence (open circle), Temperate (solid triangle), SubAntarctic (solid square), Broadly Antarctic (half-solid square), Antarctic (open square). **C.** MDS plot of species distribution affinities with genera and subgenera superimposed. *Electrona* (solid triangle), *Metelectrona* (half-solid triangle), *Hierops* (open circle), *Protomyctophum* (solid circle), *Krefftichthys* (half-solid square), *Gymnoscopelus* (open square), *Nasolychnus* (solid square)

Group B1 consisted of Widespread, Tropical and Subtropical Pattern species, except for *D. meadi* and *D. ostenfeldi*. Group B2 consisted of Convergence and Bitemperate Pattern species, except for *Symbolophorus barnardi*. Group C encompassed Antarctic Pattern species. Group D1 consisted of Broadly Antarctic Pattern species, except for *Electrona carlsbergi* and *Lampanyctus achirus*. Group D2 comprised SubAntarctic

Pattern species, except for *Lampichthys procerus*. However, *L. procerus*, *D. meadi* and *D. ostenfeldi* exhibit close spatial proximity to the bulk of the Convergence Pattern species. Group D (Broadly Antarctic and SubAntarctic Patterns) encompassed *Electrona*, *Krefftichthys*, *Metelectrona* and all subgenera of both *Protomyctophum* and *Gymnoscopelus* (Fig. 2C).

Two major groups of stations could be identified in the MATRIX-II series (not figured): one group with fishing depths between 440 m and 825 m, the other between 295 m and 450 m. This indicated a transition in relative abundances in the 440-450 m depth horizon, depths characterized by the top of the North Atlantic Deep Water (NADW) layer during the cruise. SIMPER analysis suggested that *P. bolini* appeared to be a good indicator species for the shallow group of stations. *Gymnoscopelus braueri* and *E. antarctica* appeared to be the indicator species for the deeper group. A higher average abundance for *E. antarctica* was associated with depths below 440-450 m.

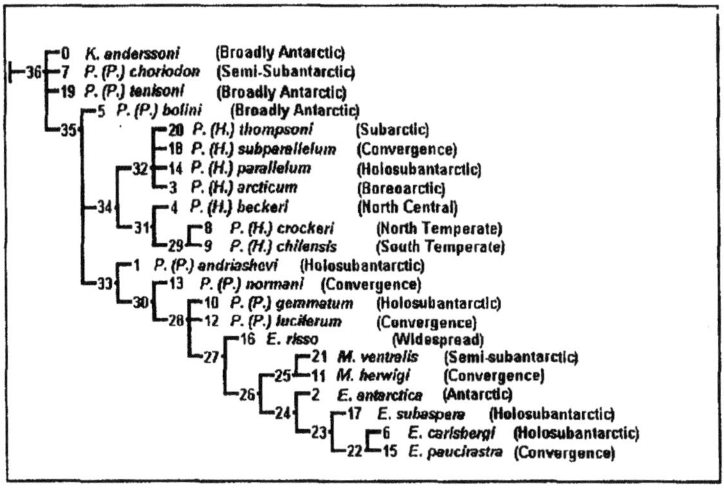

Fig. 3. Cladogram for Electronini. Pattern types after White [8] and Hulley [14]

The outputs from the cladistic analysis found two cladograms (180 steps long; consistency index 0.89). The Nelson consensus tree is given in Fig. 3. Broadly Antarctic Pattern species were restricted to the base of the cladogram. HolosubAntarctic, SemisubAntarctic and Convergence Pattern species were represented at various hierarchical levels within the cladogram. The sole Antarctic Pattern species (*E. antarctica*) appeared as one of the most highly evolved members of the tribe.

Discussion

MATRIX-I data indicate that station affinities could be correlated with the water temperature at 200 m (Fig. 1). The use of isotherms at this depth appears to be particularly suitable in the interpretation of lanternfish catch data from large, nonclosing, mesopelagic samplers [14,16]. The absolute value of the temperature should be viewed only as an index of the integrated quality of the water column. Latitudinal banding of the station groups from north to south is suggested by Fig. 1A. With some exceptions (Fig. 2A,B), the distribution pattern types (generalized distribution tracts) correspond to those given in the literature [3,6,16,17].

Species of the genera *Electrona, Krefftichthys, Metelectrona* and *Protomyctophum* (tribe Electronini) and *Gymnoscopelus* (tribe Gymnoscopelini) are considered to be members of the cold-water complex of species [18]. Except for *Metelectrona*, all occur only in one pattern type, namely the Broadly Antarctic Pattern (Fig. 2C). This suggests that Broadly Antarctic species (i.e. species attaining sexual maturity throughout their distribution range between the STC and the Antarctic continental shelf break, and thereby demonstrating adaptation to a wide range of environmental conditions) may be most ancestral.

Because of the limited number of characters (Table 1), the cladogram (Fig. 3) should be regarded only as a preliminary result. Greater resolution of the various subgroupings is required. However in support of the above hypothesis, the following should be noted: (1) the ancestor species (*K. anderssoni*) possesses a Broadly Antarctic Pattern; and (2) other species possessing this pattern type are restricted to the base of the cladogram (i.e. the most ancestral species). The pattern type does not occur again in the phylogeny of the tribe.

Such interpretation differs only slightly (and mainly in nomenclature) from that proposed by White [8]. On the basis of surface water-mass cladograms, White [8: Figs. 3-6] envisages a subdivision of the ancestral Austral region into South Temperate and South Polar regions, brought about by the deepening of the Drake Passage and the establishment of the Circumpolar Current (Early Miocene). The species cladogram (Fig. 3) indicates a somewhat more complex evolution in the distribution of the Electronini than White's [8: Fig. 6] proposal of a subsequent subdivision of the South Polar region into distinct Antarctic and subAntarctic biogeographic provinces (Early-to-Middle Miocene). Species of the monophyletic subgenus *Hierops* which occur outside the Southern Ocean region appear to have evolved along two separate lines. Each possesses Convergence or South Temperate representatives. *Protomyctophum thompsoni* (subArctic: North Pacific) and *P. arcticum* (Boreoarctic: North Atlantic) are grouped with the Convergence / HolosubAntarctic species *P. subparallelum* and *P. parallelum*. *Protomyctophum beckeri* (North Central: Pacific) and *P. crockeri* (North Temperate: North Pacific) are grouped with South Temperate *P. chilensis*. Greater resolution of these relationships is required. It is possible that the history of these species includes both a northern dispersion and subsequent allopatric speciation. Such a scenario may have occurred more than once, as is indicated by the later evolution and subsequent northern distribution of *E. risso* (Fig. 3).

The phylogeny of species of the subgenus *Protomyctophum* also requires greater resolution. Their phylogeny, and those of *Electrona* (with the exception of *E. antarctica*) and *Metelectrona*, appear to be associated with an evolution from ancestral SubAntarctic and Convergence Pattern species in association with possible vicariance events. The multiplicity of HolosubAntarctic and Convergence Pattern pairs suggests that such events may have also been operational on more than one occasion. The evolution of the only Antarctic Pattern species, *E. antarctica*, appears late in the phylogeny (Fig. 3). A proposed model for the evolution of the tribe (Fig. 4) suggests the following possible stages:

Stage 1: The opening of the Drake Passage to deep water, together with the northern migration of Australia during the Early Miocene (22 Ma) led to the development of the Circumpolar Antarctic Current and the STC, and to the restriction of Tethys circulation [7,19]. In association with continuing climatic cooling from the beginning of the Oligocene, this resulted in the commencement of gyral decoupling and the subsequent isolation of those Tethyan (Panthalassian) species living in high southern latitudes. The derivatives of these Austral species (*sensu* White [8]) are represented by lanternfish species which have a Broadly Antarctic Pattern (*sensu* Hulley [14]). In the Electronini such species include *K. anderssoni, P. tenisoni* and *P. bolini*. Recent unpublished but fragmentary data suggest that *P. choriodon* may have to be reassigned to this pattern type.

Stage 2: During the Middle Miocene (14-12 Ma), continued gyral decoupling and the formation of the Antarctic ice sheets were evident [7]. Sinking of the Iceland Ridge below sea level created fundamental changes in deep water circulation [19]. In the Southern Ocean, these changes led to development of a system of hydrographic fronts and consequent partitioning of the environment. Isolation of the populations gave rise to

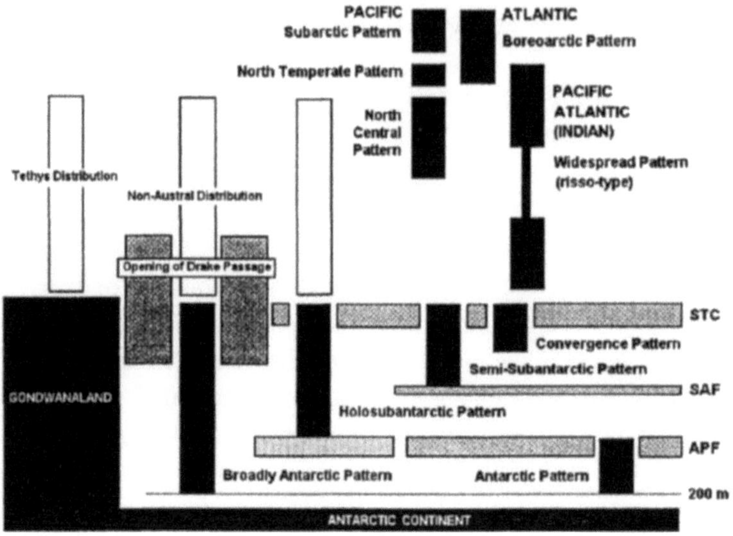

Fig. 4. Model for evolution of Electronini

HolosubAntarctic, Semi-SubAntarctic and Convergence species. This initial process of partitioning and isolation may have been repeated by numerous similar vicariance events brought about by latitudinal shifts in the position of the APF, changes in the temperatures of water masses and sea-ice cover from the Middle Miocene to the Early-to-Middle Pliocene, since the phylogeny of the species concerned is complex (Fig. 3).

Stage 3: The Terminal Miocene Event was marked by a northern shift in the siliceous biogenic belt, intensification of oceanic circulation, particularly in upwelling systems, and increased productivity in all oceans [7,19]. This resulted in the northward dispersion of ancestral *Hierops* species living in the SubAntarctic region, and was facilitated by the increased food supply in shallow water depths in tropical regions, and by equatorial transgression due to increased upwelling in eastern boundary current systems. Such dispersion and subsequent allopatric speciation also appears to have occurred more than once, so that these processes probably spanned the Late Miocene to Early Pliocene period.

Stage 4: Although the APF had become established by the Pliocene, its influence on the evolution of Antarctic Pattern species may have occurred only in the Late Pliocene (3 Ma), when stratification of the water column was sufficient either to accrue an energetic advantage to vertically migrating fishes, or to provide a refuge in the austral winter. As has been shown above, the distribution of modern representatives of these species (e.g. *E. antarctica*) appears to be closely correlated with the NADW. The onset of Northern Hemisphere ice-sheet formation in the Late Pliocene has been related to intensification of the Gulf Stream system due to the final closure of the Middle American Seaway [19]. The effect was to increase precipitation in high northern latitudes. When coupled with increased glaciation, this would have resulted in an intensification of the thermo-haline circulation and the increased production of NADW.

Two data sets lend support to the model. Firstly, fossil otoliths of *P. (H.) thompsoni* have been found only in Late Pliocene and Pleistocene deposits of California. The species is absent from the Miocene deposits [20]. This is in accordance with Stage 3, which postulates a northern dispersion and speciation event dating from after the Terminal Miocene Event. Secondly, Schwarzhans [10,21] has reported fossil otoliths of *E. risso* from the Lower Pliocene of Europe, and Fitch [20] has recorded the species from the Pleistocene of California. These data again concur with an Early Pliocene distribution for the species as proposed in Stage 3.

Finally, the model predicts that in the subgenus *Gymnoscopelus*, the only Antarctic Pattern species (*G. opisthopterus)* should be the most derived species, while the two Broadly Antarctic species (*G. braueri, G. nicholsi*) should represent the modern derivatives of the most ancestral species. The testing of these predications must await further investigation. Unfortunately, *Gymnoscopelus* has no fossil record because *G. fitchi* from the Lower Pliocene of Italy is now considered to belong to the genus *Myctophum* [10]. However, the otoliths of extant species of *Gymnoscopelus* closely resemble those of *Lampanyctodes* [10]. On osteological grounds, Paxton [22] considered the latter to be the most primitive genus of the Tribe Gymnoscopelini. The monotypic *L. hectoris* is a pseudoceanic species found only on the continental shelves and slopes of Chile, South Africa and southern Australia [6,17]. The suggestion therefore is that while recent members of Electronini may have arisen from a pelagic ancestral form living in high southern latitudes, recent species of the genus *Gymnoscopelus* may represent derivatives of a high southern latitude epibenthic ancestor subjected to the same vicariance events.

Acknowledgments

I am indebted to the captains and crew of the "Walther Herwig" and "Polarstern"; to my ship-board colleagues; and to Dr. B. Cook (South African Museum, Cape

Town) and Dr. M. Miya (Chiba Natural History Museum, Japan) for their advice and unpublished data respectively.

References

1. Hulley PA (1994) Lanternfishes. In: Paxton JR, Eschmeyer W (eds) Encyclopedia of fishes. University of New South Wales Press, Sydney, pp 127-128
2. Stiassny MLJ (1996) Basal ctenosquamate relationships and interrelationships of the Myctophiform (Scopelomorph) fishes. In: Stiassny MLJ, Parenti LR, Johnson DR (eds) Interrelationships of fishes. Academic Press, San Diego, London, pp 405-423
3. Hulley PA (1990) Family Myctophidae. In: Gon O, Heemstra PC (eds) Fishes of the Southern Ocean. JLB Smith Institute of Ichthyology, Grahamstown, pp 146-178
4. Hulley PA, Kock K-H, Pusch C, White MG. Mesopelagic fishes. In: Cruise report of Cruise ANT XIV/2 of RS 'Polarstern'. Alfred Wegener Institute, Bremerhaven (in press)
5. McGinnis RF (1977) Evolution within pelagic ecosystems: aspects of the distribution and evolution of the family Myctophidae. In: Llano G (ed) Adaptations within Antarctic ecosystems. Smithsonian Institution, Washington, pp 547-556
6. McGinnis RF (1982) Biogeography of lanternfishes (Myctophidae) south of 30°S. Ant Res Ser 35:1-110
7. Eastman JT (1993) Antarctic fish biology: evolution in a unique environment. Academic Press, San Diego London
8. White BN (1994) Vicariance biogeography of the open-ocean Pacific. Prog Oceanogr 34:257-284.
9. Brzoboharty R (1981) Zur Palökologie der fossilen Myctophiden (Myctophidae, Teleostei). Zap Karp ser paleon 6:31-48
10. Schwarzhans W (1980) Die tertiäre Teleosteer-fauna Neuseelands, rekonstruiert anhand von Otolithen. Berliner geowiss Abh (A)26:1-211
11. Fraser-Brunner A (1949) A classification of the fishes of the family Myctophidae. Proc Zool Soc Lond 118:1019-1107
12. Bolin RL (1939) A review of the myctophid fishes of the Pacific coast of the United States and of Lower California. Stan Ichthyol Bull 1:89-156
13. Andriashev AP (1965) A general review of the Antarctic fish fauna. In: van Oye, van Miegham J (eds) Biology and ecology in Antarctica. Monogr Biol 15. Junk, The Hague, pp 491-550
14. Hulley PA (1981) Results of the research cruises of FRV 'Walther Herwig' to South America. LVIII. Family Myctophidae (Osteichthyes, Myctophiformes). Arch FischWiss 31:1-300
15. Farris JS (1988) Hennig86 reference manual. Mimeograph. 22 pp (unnumbered)

16. Hulley PA, Krefft G (1985) A zoogeographic analysis of the fishes of the family Myctophidae (Osteichthyes, Myctophiformes) from the 1979-Sargasso Sea Expedition of R.V. 'Anton Dohrn'. Ann S Afr Mus 96:19-52
17. Hulley PA (1986) Family 86: Myctophidae. In: Smith MM, Heemstra PC (eds) Smith's sea fishes. Macmillan, Johannesburg, pp 282-321
18. Becker VE (1964) On the temperate-cold water complex of myctophids (Myctophidae, Pisces). Okeanol 4(3):469-475
19. Kennett JP (1982) Marine geology. Prentice-Hall, Englewood Cliffs
20. Fitch JE (1969) Fossil lanternfish otoliths of California, with notes on fossil Myctophidae of North America. Contr Sci Los Ang Cty Mus Nat Hist 173:1-20
21. Schwarzhans W (1978) Otolithen aus dem Unter-Pliozän von Süd-Sizilien und aus der Toscana. Berliner geowiss Abh (A)82:1-52
22. Paxton JR (1972) Osteology and relationships of the lanternfishes. Sci Bull Nat Hist Mus Los Ang Cty 13:1-81

Diversity of Digenea, Parasites of Fishes in Various Areas of the Antarctic

Krzysztof Zdzitowiecki

W. Stefański Institute of Parasitology, Polish Academy of Sciences, Twarda 51/55, 00-818 Warszawa, Poland

A critical list of the Antarctic fish Digenea recently published [1] concerns 36 species. However, a genuine occurrence of one species, *Lecithocladium falklandicum*, in the Antarctic is doubtful. A validity of another one, *Gonocerca lobata*, is questionable. These two species are omitted from the present review. On the other hand, new material collected in the Weddell Sea has recently been reported by the present author and a further ten undescribed species were recognized. Their descriptions have been recently published or are in press [2-7]. Thus 44 species are taken into account. This is not a final number of digeneans occurring in fish in the Antarctic, as strongly contracted and/or damaged undeterminable (even to generic level) specimens of not less than two further species are in the present author's collection. The presence of further undescribed species is very possible, as only limited data are available for parasites of Antarctic fishes of such important families as Liparididae, Zoarcidae, Murenolepididae and Macrouridae, as well as some Notothenioidei like Artedidraconidae and Bathydraconidae occurring in deep waters. It should be noted that only one species, *Lecithochirium whitei*, is known as a parasite specific to a bathypelagic fish, *Bathylagus antarcticus*. All others normally occur in benthic and bentho-pelagic fishes; only a few species and specimens have been reported also from pelagic fishes.

Data from five areas within the Antarctic Region - the Ross Sea (R), the Indian sector of the Southern Ocean (I), the Weddell Sea (W), the South Shetland Islands area (SS) and the South Georgia area (G), as well as from the SubAntarctic (SA) are taken into account. A few species are known as cosmopolitan or bipolar (C). These abbreviations are used in Tables 1-4.

Limited data from areas of the South Orkney Islands and the western side of the Antarctic Peninsula are restricted to species occurring in the South Shetland Islands area and are not separately given. Fish classification given by Gon and Heemstra [8] is used.

Representatives of all the higher digenean taxa occurring in Antarctic fishes are common in the whole World Ocean.

G. di Prisco, E. Pisano, A. Clarke (Eds)
Fishes of Antarctica. A biological overview
© Springer-Verlag Italia 1998

The first high taxon taken into account is the superfamily Hemiuroidea (Table 1). Out of 12 species belonging to five families recorded in the Antarctic, nine species normally occur in the stomach; the remaining occur in the intestine and the only parasite of elasmobranchs (Rajidae), *Otodistomum cestoides*, occurs in an immature stage (metacercaria) in cysts in the body cavity of teleosts. Such metacercariae were found in the Weddell Sea. Apart from *O. cestoides*, only *Lecithochirium whitei* and probably *Lecithophyllum champsocephali* have narrow specificity to final hosts. Presently nothing may be said about the specificity of *Aphanurus* sp., as only a single specimen was found. The remaining eight species occur in fishes belonging to 2-7 orders. Four species are cosmopolitan or bipolar. Probably three of them may be found in all areas of the Antarctic, if sufficient numbers of species and specimens of fishes had been examined. The fourth species, *Derogenes varicus*, the most abundant

Table 1. Records of Hemiuroidea in Antarctic fishes

	Parasite	Site	Hosts	Areas of records						
				R	I	W	SS	G	SA	C
1.	*Otodistomum cestoides*	stomach	Rajidae				+		+	+
	O. cestoides metacercariae	cysts	Notothenioidei			+				
2.	*Genolinea bowersi*	stomach	Notothenioidei, Macrouridae[a]	+	+	+	+	+	+	
3.	*Aphanurus* sp.	intestine	Zoarcidae			+				
4.	*Derogenes johnstoni*	stomach	Notothenioidei, Liparididae[a]	+	+	+				
5.	*D. varicus*	stomach	various				+		+	+
6.	*Gonocerca phycidis*	stomach	Notothenioidei, Gadiformes, Stromateidae	+		+	+	+	+	+
7.	*Elytrophalloides oatesi*	stomach	various	+	+	+	+	+	+	
8.	*Glomericirrus macrouri*	stomach	Notothenioidei, Gadiformes			+	+		+	+
9.	*Lecithochirium whitei*	stomach	Bathylagidae		+			+		
10.	*Lecithaster macrocotyle*	intestine	Notothenioidei, ?Muraenolepididae, ?Achiropsettidae	+			+	+	+	
11.	*L. micropsi*	intestine	Notothenioidei, Muraenolepididae					+	+	
12.	*Lecithophyllum champsocephali*	stomach	Channichthyidae				+			

[a]Accidental host

digenean species in the World Ocean, is replaced in the high Antarctic by *Derogenes johnstoni*. These species are closely related, but are distinguishable on the base of dimensions and morphology of the egg shell. The second pair of closely related species are *Lecithaster macrocotyle* and *L. micropsi*. Their distribution is restricted to the southern hemisphere. The former occurs in the low Antarctic, in the SubAntarctic and rarely in the Ross Sea; the latter occurs in the SubAntarctic and off South Georgia. Specimens of both pairs of parasites occur partly in the same fish species, for example *Derogenes* spp. in *Trematomus hansoni*, and *Lecithaster* spp. in *Chaenocephalus aceratus*. It indicates that their presence or absence is not associated with the distribution of final hosts, but rather with the presence or absence of the particular molluscan hosts of the parthenogenetic generations.

All other digenean species recorded in Antarctic fishes are intestinal parasites. Fourteen species of the family Opecoelidae, accompanied by one species endemic to the Kerguelen subregion, are listed in Table 2. Eight species occur in the Eastern Antarctic, but only two of them also off the South Shetland Islands. Two other species occur off South Georgia and in the SubAntarctic. The specificity to final hosts is narrower than that of Hemiuroidea. Eight species occur mainly or entirely in Notothenioidei, three appear to be specific to the Muraenolepididae (two of them were found in Nototheniidae, but all parasites were immature), one species is specific to Zoarcidae, three species normally occur in both Nototheniidae and either Zoarcidae or Congiopodidae. It should be noted, that three species of parasites normally occurring in Notothenioidei rarely occur in Liparididae and another one in Zoarcidae. Probably, we are observing the moment of acquisition of new hosts and new species may be formed in future. Some closely related but distinct species occur in different areas. For example, *Macvicaria pennelli* occurs in the Ross Sea and South of the Indian Ocean, *M. georgiana* - off the South Shetland Islands and South Georgia, *M. antarctica* - in theSubAntarctic and rarely off South Georgia. Similar digeneans occurring in the Weddell Sea appear to be morphologically intermediate between typical *M. pennelli* and *M. georgiana*. They are tentatively assigned to the latter species. Similarly, *Neolebouria antarctica* occurs in the Western Antarctic, whereas *N. terranovaensis* occurs in the Eastern Antarctic. *Stenakron glacialis* was recorded in the subcontinental waters of the Antarctic, whereas *S. kerguelense* was found in the Kerguelen subregion of the SubAntarctic. In contrast, two representatives of the genus *Discoverytrema* occur in the same host species (*Muraenolepis microps*), habitat (posterior part of the small intestine) and locality (off South Georgia). The same is true for *Macvicaria georgiana* and *M. microtestis* in the Weddell Sea. No one opecoelid species occurring in subcontinental waters was recorded in the

Table 2. Records of Opecoelidae in Antarctic fishes

	Parasite	Hosts	Areas of records					
			R	I	W	SS	G	SA
1.	*Discoverytrema markowskii*	Muraenolepididae					+	+
	D. markowskii juv.	Nototheniidae*					+	
2.	*D. gibsoni*	Muraenolepididae					+	
	D. gibsoni juv.	Nototheniidae*					+	
3.	*Macvicaria pennelli*	Notothenioidei, Zoarcidae*	+	+				
4.	*M. georgiana*	Notothenioidei			+	+	+	
5.	*M. antarctica*	Nototheniidae					+	+
6.	*M. microtestis*	Notothenioidei			+			
7.	*M. ophthalmolyci*	Zoarcidae, Nototheniidae				+		
8.	*M. longibursata*	Zoarcidae			+			
9.	*M. muraenolepidis*	Muraenolepididae					+	
10.	*Neolebouria terranovaensis*	Notothenioidei, Liparididae*	+	+	+			
11.	*N. antarctica*	Notothenioidei, Liparididae*				+	+	
12.	*Helicometra antarcticae*	Nototheniidae, Zoarcidae	+					
13.	*H. rakusai*	Nototheniidae		+	+			
14.	*Stenakron glacialis*	Notothenioidei, Liparididae*		+	+	+		
15.	*S. kerguelense*	Notothenioidei, Congiopodidae						+

*Accidental host

SubAntarctic. The South Georgia area represents an intermediate position, as both Antarctic and SubAntarctic opecoelids occur there.

The third higher taxon of the Antarctic fish digeneans is the family Lepocreadiidae and 15 species are listed in Table 3. Most of the species are specific either to Notothenioidei or even only to Nototheniidae (five specific species and one additionally recorded in a zoarcid) or to Gadiformes - families Muraenolepididae and/or Macrouridae (eight species).

Only one species is specific to Zoarcidae. Distribution is usually restricted either to exclusively subcontinental waters (seven species) or to subcontinental waters and the South Georgia area (three species). Two species were recorded only off South Georgia, but their endemicity is doubtful due to lack of data from the same hosts occurring in other areas. Three species occur in both Antarctic and SubAntarctic regions. *Lepidapedon garrardi* is common in all areas investigated. The occurrence

Table 3. Records of Lepocreadiidae in Antarctic fishes

	Parasite	Hosts	Areas of records					
			R	I	W	SS	G	SA
1.	*Lepidapedon garrardi*	Nototheniodei	+	+	+	+	+	+
2.	*L. balgueriasi*	Nototheniidae			+			
3.	*L. notogeorgianus*	Nototheniidae				+	+	
4.	*L. tertius*	Zoarcidae				+		
5.	*L. paralebouri*	Muraenolepididae					+	
6.	*L. brayi*	Macrouridae			+			
7.	*L. ninae*	Macrouridae			+			
8.	*Neolepidapedon magnatestis*	Nototheniidae				+	+	+
9.	*N. trematomi*	Nototheniodei, Zoarcidae[a]		+	+	+	+	
10.	*N. macquariensis*	Nototheniodei		+				+
11.	*Paralepidapedon antarcticum*	Macrouridae		+				
12.	*P. dubium*	Macrouridae		+				
13.	*P. awii*	Macrouridae			+			
14.	*Muraenolepitrema magnatestis*	Muraenolepididae					+	
15.	*Postlepidapedon opisthobifurcatus*	Muraenolepididae, Macrouridae			+		+	

[a]Accidental host

of any Antarctic lepocreadiid species further north than in the SubAntarctic is not known. Three species belong to two other higher taxa - families Monorchiidae (one species) and Fellodistomidae (two species) (Table 4). *Postmonorchis variabilis* is common in its proper hosts (Notothenioidei) off South Georgia and in the SubAntarctic, and absent in subcontinental waters. Both fellodistomid species were recognized for the first time in 1996 in the Weddell Sea and recently described as new [6]. According to

Table 4. Records of Monorchiidae and Fellodistomidae in Antarctic fishes

	Parasite	Hosts	Areas of records		
			W	G	SA
	Monorchiidae				
1.	*Postmonorchis variabilis*	Nototheniidae, Harpagiferidae		+	+
	Fellodistomidae				
1.	*Steringophorus arntzi*	Bathydraconidae, Artedidraconidae	+		
2.	*S. liparidis*	Liparididae	+		

present data, each is specific to one higher taxon of fishes, the first one to the superfamily Notothenioidei, the second to the family Liparididae. Other representatives of Monorchiidae and Fellodistomidae (not listed here) were recorded in the SubAntarctic.

Data on the diversity of Digenea of fishes in different areas of the Antarctic (Table 5) should be regarded as preliminary ones, as they are strongly dependent upon the extent of research. For example, the smallest number of digenean species (9) has been reported from the Ross Sea, but almost all data are based on examinations of fishes caught in relatively shallow waters (down to ca. 200 m) with the only exception being one zoarcid species occurring at a depth of ca. 600 m. The greatest number of species (22) has been reported from the Weddell Sea, where a lot of deep water fish species have been investigated, including many others than Notothenioidei. More digenean species were recorded in the South Georgia area than off the South Shetland Islands. It is clearly dependent upon the research on muraenolepidids in the former area, as numbers of parasites of Notothenioidei are similar in both areas. Data from the Indian sector of the Southern Ocean are considerably less numerous than those from the Weddell Sea. They are based mainly on material collected over 60 years ago and elaborated 40 years later [9], with additional data based on a limited number of fishes received by the present author from Russian, Italian and French ichthyologists. Thus, a low number of parasite species appears to be associated with too poor investigations. It should be noted that intensities of digenean infections are greater in coastal zones (especially in fjords) in the Western Antarctic, reaching to more than 1000 parasites in some *Notothenia neglecta* and *N. rossii*, than in the open sea and deep waters, especially in the Eastern Antarctic (normally several, rarely more than ten parasites in one fish).

Table 5. Species diversity and specificity of fish Digenea in different areas of the Antarctic

Area and specificity	Number of species		
	Parasitic in Notothenioidei	Parasitic in other fishes	Total
Ross Sea	9	2	9
Indian sector	10	6	13
Weddell Sea	13 (14)	8	21 (22)
South Shetlands	15	4	17
South Georgia	13 (15)	9	20
Specific parasites	15	17 (14)	32 (29)
Nonspecific parasites	12 (15)	12 (15)	12 (15)
Total	27 (30)	29	44

Data from areas and hosts out of the Antarctic are omitted. Numbers including immature forms are in parentheses

Taking into account the data from all Antarctic areas together, probably all (or almost all) digenean species occurring commonly in fishes living in shallow waters (down to 200 m) have been recognized. Unfortunately, our knowledge of species occurring in fishes living at deeper parts of the shelves and at the continental slope is limited. Examinations of fishes other than Notothenioidei, including deep water artedidraconids and bathydraconids, are needed in all areas. In total, 15 digenean species appear to be specific parasites of Notothenioidei in the Antarctic and 17 are specific parasites of other fishes; 12 digenean species occur in the Antarctic in both Notothenioidei and other fishes. However, three further species, *G. bowersi, D. varicus* and *L. macrocotyle,* found in the Antarctic only in Notothenioidei have been recorded in the SubAntarctic in other fishes. These results should be regarded as preliminary ones. The final conclusions of this review are as follows:

1. Out of 44 digenean species, 43 normally occur in bottom fishes and one in bathypelagic ones.
2. Intensities of infections with Digenea are greater in coastal zones, but the species diversity is greater in the open sea.
3. Four species are cosmopolitan or bipolar, 30 appear to be endemic to the Antarctic (four hemiuroids, 12 opecoelids, 12 lepocreadiids and two fellodistomids).
4. Numbers of digenean species known as occurring in each area are not final; those from the Ross Sea and Indian sector of the Southern Ocean are especially far from the reality.
5. Further investigations, especially on deep water fishes in the high Antarctic, are necessary and findings of undescribed species are expected.

Acknowledgments

The present review is based on material collected by the author during three expeditions of the Polish Academy of Sciences headed by Professor S. Rakusa-Suszczewski, the Expedition ANT XIII/3 organized by the Alfred Wegener Institute for Polar and Marine Research headed by Professor W. Arntz, additional materials received from Dr. M. Vacchi, Dr. E. Pisano, Dr. S. Palladino, Dr. C. Ozouf-Costaz, Dr. M. White, Dr. A. Balushkin, Professor A. Kompowski and Mr. T. Zadrózny, loans from the Natural History Museum in London and Dr. M. Ostrowski de Nunez, and wide literature data listed by Zdzitowiecki [1].

References

1. Zdzitowiecki K (1996) Antarctic Digenea, parasites of fishes. Synopses of the Antarctic Benthos 8. Koeltz Scientific Books, Koenigstein (in press)

2. Zdzitowiecki K, Cielecka D (1997) Digenea of fishes of the Weddell Sea. I. Parasites of *Macrourus whitsoni* (Gadiformes: Macrouridae). Acta Parasit 42: 23-30
3. Zdzitowiecki K, Cielecka D (1997) Digenea of fishes of the Weddell Sea. II. The genus *Macvicaria* (Opecoelidae). Acta Parasit 42: 77-83
4. Zdzitowiecki K, Cielecka D (1997) Digenea of fishes of the Weddell Sea. III. The Lepocreadiidae (genera *Neolepidapedon* and *Lepidapedon*), parasites of Notothenioidea. Acta Parasit 42: 84-91
5. Zdzitowiecki K (1997) Digenea of fishes of the Weddell Sea. IV. Three opecoelid species of the genera *Neolebouria, Helicometra* and *Stenakron*. Acta Parasit 42: 138-143
6. Zdzitowiecki K (1997) Digenea of fishes of the Weddell Sea. V. Two new species of the genus *Steringophorus* (Fellodistomidae). Acta Parasit 42: 144-148
7. Zdzitowiecki K (1997) Digenea of fishes of the Weddell Sea. VI. The superfamily Hemiuroidea. Acta Parasit 42 in press
8. Gon O, Heemstra PC (1990) Fishes of the Southern Ocean. JLB Smith Institute of Ichthyology, Grahamstown
9. Prudhoe S, Bray RA (1973) Digenetic trematodes from fishes. Rep BANZ Antarctic Res Exped, ser B 8: 195-225

Part 2

Physiology and Biochemistry

Cold Adaptation and Stenothermy in Antarctic Notothenioid Fishes: What Has Been Gained and What Has Been Lost?

George N. Somero[1], Peter A. Fields[1], Gretchen E. Hofmann[2], Randi B. Weinstein[3], and Helena Kawall[4]

[1]Hopkins Marine Station, Stanford University, Pacific Grove, CA 93950, USA
[2]Department of Biology, University of New Mexico, Albuquerque, NM 87131, USA
[3]Department of Biochemistry, University of Arizona Health Sciences Center, Tucson, AZ 85724, USA
[4]Department of Marine Science, University of South Florida, St. Petersburg, FL 33701, USA

Introduction

Antarctic notothenioid fishes inhabit the coldest and most thermally stable waters in the ocean. During their evolutionary histories, these fishes have adapted to the threats posed by potentially freezing temperatures as well as the decelerating effects of reduced temperatures on rates of physiological processes. Cold adaptation of numerous biochemical systems has been a major feature of the evolutionary development of these species [1-4].

The extremely stable temperatures found in the Antarctic Ocean also appear to have influenced profoundly the characteristics of these fishes, by allowing the development of an extraordinary degree of stenothermy. A study of the thermal tolerance limits of several species of nototheniid fishes from McMurdo Sound, where temperatures are virtually constant both seasonally and with depth, found that heat death occurs at only a few degrees above 0 °C [5]. Long-term acclimation of McMurdo Sound nototheniids is possible only at temperatures below 5 °C. Thus, these fishes have the narrowest range of thermal tolerance of any animal species yet studied.

The acquisition of cold-adapted biochemical systems enables Antarctic notothenioids to perform physiological tasks, such as mitochondrial ATP synthesis [6], muscle contraction [7], and neural transmission [3], at temperatures at which less cold-adapted ectotherms would cease to metabolize. The gain of these remarkable capacities for function at near-

G. di Prisco, E. Pisano, A. Clarke (Eds)
Fishes of Antarctica. A biological overview
© Springer-Verlag Italia 1998

freezing temperatures has, however, been paired with the loss of abilities to withstand, and to acclimatize to, temperatures much above 0 °C. Although cold adaptation and stenothermy have developed in parallel in Antarctic notothenioids, they are not necessarily linked in any mechanistic sense. That is, there is no a priori basis for concluding that adaptation to low temperatures should necessarily restrict a species' ability to tolerate high temperatures. Thus, as we examine cold adaptation and stenothermy in Antarctic notothenioids, it is important to keep in mind that these two most striking characteristics may be the consequence of quite different causal factors.

Metabolic Compensation to Low Temperature

A long-standing debate in polar biology concerns whether or not the metabolic (respiratory) rates of polar ectotherms, including fishes, are rate-compensated for function at low temperatures [8]. Do the respiratory rates of notothenioid fishes exhibit a temperature-compensatory upwards adjustment, such that, at the low body temperatures of these species, their rates of oxygen consumption are similar to those of more warm-adapted species at their higher habitat (body) temperatures? Initial studies were interpreted as providing evidence for a substantial degree of metabolic cold adaptation in Arctic fishes [9] and Antarctic nototheniid fishes (albeit not in an Antarctic zoarcid) [10]. However, further study of polar and temperate species, coupled with critical reexamination of earlier data [8,11,12] undermined the generalization that polar fishes are highly cold-adapted in their respiratory rates. As emphasized by Holeton [11] and Clarke [8], the use of whole-organism oxygen consumption rates to study metabolic cold adaptation is fraught with logical and experimental pitfalls. For instance, comparisons of active and sluggish species, especially species belonging to distantly related families, may yield more information about the effects of "mode of life" (Zimmermann and Hubold, this Vol.) on metabolic rate than about adaptation to temperature per se. The time period during which a fish is allowed to acclimate to the experimental chamber, and the dietary status and reproductive state of the fish also may greatly affect its rate of oxygen consumption. Because of the difficulties associated with studies of whole-organism respiration, we sought to find a less ambiguous index of metabolic temperature compensation, one that would allow meaningful comparisons to be made between warm- and cold-adapted fishes belonging to widely different families and having different modes of life.

Our approach to resolving the question about the degree to which metabolic rates of Antarctic fishes are cold adapted involves measurement of in vitro activities of enzymes of ATP-generating pathways. There are several reasons why in vitro enzymatic activities can be expected to yield insights into metabolic compensation and to avoid the pitfalls associated with studying whole-organism respiration. First, in vitro enzymatic activities are not subject to artifacts arising from the response of a fish to confinement in a metabolic chamber. Second, at least for certain organs, enzymatic activities are likely to be independent of nutritional and reproductive states. Third, in some organs, the fish's mode of life, for example, its locomotory habit and general activity level, should not be a strong determinant of ATP-generating ability. The organ in which enzymatic activities seem most likely to provide a clear indication of metabolic compensation to temperature is brain [13]. We hypothesize that the metabolic activity of brain should be largely independent of a fish's mode of life and its reproductive and nutritional status. The activities of enzymes in brains of temperate zone fishes with widely different modes of life, for instance, pelagic versus benthic existence, are remarkably similar [14]. Long-term food deprivation led to significant changes in enzymatic activities in muscle but not in brain of the scorpaenid fish *Sebastolobus alascanus* [15]. Size-dependent effects on enzymatic activity have been observed in brain and other tissues, but these effects can readily be taken into account in analyses [16]. Thus, if temperature compensation of ATP production is a critical feature of adaptation to temperature, it should be manifested especially clearly in brain.

Our comparisons of the in vitro activities (International Units per gram fresh (wet) weight of tissue) of two enzymes involved in ATP generation, citrate synthase (CS) of the citric acid cycle and lactate dehydrogenase (LDH) of the glycolytic pathway, in homogenates of brains from several Antarctic notothenioid and tropical fishes provided clear evidence for partial temperature compensation of brain ATP-generating capacity [13]. Rates of CS and LDH activity were measured at a common temperature, 10 °C. A Q_{10} of 2.0 was used to adjust the 10 °C rates to the rates expected at each species' normal physiological temperatures (-1.9 °C for Antarctic fish and 25-30 °C for tropical species). Rather than observing the approximately 8-fold difference in activity at physiological temperatures that would be predicted by Q_{10} relationships if no compensation were present, LDH activity was only 2.6-fold higher and CS 4.9-fold higher in the tropical species. The factors that underlie this incomplete temperature compensation of brain ATP-generating capacity remain to be discovered.

It seems possible that the energy costs of maintaining ion disequilibrium might be reduced at low temperatures, but this hypothesis has not been tested [17]. In contrast to brain tissue, CS and LDH activities in locomotory muscle varied widely among Antarctic and tropical species in concert with differences in locomotory habits: active swimmers had high activities of both enzymes compared to sluggish species (Kawall and Somero, unpublished observations). As would be predicted by the analyses of Holeton [11] and Clarke [8], no apparent temperature-related differences in muscle ATP-generating capacity were observed (Kawall and Somero, unpublished observations).

Cold Adaptation and Stenothermy in Mitochondrial Function

The ability to respire - regardless of rate - at extremely low temperatures is dependent upon mitochondrial function, a highly temperature-sensitive process [18,19,6] that, in eurythermal species, is subject to the effects of acclimation [20]. There are several properties of mitochondrial respiration that provide insights into the thermal tolerance limits of mitochondrial ATP generation. These properties include 1) the effectiveness of coupling of electron transport to phosphorylation of ADP (as measured by the Acceptor Control Ratio (ACR), the degree to which respiration is stimulated by introducing ADP into the respiration medium), and 2) the temperatures at which sharp discontinuities ("breaks") in slope are observed in Arrhenius plots of log respiration rate *versus* the reciprocal of absolute temperature [Arrhenius Break Temperatures (ABTs)]. We studied these two properties of mitochondria isolated from livers of a nototheniid fish, *Trematomus bernacchii,* and several temperate and tropical species (Weinstein and Somero, in preparation).

Our results provide further illustrations of the degree of cold adaptation present in the notothenioids and the restricted range of temperatures over which their physiological systems can function. Mitochondria from *T. bernacchii* exhibited a high capacity for ATP synthesis (high ACR values) only at temperatures below approximately 18 °C; at higher temperatures electron transport rapidly became uncoupled to ATP production. The temperatures at which sharp breaks in the slopes of Arrhenius plots occurred also reflected the low adaptation temperature of the Antarctic fish (Fig. 1).

The Arrhenius Break Temperature for mitochondria from *T. bernacchii* is the lowest reported for any organism, and falls well below the value predicted by the regression equation based on ABT data from a wide range of vertebrates and invertebrates (Fig. 1) [18-20]. Moreover, 2 weeks of acclimation to 4 °C led to no change in the ABT, a result which contrasts with the plasticity in this trait found in more eurythermal animals ([20], Weinstein and Somero, in preparation).

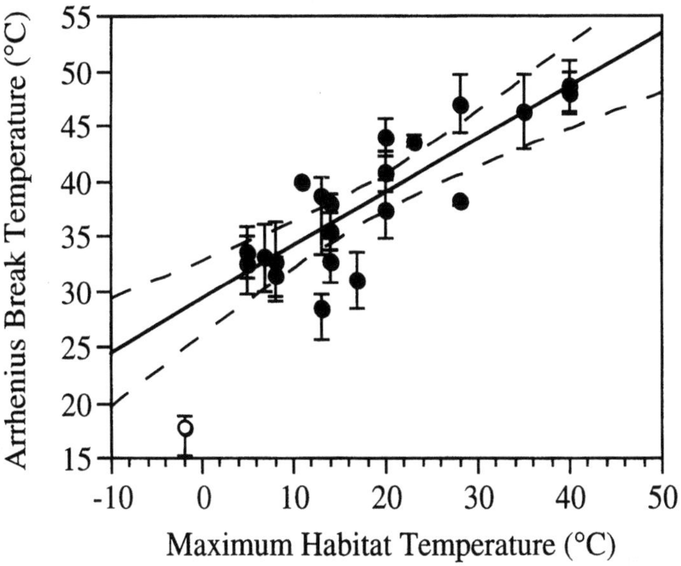

Fig. 1. The relationship between the Arrhenius Break Temperature (ABT) for uncoupled respiration of mitochondria and maximum habitat temperature. Filled symbols represent ABT values determined for a wide variety of marine invertebrates [18,20] and marine fishes (Weinstein and Somero, in preparation). The regression line is based on all data shown by filled symbols

In the context of metabolic compensation to temperature, it should be noted that the respiratory rates of mitochondria of the Antarctic fish *Notothenia coriiceps*, when expressed on the basis of oxygen consumed per unit mitochondrial protein, do not exhibit temperature compensation [6]. Thus, compensation to temperature in rates of mitochondrial ATP generation, as evidenced by elevated levels of CS activity in brain tissue of Antarctic fish, are likely to be due to increased numbers of mitochondria rather than more efficient mitochondria.

In summary, mitochondrial function in Antarctic fishes provides further evidence in these species for a high level of adaptation to cold temperatures, a low upper temperature limit for function, and a lack of ability to acclimate to higher temperatures.

Is the Heat Shock Response Missing in *Trematomus bernacchii*?

A further manifestation of the high degree of stenothermy found in notothenioid fishes was discovered in studies of the heat shock response in

T. bernacchii (Hofmann and Somero, in preparation). The heat shock response entails the induction of synthesis of one or more types of molecular chaperones, proteins that assist in the folding and compartmentalization of other proteins [21]. All cells examined to date have been found to contain several types of molecular chaperones which are essential for facilitating protein synthesis and maturation under normal (nonstressful) cellular conditions. In addition to these constitutively expressed chaperones, stress-induced chaperones ("heat shock" or "stress" proteins) also have been found in all organisms examined. Stress-induced chaperones prevent aggregation of unfolded proteins and, therefore, can be viewed as an important "rescue" mechanism for helping to restore the native structure of proteins damaged by environmental stress, such as that resulting from exposure to high temperatures.

To investigate the heat shock response in *T. bernacchii*, we employed an in vivo protein synthesis protocol that has been used successfully to study the induction of heat shock proteins in several other species of fishes [22,23]. This method involves the injection of specimens with a mixture of ^{35}S-labeled methionine and cysteine, followed by exposure of the animals to a range of temperatures, including temperatures predicted to be high enough to induce the heat shock response. Following 2 h of exposure to test temperatures, the animals are returned to a normal physiological temperature and protein synthesis is allowed to occur for an additional hour. At the end of this period, animals are sacrificed and the newly synthesized proteins are visualized using SDS-PAGE and autoradiography. When used in studies of temperate and tropical fishes, this method has always yielded clear evidence for the induction of synthesis of heat shock proteins, as shown for the eurythermal goby fish *Gillichthys mirabilis* in Fig. 2.

Heat shock proteins of two size classes, 70 kDa and 90 kDa, are strongly expressed in the gill tissue of this goby at 37 °C, but not at 25 °C. In contrast to the response seen in *G. mirabilis* and several other marine teleosts [23], we found no evidence of any induction of heat shock protein synthesis in gill tissue (Fig. 2), brain, liver, spleen or muscle of *T. bernacchii* exposed to high temperatures (6 °C, Fig. 2, or 9 °C, data not shown). These negative findings represent the first case in which the heat shock response has failed to be observed in an animal. Western analysis showed that tissues of *T. bernacchii* contain constitutively expressed molecular chaperones, in common with all other species, but our failure to detect induction of stress-induced chaperones, i.e., heat shock proteins, suggests that the capacity to increase chaperone synthesis in response to

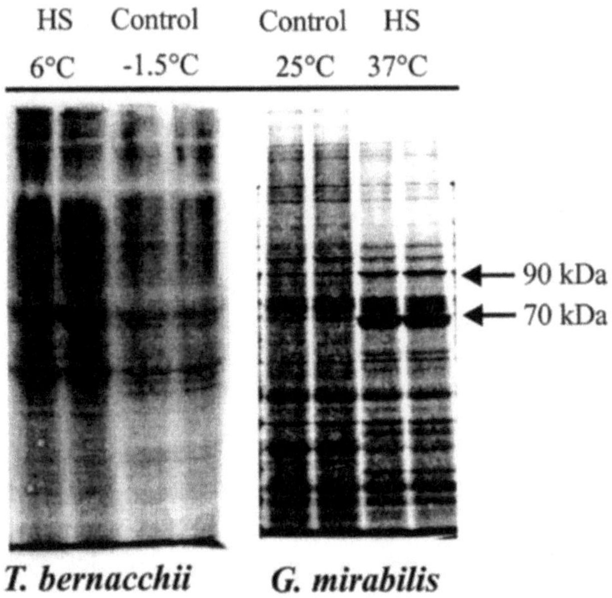

Fig. 2. Protein synthetic patterns in gill tissue from control and heat-shocked (HS) *Trematomus bernacchii* and *Gillichthys mirabilis*. Bands shown on these autoradiographs denote new protein synthesis. For *G. mirabilis*, the bands corresponding to heat shock proteins of the 70 kDa and 90 kDa classes are indicated. These bands are absent in the lanes for *T. bernacchii*. Data are from Hofmann and Somero (unpublished)

thermal stress has been lost in these fishes during tens of millions of years of evolution in extremely cold and stable waters. Whether stress-induced synthesis of chaperones occurs in response to other factors known to trigger synthesis of heat shock proteins, such as heavy metals, remains to be determined. It will be important to examine notothenioid fishes further to see whether the genes encoding stress-induced chaperones have been lost or rendered nonfunctional, as has been found for certain of the genes for oxygen-binding proteins in members of the family Channichthyidae [24,25]. Furthermore, it will be necessary to examine in more detail the possibility that mRNA encoding one or more stress-induced chaperones is transcribed ([26], Carratù and Maresca, this Vol.), but not translated in these fishes. Taken at face value, the failure to observe heat-induced expression of any heat shock protein in any tissue of *T. bernacchii* is another manifestation of the extent to which the notothenioid fishes are adapted to low and extremely stable water temperatures.

Adaptation in Enzyme Kinetic Properties: Lactate Dehydrogenase-A Orthologs of Notothenioid Fishes

Many structural and functional properties of proteins reflect adaptation to temperature [27-29]. When orthologous homologs (orthologs) of a protein from differently thermally adapted species are compared, their thermal stabilities typically vary directly with adaptation temperature. Furthermore, temperature-sensitive kinetic properties such as Michaelis-Menten constants (K_m values) are highly conserved among species at their physiological temperatures. K_m values generally provide an index of the ability of an enzyme to bind its substrate(s) and cofactor(s). Rapid increases in K_m with changing temperature are a sign that the enzyme is losing its ability to bind ligands and is becoming impaired in function. K_m values generally increase with rising temperature [29], although for some enzymes steep rises in K_m with falling temperature have been observed [4]. The enzymes of eurythermal species typically have K_m values that are less perturbed by temperature than those of enzymes from stenothermal species [30,31].

To obtain further insights into the biochemistry of cold adaptation and stenothermy in Antarctic notothenioid fishes, we compared the effects of temperature on the K_m of pyruvate for A-type (muscle isoform) lactate dehydrogenases (LDH-As) from six nototheniid and channichthyid species (Fig. 3; Fields and Somero, unpublished observations).

All of the notothenioid species' LDH-As exhibited similar responses of K_m of pyruvate to temperature. At temperatures near 0 °C, the K_m of pyruvate for all orthologs from the notothenioids was in the range of 0.15 to 0.25 mM pyruvate, i.e., within the range of K_m values at physiological temperatures typical of LDH-As of other vertebrates [29]. (Note that the box enclosing the K_m values for LDH-As of the notothenioid species includes the 95% confidence limits around the K_m values.) At temperatures only a few degrees above 0 °C, however, the K_m of pyruvate began to rise sharply, an index of the stenothermy of the LDH-A orthologs of the notothenioids. In contrast, the slope of the K_m of pyruvate versus temperature plots for the LDH-As of two goby fishes, the extremely eurythermal species *Gillichthys seta*, which encounters habitat eurythermal congener, *Gillichthys mirabilis*, is relatively shallow, illustrating that the eurythermality observed at the level of the whole organism may also be reflected in the kinetic properties of individual enzymes.

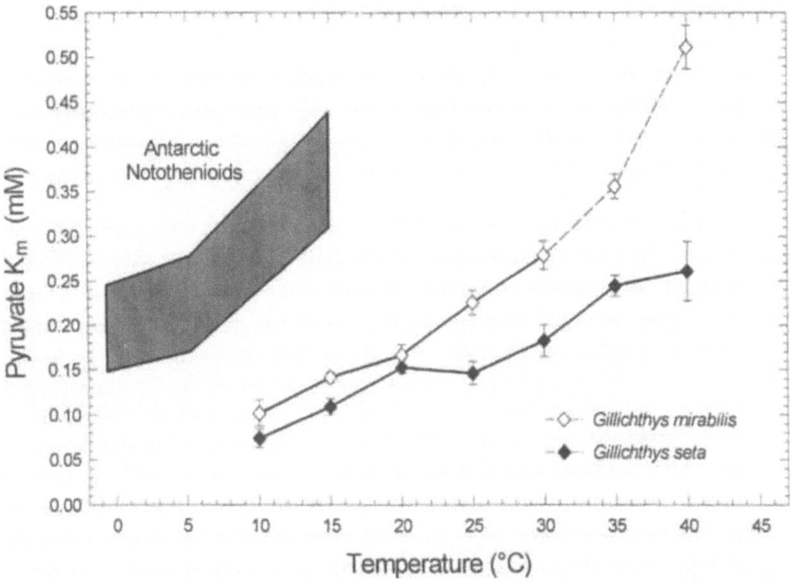

Fig. 3. The effect of temperature on the Michaelis-Menten constant (K_m) of pyruvate for LDH-A orthologs of six Antarctic notothenioid fishes (*Chionodraco rastrospinosus, Chaenocephalus aceratus, Champsocephalus gunnari, Lepidonotothen nudifrons, Notothenia coriiceps,* and *Parachaenichthys charcoti* (shaded box)) and two species of goby, *Gillichthys mirabilis* and *Gillichthys seta.* Data on LDH-As of *Gillichthys* congeners are from Fields and Somero [31] and data for LDH-As of notothenioids are unpublished data of Fields and Somero. The box enclosing the data for the notothenioid species includes the 95% confidence intervals for the estimates of the K_m of pyruvate. The dark line segments for the *Gillichthys* congeners indicate approximate ranges of habitat (body) temperatures

The loss of substrate binding ability at temperatures only slightly above 0 °C has also been observed in other enzymes from Antarctic fishes, including pyruvate kinase of *T. bernacchii* [32] and acetylcholine esterase of *Pagothenia borchgrevinki* [33]. The very steep rise in K_m of acetylcholine for the latter enzyme at temperatures only slightly above 0 °C is paired with a rapid increase in the rate at which quanta of acetylcholine are released at neuromuscular junctions in *P. borchgrevinki* [3]. A rapid increase in the rate of release of neurotransmitter, coupled with a failure of the transmitter-degrading enzyme to bind its substrate, might lead to severe impairment of neuromuscular function, and may explain in part the very low upper lethal temperatures of notothenioid fishes [4]. Stenothermy of enzyme function thus may play a critical role in governing the thermal tolerances of these fishes.

Conclusions

The biochemical and physiological characteristics of Antarctic notothenioid fishes reflect adaptation to the low and extremely stable temperatures found in the Antarctic Ocean. Antifreeze proteins help to protect these fishes from freezing (Cheng and DeVries, this Vol.), and other biochemical adaptations facilitate metabolic function at near-freezing temperatures. The extent to which the respiration rates of Antarctic fishes are cold-adapted remains unclear. Indeed, as Clarke [8] has cogently argued, it seems unlikely that metabolic compensation to temperature can be studied meaningfully at the level of whole-organism respiration. However, our data on activities of LDH and CS in brain indicate that partial compensation to temperature exists in ATP-producing pathways in this organ. The finding that compensation, while substantial, is still only partial raises the interesting question as to why compensation in ATP production is not complete. Future studies must examine the energy costs of maintaining neural function, for instance, transmembrane ion gradients, in nervous tissue at different temperatures [17].

The stenothermy of Antarctic fishes noted at the whole organism level [5] is also evident in many of the biochemical characteristics of these species. The mitochondria of *T. bernacchii* have the lowest tolerance of high temperature reported for any organism. The temperature at which the Acceptor Control Ratio decreases rapidly, and the extremely low Arrhenius Break Temperature for mitochondrial respiration both reflect what appears to be a general characteristic of the thermal responses of the physiological systems of Antarctic notothenioids: a pronounced ability to function well at extremely low temperatures, but the inability to function at, and to acclimate to, temperatures much above 0 °C.

The lack of induction of heat shock proteins in *T. bernacchii* may be another illustration of the loss of function at high temperatures that has occurred during the evolution of the notothenioids. These fishes provide striking examples of what might be termed the "blind cave fish phenomenon," that is, the loss of traits that have ceased to be important for the organism in its new environment. The absence of hemoglobin and myoglobin in members of the Channichthyidae may be indicative of what has happened to other biochemical systems in the notothenioids, for which low temperatures or stable temperatures have eliminated the *raison d'être* for the protein(s) encoded by the gene(s). The loss of ability to express heat shock proteins seems consistent with evolution for many millions of years in an environment having only minimal variation in temperature.

Thus, evolution of the notothenioids has involved concurrent acquisition of biochemical adaptations essential for survival at low temperatures and the loss of other biochemical capacities that are unnecessary in the context of life at extremely stable temperatures.

Acknowledgments

Portions of this work were supported by National Science Foundation grants IBN92-06660 to GNS and OPP93-17696 to Dr. D. Manahan for support of the McMurdo Biology Course. We thank Dr. Manahan for his encouragement in many phases of these studies, and Dr. B. Sidell for his generous provisions of specimens of Antarctic fishes.

References

1. Clarke A, Johnston IA (1996) Evolution and adaptive radiation of Antarctic fishes. Trends Ecol Evol 11:212-218
2. Eastman J (1993) Antarctic fish biology: evolution in a unique environment. Academic Press, San Diego
3. Macdonald JA, Montgomery JC, Wells RMG (1988) The physiology of McMurdo Sound fishes: current New Zealand research. Comp Biochem Physiol 90B:567-578
4. Somero GN (1991) Biochemical mechanisms of cold adaptation and stenothermality in Antarctic fish. In: di Prisco G, Maresca B, Tota B (eds) Biology of Antarctic fish, Springer-Verlag, Berlin, pp 232-247
5. Somero GN, DeVries AL (1967) Temperature tolerance of some Antarctic fishes. Science 156:257-258
6. Johnston IA, Guderley HE, Franklin CE, Crockford T, Kamunde C (1994) Are mitochondria subject to evolutionary temperature adaptation? J Exp Biol 195:293-306
7. Johnston IA, Ball D (1997) Thermal stress and muscle function in fish. In: Wood CM, McDonald DG (eds) Global warming: implications for freshwater and marine fish. Soc Exp Biol Sem Ser 61, Cambridge University Press, Cambridge, pp 79-104
8. Clarke A (1991) What is cold adaptation and how should we measure it? Am Zool 31:81-92
9. Scholander PF, Flagg W, Walters V, Irving L (1953) Climatic adaptation in arctic and tropical poikilotherms. Physiol Zool 26:67-92
10. Wohlschlag DE (1964) Respiratory metabolism and ecological characteristics of some fishes in McMurdo Sound, Antarctica. In: Lee MO (ed) Biology of the Antarctic seas, vol 1. Am Geophy Union, Washington, DC, pp 33-6230
11. Holeton GF (1974) Metabolic cold adaptation of polar fish: fact or artefact? Physiol Zool 47:137-152

12. Wells RMG (1978) Respiratory adaptation and energy metabolism in Antarctic nototheniid fishes. NZ J Zool 5:813-815
13. Kawall HG, Somero GN (1997) Temperature compensation of enzymatic activities in brain of Antarctic fishes: evidence for metabolic cold adaptation. Antarct J (in press)
14. Sullivan KM, Somero GN (1980) Enzyme activities of fish skeletal muscle and brain as influenced by depth of occurrence and habits of feeding and locomotion. Mar Biol 60:91-99
15. Yang T-H, Somero GN (1993) Effects of feeding and food deprivation on oxygen consumption, muscle protein concentration and activities of energy metabolism enzymes in muscle and brain of shallow-living (*Scorpaena guttata*) and deep-living (*Sebastolobus alascanus*) scorpaenid fishes. J Exp Biol 181:213-232
16. Somero GN, Childress JJ (1980) A violation of the metabolism-size scaling paradigm: activities of glycolytic enzymes in muscle increase in larger-size fish. Physiol Zool 53:322-337
17. Hochachka PW (1988) Channels and pumps - determinants of metabolic cold adaptation strategies. Comp Biochem Physiol 90B:515-519
18. Dahlhoff E, O'Brien J, Somero GN, Vetter RD (1991) Temperature effects on mitochondria from hydrothermal vent invertebrates: evidence for adaptation to elevated and variable habitat temperature. Physiol Zool 64:1490-1508
19. Guderley HE, St. Pierre J (1996) In: Johnston IA and Bennett AF (eds) Animals and temperature: phenotypic and evolutionary adaptation. Soc Exp Biol Sem Ser 59, Cambridge University Press, Cambridge, pp 127-152
20. Dahlhoff E, Somero GN (1993) Effects of temperature on mitochondria from abalone (genus *Haliotis*): adaptive plasticity and its limits. J Exp Biol 185:151-168
21. Hartl FU (1996) Molecular chaperones in cellular protein folding. Nature (Lond) 381:571-580
22. Dietz TJ, Somero GN (1992) The threshold induction temperature of the 90-kDa heat shock protein is subject to acclimatization in eurythermal goby fishes (genus *Gillichthys*). Proc Natl Acad Sci USA 89:3389-3393
23. Dietz TJ, Somero GN (1993) Species- and tissue-specific synthesis patterns for heat-shock proteins hsp70 and hsp90 in several marine teleosts. Physiol Zool 66:863-880
24. Cocca E, Ratnayake-Lecamwasam M, Parker SA, Camardella L, Ciaramella M, di Prisco G, Detrich HW (1995) Genomic remnants of α-globin genes in the hemoglobinless Antarctic icefishes. Proc Natl Acad Sci USA 92:1817-1821
25. Sidell BD, Vayda ME (1998) Physiological and evolutionary aspects of myoglobin expression in the hemoglobinless Antarctic icefishes. In: Playle R, Pörtner HO (eds) Cold ocean physiology, Cambridge University Press, Cambridge, in press

26. Maresca B, Patriarca E, Goldenberg C, Sacco M. (1988) Heat shock and cold adaptation in Antarctic fishes: a molecular approach. Comp Biochem Physiol 90B:623-629
27. Arpigny JL, Feller G, Davail S, Génicot S, Narinx E, Zekhnini Z, Gerday C (1994) Molecular adaptations of enzymes from thermophilic and psychrophilic organisms. Adv Comp Envir Physiol 20: 270-
28. Jaenicke R (1991) Protein stability and molecular adaptation to extreme conditions. Eur J Biochem 202:715-728
29. Somero GN (1995) Proteins and temperature. Annu Rev Physiol 57:43-68
30. Coppes Z, Somero GN (1990) Temperature adaptive differences between the M_4-lactate dehydrogenases of stenothermal and eurythermal sciaenid fishes. J Exp Zool 254:127-131
31. Fields PA, Somero GN (1997) Amino acid sequence differences cannot fully explain interspecific variation in thermal sensitivities of gobiid fish A_4-lactate dehydrogenases (A_4-LDHs). J Exp Biol 200:1839-1850
32. Somero GN, Hochachka PW (1978) The effect of temperature on catalytic and regulatory functions of pyruvate kinases of rainbow trout and the Antarctic fish *Trematomus bernacchii.* Biochem J 110:395-400
33. Baldwin J (1971) Adaptation of enzymes to temperature: acetylcholinesterases in the central nervous system of fishes. Comp Biochem Physiol 40:181-187

Do Antarctic Fish Respond to Heat Shock?

Luisella Carratù[1], Andrew Y. Gracey[2], Stefania Buono[1], and Bruno Maresca[1]

[1]International Institute of Genetics and Biophysics, CNR, Via Marconi 12, 80125 Napoli, Italy
[2]University of Liverpool, Department of Environmental and Evolutionary Biology, Liverpool L69 3BX, UK

Introduction

Temperature is a major environmental factor requiring adaptive responses and it is a central selective element in speciation. One of the best known and complex mechanisms that is primarily involved in protecting cells from various forms of stresses, such as temperature, is the stress (heat shock) response [1]. For poikilotherms, fluctuations in environmental temperatures can be fatal. Sudden drops in temperature can lead to a reduction in membrane fluid state that, in turn, causes cessation of normal functions. Unless rapidly corrected, such alterations will lead to physiological damage and, ultimately, to death. Several essential cellular activities depend on proper membrane functionality [2]. For elevated environmental temperatures, some of the problems encountered during chilling are also relevant. There still exist the dual needs of being able to "sense" the elevated environmental temperature and to couple this detection to the induction of gene expression - such as for heat shock genes. Recently, there have been several lines of evidence which suggest that changes in response to an abrupt rise in the environmental temperature may have analogies with chilling adaptation. Thus, membrane lipid composition, as well as the dynamic state of membrane lipids, represent the basic elements for membrane functionality. Membranes, indeed, have multiple physical properties which permit the cell to sense environmental changes. Several recent studies from our and other laboratories now indicate that an immediate change in temperature causes a reorganization of membrane components [3,4]. In addition, abrupt changes occurring in the environment during pathological states or during the aging process can cause modifications in membrane structure, membrane protein activity, signal transduction and gene expression with consequent loss of physiological functions [5,6]. Recently, it has been shown that some forms of heat injury are due to the reorganization of lipid protein complexes of the membrane at the time of heat shock [7,8]. It has also been shown that a

G. di Prisco, E. Pisano, A. Clarke (Eds)
Fishes of Antarctica. A biological overview
© Springer-Verlag Italia 1998

primary heat shock sensor (HSS) responsible for monitoring sudden temperature variations is membrane-associated. Furthermore, we have shown that modification of membrane fluid state and/or loss of bilayer integrity, when the temperature exceeds the physiological range (during heat shock), influences the level of expression of heat shock genes by setting the temperature at which optimum heat shock response occurs [3].

Fish are ectotherms whose body temperature is determined by the water temperature. Antarctic fish represent a particular class of poikilothermic animals since they live in a very stable environment, and, therefore, are more stenothermal than other fish species. Antarctic fish live in the Southern Ocean at temperatures near the freezing point of sea water. Furthermore, water temperature has been remarkably stable over the past 20 million years, exerting a strong and constant selective pressure on biochemical as well as physiological mechanisms of adaptation of the marine organisms. Antarctic fish can live at -1.9 °C, with seasonal fluctuations of the order of only 0.2 °C [9,10]. Broadly speaking, two classes of Antarctic fish can be distinguished based on their upper acclimation temperature. This temperature, termed "ultimate incipient lethal temperature," is 4-5 °C in the ice fish family (e.g. *Chionodraco hamatus*) and around 12-14 °C in other Antarctic fish (e.g. *Notothenia rossii*). Profound evolutionary changes have occurred in several enzymatic and structural proteins of Antarctic fish that have allowed these organisms to maintain roughly the same metabolic and locomotory activities of other fish [2]. So far, very little is known on the mechanisms of regulation of gene expression and how these have contributed to the adaptation to these extreme temperatures.

The question that arose several years ago concerned the possibility that stenothermal species may have lost the ability to respond to heat shock as a result of the extreme environmental condition present in the waters of the Antarctic Ocean [11]. To address this issue, we cloned *hsp70* gene from the Antarctic fish *C. hamatus*. This gene codes for the most conserved heat shock protein throughout evolution, HSP70. Moreover, in the same species a parallel project regarding the identification and analysis of expression of Δ^9-desaturase, whose product is involved in adaptation of membranes to cold, is in progress in our lab. Understanding the regulation of the expression of these two classes of genes in organisms such as Antarctic fish living under the selective pressure of extreme climatic environments, will give an insight into the molecular mechanisms of adaptation, while an analysis of the gene structure will reveal how these sequences have evolved in these organisms.

Physical Analysis of *C. hamatus hsp70* Promoter

We have cloned and sequenced *hsp70* gene of the Antarctic fish *C. hamatus*. The deduced protein sequence revealed a high degree of homology with most of the heat inducible HSP70 proteins. Moreover, the functional domains characteristic among nearly all HSP70s are conserved in the Antarctic protein, particularly in the $-NH_2$ terminal region that interacts with proteins during the transport or the folding/unfolding processes, and in the -COOH terminal part, exerting ATPase activity [12].

Special attention was given to the characterization of the promoter region of *hsp70* gene. In all promoters of eukaryotes, despite the evolutionary distance between organisms, typical *cis*-acting elements (HSE, heat shock element) have been found [13]. HSEs are regions for specific binding of heat shock transcription factor (HSF) which is essential to induce transcriptional activation [14]. Most heat shock promoters contains several HSEs located both proximal and distal to the initiation of transcription start site. The HSE can be described as a series of pentameric units arranged as inverted adjacent arrays of the sequence 5'-nGAAn-3' [15-17].

A functional HSE is composed of a minimum of three pentamers with additional reiteration resulting in higher affinity interactions between the HSF and the HSE [18-20].

Distal HSEs can function as enhancer elements and may be crucial for transcriptional activity [21,22]. Multiple HSEs may have a synergistic effect on transcriptional activity, while rotational alignment confers higher transcriptional activity when two HSEs are within 80 bp apart [23,24].

Amin and co-authors [23] have also conducted experiments suggesting that HSEs does not need to be located at a specific distance from a downstream sequence such as the TATA box. Figure 1 summarizes the results of sequence analysis performed on *C. hamatus hsp70* gene promoter.

We identified six HSEs, three of which match exactly with two 5 bp units nGAAn. Moreover, typical upstream elements found in most eukaryotic promoters of genes transcribed by RNA polymerase II [25] are present also in *hsp70* promoter of *C. hamatus*. This is the case of the TATA box and the CAAT box, usually located respectively at about 20 and 80 nt upstream of the start point of transcription. At the start point there is a tendency for the first base of mRNA to be A, flanked on either side by pyrimidines spanning the positions between -3 and +5 [25]. The promoter under analysis presents all those features, with multiple possible

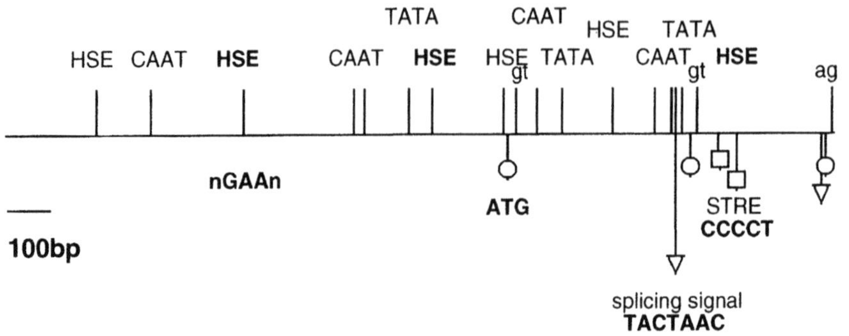

Fig. 1. *Chionodraco hamatus hsp70* gene promoter (2430 bp)

sites of initiation of transcription, as we measured experimentally by primer extension analysis.

Translation may start at either one of the three ATG codons depicted in Fig. 1 without affecting the final product, HSP70. The splicing sites found (GT-AG) follow the GT-AG rule [25] and both introns contain an internally conserved sequence that resembles the TACTAAC consensus sequence of *Saccharomyces cerevisiae* needed for branch site formation and involved in splicing events [26]. Alternative splicing events in different tissues is under investigations using the reverse transcriptase-polymerase chain reaction (RT-PCR) technique.

An additional element found in the *C. hamatus hsp70* promoter is called STRE (stress element) that has been shown to mediate transcription activation by exposure to hydrogen peroxide or increases in osmolarity [27]. STREs have been found in several promoters of yeast genes, and among those some are heat shock genes. Both HSEs and STREs can be involved in the induction of stress resistance in organisms [28].

Expression of *hsp70* Gene

We have initiated the study of the expression of *hsp70* gene with RNAs extracted from spleen of *Trematomus bernacchii* since during the past expeditions to Antarctica very few samples of *C. hamatus* were caught and heat shock treatments could not be performed. However, *T. bernacchii* is a stenothermal fish species like *C. hamatus*. Fish samples were caught off McMurdo Sound by rod and line. The upper lethal temperature of all these species, adapted at -1.9 °C, is approximately 6 °C [29]. Fish were kept in aquaria at -1.2±0.5 °C for 3 weeks, killed and isolated spleens were subjected to heat shock at 5 °C for 4h.

Total RNA was purified also from spleen excised from fish adapted at -1.2 °C (C_0) and from spleen incubated for 4 h at -1.2 °C in saline solution (C_4). Normalized Northern blots probed with a fragment of the cloned *C. hamatus hsp70* gene revealed an increase of about 15% in the level of *hsp70* mRNA in C_4 compared to C_0, suggesting that handling/extraction of the spleen and the subsequent incubation in saline solution was stressful enough to induce an increase in hsp70 transcription. After 4 h of incubation at 5 °C, level of transcription increased about 50% compared to C_0 and about 10% compared to C_4. Possibly, the heat shock response was masked by an initial induction of stress proteins in response to handling conditions.

The level of *hsp70* transcript from *T. bernacchii* was compared with RNA purified from *N. rossii*, a more eurythermal species that experiences significant seasonal temperature variations. In this fish, the optimal temperature of induction of HSPs is 8 °C with an increase in the level of *hsp70* transcript of about 100% compared to the sample heat shocked at 5 °C [30]. The reason why the heat shock response is different in these two species may be due to the difference in the temperature stability of the environment that they live in. It is possible to reason that fish like *T. bernacchii* adapted to a constant water temperature without seasonal temperature variations have a poor capacity to respond to heat stresses, while for species like *N. rossii* the ability to induce the heat shock response is a constraint for survival.

Discussion

Differences in *hsp70* mRNA induction in different species of Antarctic fish open new molecular and evolutionary questions that need to be addressed. Differences in regulation of expression may reside in differences in the promoters of *hsp70* genes of these fish species. The lack of expression of *hsp70* in *T. bernacchii* and the results of the preliminary experiments of primer extension make it very likely that HSEs present in the *T. bernacchii hsp70* promoter may not be functional. This hypothesis is confirmed by our experiments of heat shock on spleen and liver of *N. rossii* in which we found transcription of *hsp70* gene. In particular, *N. rossii* should have functionally active tandem arrays of nGAAn units and HSF could induce transcription by binding to these *cis*-elements. To compare differences in the promoter organization and in their activity, we cloned *N. rossii* and *T. bernacchii hsp70* promoters from genomic DNA and sequencing is in progress.

Other and more complex levels of regulation of the heat shock gene expression may be exerted by the HSF itself, in particular the HSF1, that in all vertebrates activates transcription of heat shock genes after temperature shock, and HSF2 that induces heat shock response during cellular differentiation, spermatogenesis and embryonic development [31]. In chicken it has been shown that low levels of HSF2 mRNA in liver and blood cells correlate with a very low or absent induction of the hsp70 mRNA after heat shock [32]. Therefore, it is reasonable at this stage of investigation to postulate that different levels of HSFs could be present in different tissues of Antarctic fish and also in different fish species, which could explain the lack of expression of *hsp70* in *T. bernacchii*.

Recent evidence suggests that the temperature at which expression of heat shock genes occurs is influenced by the membrane fluid state which in turn is regulated by the activity of the Δ^9-desaturase gene [3]. For this reason it will be crucial to study in detail Δ^9-desaturase gene regulation and determine fatty acid content of membranes from different fish tissue to correlate it to the capacity to induce gene expression.

Our experiments on the differential expression of *hsp70* gene confirm the notion that stenothermal Antarctic species have conserved a low thermal resistance and suggest that the constant low temperature has modified the regulation of expression of these highly conserved genes. This represents, to our knowledge, the first example of a high eukaryotic organism that has lost the heat shock gene induction in response to the evolutionary adaptation to the Antarctic Ocean.

Acknowledgments

This work was supported by a Grant from the Italian Antarctic Programme (PNRA) and by a Grant from CNR Progetto Strategico Stress Cellulare.

References

1. Lindquist S, Craig EA (1988) The heat shock proteins. Ann Rev Genet 22: 631-677
2. Cossins AR, Bowler K (1987) Temperature biology of animals. Chapman & Hall, New York
3. Carratù L, Franceschelli S, Pardini CL, Kobayashi GS, Horvath I, Vigh L, Maresca B (1996) Membrane lipid perturbation modifies the set point of the temperature of heat shock response in yeast. Proc Natl Acad Sci USA 93: 2120-2125
4. Vigh L, Los DA, Horwath I, Murata N (1993) The primary signal in the biological perception of temperature: Pd-catalyzed hydrogenation of membrane lipids stimulated the expression of the *des*A gene in *Synechocystis* PCC6803. Proc Natl Acad Sci USA 90: 9090-9094
5. Hegner D (1980) Age-dependence of molecular and functional changes in biological membrane properties. Mech Ageing Dev 12: 207-214

6. Dimri GP, Campisi J (1994) Altered profile of transcription factor-binding activities in senescent human fibroblasts. Exp Cell Res 212: 116-124
7. Mejia R, Gomez Eichelmann MC, Fernandez MS (1995) Membrane fluidity of *Escherichia coli* during heat-shock. Biochim Biophys Acta 1223: 195-200
8. Revathi CJ, Chattopadhyay A, Srinivas UK (1994) Change in membrane organization induced by heat shock. Biochem Mol Biol Int 16: 925-950
9. Eastman JT (1993) Antarctic fish biology. Evolution in unique environment. Academic Press, San Diego
10. Somero GS (1991) Biochemical mechanisms of cold adaptation and stenothermality in Antarctic fish. In: di Prisco G, Maresca B, Tota B (eds) Biology of Antarctic fish. Springer-Verlag, Heidelberg. pp 216-231
11. Maresca B, Patriarca E, Goldenberg C, Sacco M (1988) Heat shock and cold adaptation in Antarctic fishes: a molecular approach. Comp Biochem Physiol 90B: 623-614
12. Freeman BC, Myers MP, Schumacher R, Morimoto RI (1995) Identification of a regulatory motif in HSP70 that affects ATPase activity, substrate binding and interaction with HDJ-1. EMBO J 14: 2131-2142
13. Bienz M, Pelham HRB (1987) Mechanisms of heat-shock gene activation in higher eukaryotes. Advances in Genetics 24: 31-72
14. Lis J, Wu C (1993) Protein traffic on the heat shock promoter: parking, stalling, and trucking along. Cell 74: 1-4
15. Pelham HRB, Bienz M (1982) A synthetic heat shock promoter element confers heat inducibility on the herpes simplex virus thymidine kinase gene. EMBO J 1: 1313-1317
16. Xiao H, Lis JT (1988) Germline transformation used to define key features of the heat shock response element. Science 223: 1123-1126
17. Amin J, Ananthan J, Voellmy R (1988) Key features of heat shock regulatory elements. Mol Cell Biol 8: 2161-2169
18. Perisic O, Xiao H, Lis JT (1989) Stable binding of *Drosophila* heat shock factor to head-to-head and tail-to-tail repeats of a conserved 5 bp recognition unit. Cell 59: 797-806
19. Xiao H, Perisic O, Lis JT (1991) Cooperative binding of *Drosophila* heat shock factor to arrays of a conserved 5 bp unit. Cell 64: 585-593
20. Kroeger PE, Sarge KD, Morimoto RI (1993) Mouse heat shock transcription factors 1 and 2 prefer a trimeric binding site but interact differently with the HSP70 heat shock element. Mol Cell Biol 13: 1610-1623
21. Bienz M, Pelham HRB (1986) Heat shock regulatory elements function as an inducible enhancer in *Xenopus hsp70* gene and when linked to a heterologous promoter. Cell 14: 753-760
22. Chen J, Pederson DS (1993) A distal heat shock element promotes the rapid response to heat shock of the *hsp26* gene in the yeast *Saccharomyces cerevisiae*. J Biol Chem 268: 7132-7122
23. Amin J, Fernandez M, Ananthan J, Lis JT, Voellmy R (1994) Cooperative binding of heat shock transcription factor to the *hsp70* promoter in vivo and in vitro. J. Biol. Chem 269: 1604-1611

24. Cohen RS, Meselson M (1988) Periodic interactions of heat shock transcriptional elements. Nature 172: 856-858
25. Lewin B (1994) Genes V. Oxford University Press
26. Langford CJ, Klniz FJ, Donath C, Gallwitz D (1984) Point mutations identify the conserved intron-contained TACTAAC box as an essential splicing signal sequence in yeast. Cell 20: 614-653
27. Marchler G, Schüller C, Adam G, Ruis H (1993) A *Saccharomyces cerevisiae* UAS element controlled by protein kinase A activates transcription in response to a variety of stress conditions. EMBO J 12: 1997-2003
28. Mager WH, Kruijff AJJ (1995) Stress-induced transcriptional activation. Microb Reviews 5: 506-531
29. Somero GN, DeVries AL (1967) Temperature tolerance of some Antarctic fishes. Science 156: 257-258
30. Harrison P, Gargano S, Maresca B (1989) Physical analysis and regulation of the gene *hsp70* in Antarctic fish. Scienza e Cultura. Proceedings of the 1st meeting on "Biology in Antarctica"
31. Morimoto RI, Kroeger PE, Cotto JJ (1996) The transcriptional regulation of heat shock genes: a plethora of heat shock factors and regulatory conditions. In: Feige U, Morimoto RI, Yahara I, Polla BS (eds) Stress-inducible cellular responses. Birkhauser Verlag, Basel Boston Berlin. pp 123-163
32. Nakai A, Morimoto RI (1993) Characterization of a novel chicken heat shock transcription factor, heat shock factor 3, suggests a new regulatory pathway. Mol Cell Biol 13: 1983-1997

Aspects of Eco-Physiological Adaptations in Antarctic Fish

Andreas P. A. Wöhrmann

Institut für Polarökologie, Wischhofstr. 1-3, Gebäude 12, University of Kiel, 24148 Kiel, Germany

Introduction

Antarctic marine ectotherms are highly specialized to cold stenothermic conditions. At low temperatures, the rates of molecular diffusion and enzyme reactions can slow down considerably, and in absence of compensating mechanisms, physiological processes in Antarctic ectothermes ought to proceed more slowly than in specimens of temperate or tropical environments [1]. Organisms exhibit a diversity in lipid structures to fashion membranes to prevailing ambient temperatures in such a manner that they become more fluid in a cold-acclimatized state than in a warm-acclimatized state [2].

Higher amounts of unsaturated phospholipid classes increase the fluidity and hence proper functioning of biomembranes at temperatures close to freezing. Exposure to low temperatures also requires biochemical and neurophysiological adjustments protecting biomembranes from freezing. Higher proportions of polyunsaturated fatty acids [3] as well as alterations in the composition of gangliosides [4] may contribute to a proper functioning of the nervous system and brain at freezing temperatures.

In addition to low temperature the continual presence of ice in subzero water presents fishes with another physiological challenge. The presence of antifreeze peptides (AFP) and glycopeptides (AFGP) is one important adaptation for survival in freezing sea-water. Because of this Antarctic fishes tend to occupy most ecological niches of the Antarctic Ocean including the surface and midwaters which are often rich in food and also permanently ice-laden (for review see Clarke [5] and Eastman [6]).

Thus, the objective of my communication is to present a comprehensive data set on several biochemical adaptations of notothenioids to life in high-Antarctic waters with special regard to fully pelagic *Pleuragramma antarcticum* [7], benthopelagic *Trematomus eulepidotus* [8] and to the seasonality of some of the biochemical processes.

G. di Prisco, E. Pisano, A. Clarke (Eds)
Fishes of Antarctica. A biological overview
© Springer-Verlag Italia 1998

Lipid Content and Seasonal Variations

The more pelagic notothenioid species live, the more they tend to reduce heavy components such as bones and scales and the stronger the tendency to accumulate lipids [9]. These modifications are important adaptations towards neutral buoyancy, since all notothenioids lack swim-bladder. However, lipids in high-Antarctic notothenioids are important for two further features, for the development of gonads and as energy reserves for the long winter period. Figure 1 shows the total lipid content in different tissues and the gonado somatic index (GSI) of female *Trematomus eulepidotus* in spring and summer. The GSI increased from 1.8 in spring to 8.2 in summer. In this time the lipid content in percent of dry weight decreased dramatically in the liver and in the gonad from 52.3% to 19.8% and from 60.7% to 17.8%, respectively. By contrast, the lipid content in the dorsal muscle was not significantly changed. However, according to Eastman [9] the liver is not an important site for lipid storage in notothenioid fishes, in contrast e.g. to the lean cod fishes which store most of their lipid in the liver. Livers of male *T. eulepidotus* had higher lipid contents as compared to livers of females, a phenomenon also reported for the benthopelagic *Notothenia rossii marmorata*, presumably due to higher protein contents required in females for the formation of ovaries. The variability of lipid contents in the liver and the gonad may be explained by the fact that these energy stores are utilized more rapidly than other fat reserves when energy is required [10].

Sidell and coworkers emphasize the importance of lipids, particularly monounsaturated fatty acids, as fuel for aerobic energy metabolism. They determined higher lipid levels in serum and oxidative muscle of the "mesopelagic" *Trematomus newnesi* as compared to the demersal *Gobionotothen gibberifrons*, which corresponded with the higher locomotory activity of the dusky notothen [11]. They also found high lipid contents of oxidative muscles being correlated with high capacities for fatty acid catabolism, in contrast to low glycolytic enzyme activities. Furthermore, a high lipid content in oxidative tissues may accelerate oxygen transfer from capillaries to mitochondria [12].

The tendency of lipid accumulation with an increasingly pelagic life style is readily explained by energetic advantages due to improved buoyancy characteristics. DeVries and Eastman [13] claimed that in *Pleuragramma antarcticum* the lipid sacs are extracellular formations. Hence, although the sacs are largely filled with triacylglycerols, they should have no metabolic function as energy reserves and serve exclusively as flotation devices. However, the cellular structures of the

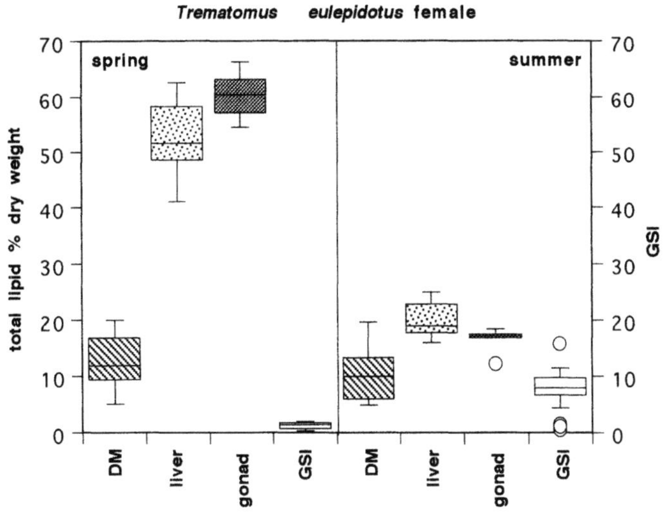

Fig. 1. Lipid contents [% dry weight (DW)] of different tissues [dorsal muscle (DM), liver and gonad] of *Trematomus eulepidotus* female specimens caught in spring and summer. Min-Max, range; x, media; Q1, lower quartil (25%); Q3, upper quartil (75%); GSI, gonado somatic index

lipid sac wall as well as adjacent capillaries indicate that the lipid stores in the sacs are available for metabolic processes [14]. The question still remains, to what extent these lipids are actually utilized as energy reserves in *P. antarcticum*, as discussed by Clarke et al. [15].

With regard to energy requirements, we tend to associate pelagic fishes with species like herring or mackerel, fast-swimming animals with high metabolic costs. Morphology, ultrastructure and blood physiology suggest, however, that pelagic Antarctic fishes like *Pleuragramma antarcticum* and *Aethotaxis mitopteryx* are rather sluggish with a low scope of activity and hence low metabolic requirements [16]. This may also have to do with the increased viscosity of sea water at near-freezing temperatures and hence higher energetic cost of swimming [7]. Reduced energy requirements due to low metabolic rates have also been indicated by starvation experiments with *Trematomus eulepidotus*, which survived extended periods (12 months) without food.

Brain Gangliosides and Neuronal Functions

Most key adaptations of Antarctic notothenioids are based on biochemical modifications [17] in metabolic pathways or important cell structures.

Since the whole metabolism is controlled by the central nervous system one expects a significant temperature correlated adaptation of the glycosphingolipid composition of the synaptic membranes. The ganglioside composition of all investigated endemic Antarctic teleosts is characterized by extreme polar brain gangliosides, which contain fractions with five or even more sialic acid residues. Brains of bathydraconids and artedidraconids contain up to 5% of pentasialogangliosides, which are the highest concentration of these highly sialylated gangliosides that have been found in any vertebrate brain. For example, European perches contain only 1% (or less) pentasialsogangliosides. This speaks in favor of a temperature correlated brain ganglioside adaptation in the CNS of Antarctic fishes [4].

Generally, in fishes there are far more polar fractions with up to 5 or more neuraminic acid residues, but the composition deviates considerably between groups of fish. Nevertheless, the pattern within related species is quite similar. This becomes evident when ranging the brain gangliosides of the various species of fish according to their composition of fractions with 3, or more, or less neuraminic acid residues (Fig. 2). Furthermore, a correlation between the polarity of the brain gangliosides and the adaptation of the species to habitats with different temperatures is evident. The brain gangliosides are composed mainly of less polar fractions in species living under warm-water conditions (lungfish, cichlid), whereas the notothenioids display mainly highly polar gangliosides. The European

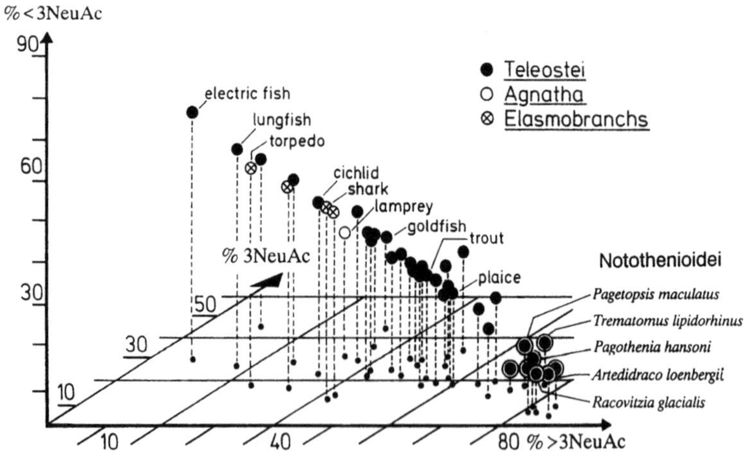

Fig. 2. Three-dimensional arrangement of brain gangliosides in relation to their content of NeuAc residues (3, more than 3 and less than 3 NeuAc residues, respectively) from 31 fish species and some Antarctic notothenioids. (Modified after Rahmann and Hilbig [18])

species living in temperate waters (carp, plaice, trout) have an intermediate position.

Gangliosides, in contrast to other membrane lipids, exhibit specific properties in the presence of calcium [19]. Gangliosides seem to be able to influence the fluidity and/or the charge (dipole) of the membrane. Due to their specific interaction with calcium, gangliosides are assumed to modulate functional membraneous proteins (ATPase, kinase, receptors).

Furthermore, calcium is known to be essential for all neuronal functions. Gangliosides, on the other hand, were shown to possess a pronounced ability to complex with calcium. Figure 3 shows the elemental mapping of

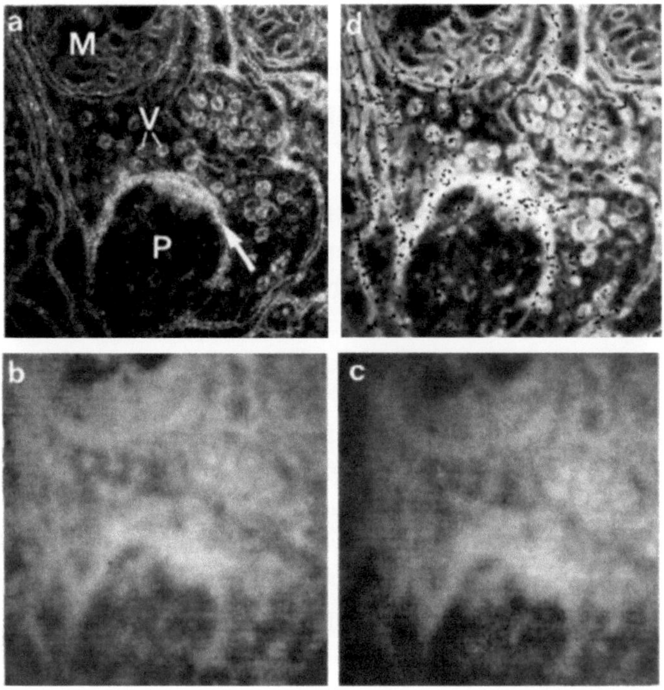

Fig. 3. Cytochemical localization of endogenous calcium in the tectum of the Antarctic silverfish *Pleuragramma antarcticum*. Precipitates are concentrated in extracellular spaces with concentrations in the synaptic cleft (arrowhead). **a** Structure-sensitive contrast produced by ESI at 250 eV energy loss. **b** ESI on the calcium edge at 360 eV. **c** Background image below the calcium adge at 330 eV. **d** Combination of structure-sensitive ESI and the most intensive calcium signals, exceeding the average grey level by more than three times the standard deviation of the grey levels. P, postsynaptic cleft; V, synaptic vesicle; M, mitochondrium. Scale bar = 0.5 μm.

calcium in the optic tectum of *Pleuragramma* after cytochemical precipitation of endogenous calcium. In contrast to other fish species in the nerve cell membrane of *Pleuragramma* highly glycosylated gangliosides are clustered at membraneous proteins, with the evidence of a high Ca^{2+} accumulation within the synaptic cleft [20].

In summary, the results support the hypothesis that the concentration and/or composition of neuronal gangliosides influence basic physico-chemical properties of neuronal membranes. As a consequence, the extremely high polarity of brain gangliosides of cold-blooded vertebrates, especially of those living in habitats near the freezing point and the polarization of the ganglioside composition may facilitate synaptic transmission. Thus, the properties of gangliosides indicate a mechanism at the molecular level that keeps the neuronal membranes functional, even under the extreme temperature conditions of the Antarctic Ocean.

Antifreezes and Their Relation to the Mode of Life

In the high-Antarctic, shallow water species are abundant above 400-450 m depth. The benthic/epibenthic *Cygnodraco mawsoni*, *Trematomus pennellii* and *Pagetopsis macropterus* are representatives of this group and they possess high amounts of both low and high molecular weight AFGP (Fig. 4). Eurybathic species are abundant over most of the depth range down to 600-700 m. *Pogonophryne scotti*, *Racovitzia glacialis*, *Trematomus lepidorhinus* and *Cryodraco antarcticus* belong to this group. The antifreeze content among these fish is moderate, and lower in active (*C. antarcticus*) than in sluggish species (*P. scotti*). Deeper shelf and upper slope species occur only at depths below 400-450 m. *Lepidonotothen kempi*, *Dacodraco hunteri* and *Muraenolepis marmoratus* in the Lazarev Sea, *Bathydraco marri* and *Trematomus loennbergi* in the Weddell Sea are examples of this group. The Lazarev Sea species have lower amounts of antifreezes which is likely related to the higher water temperature [21].

While antifreeze protection down to -2.2 °C is necessary for fishes inhabiting relatively shallow ice-laden water, most nonnotothenioid species and some notothenioids (e.g. *Neopagetopsis ionah*) live at 600 m depth and deeper, a habitat with different physical parameters and less ice. Seawater at the surface of the Weddell Sea freezes at approx. -1.86 °C. Because of the pressure effect on the freezing point (0.00753 °C 10 m^{-1}; [22]), seawater at 500 m depth with a salinity of 34.7‰ freezes in situ at -2.24 °C. Hence, based on the freezing point of serum, the freezing point

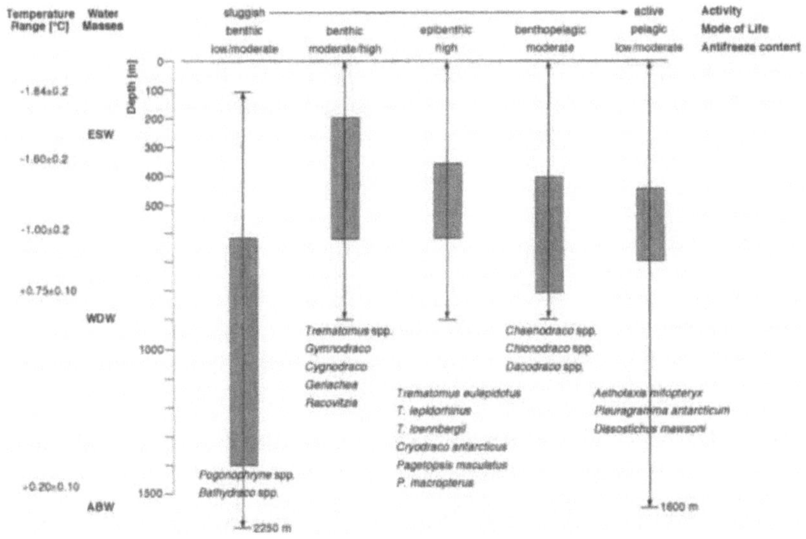

Fig. 4. Vertical distribution of investigated Antarctic fishes and their activity, mode of life and antifreeze content. Total range of distribution for each fish group is denoted by a vertical line, the bar represents the minimum and maximum depth of capture of the investigated species. ESW, Eastern Shelf Water; WDW, Warm Deep Water; ABW, Antarctic Bottom Water. (After Wöhrmann [24])

would be -1.50 °C in *Paraliparis somovi* (623 m), -2.28 °C in *Pogonophryne scotti* (830 m), -1.90 °C in *N. ionah* (799 m), and -2.36 °C in *Bathydraco antarcticus* (1400 m).

In the presence of ice, however, supercooling is not possible as the fish's gills and integument constitute no barrier to the propagation of ice in the body and even partial freezing leads to death [23]. In the Southern Weddell Sea near the Filchner Ice Shelf the water at the undersurface of ice shelves is at its equilibrium freezing point. As this water flows out from under the shelf and rises, it becomes supercooled and potentially ice formation could occur at considerable depths. Large masses of single-crystal ice platelets were discovered in this area.

Little is known about the bathymetric range of the benthopelagic fish fauna. Some notothenioids, such as *Pogonophryne macropogon* and *Bathydraco antarcticus*, as well as a number of zoarcids and liparidids have been found as deep as 2000-2600 m and could be referred to as deep-sea rather than deeper shelf and upper slope species. They do not require antifreezes due to the absence of ice crystals in deep waters. Nevertheless, these benthic and sluggish species still possess antifreezes although in small amounts. Antifreezes are the nonglycosylated AFP in the case of zoarcids and liparidids, and the AFGPs in *Pogonophryne* spp. and *Bathydraco* spp. Moreover, *Bathydraco* spp., distributed in the cold water of the Filchner Depression, synthesize additional AFP in the same concentration as the AFGP. Antifreeze activity is 0.84 °C at a peptide concentration of 40 mg ml^{-1}. So far, these peptides have not been further characterized [21].

The measurable thermal hysteresis (0.1-0.5 °C) of blood serum of several deep-living species (e.g. *Muraenolepis marmoratus*, *Paraliparis somovi*, *Neopagetopsis ionah*, *Bathydraco* spp.) suggests that the serum contains modest amounts of antifreezes, but that this does not lower the freezing point of serum sufficiently to protect these species from freezing at the usual -1.9 °C surface temperature in the Shelf Surface Water. Combined with elevated osmolality of body fluids, a widespread phenomenon in polar teleosts, the freezing point depression of serum in deep water species is only 0.98-1.38 °C below the freezing point of pure water.

However, the freezing behavior of the serum of *Paraliparis somovi* is similar to that reported for some northern boreal species [24]. The freezing points are not well defined and the ice growth in the serum progresses through periods of stops and starts as the temperature is lowered. In other nonnotothenioid species from the Lazarev Sea (e.g. *Macrourus holotrachys* and *Lycenchelys hureaui*) freezing point depression of serum could not be detected due to the extremely low concentrations of the antifreeze peptides.

Summary

The lipid content in liver and muscle of notothenioid fish decreases with the development of gonads and starvation experiments with *Trematomus eulepidotus* indicate that high-Antarctic notothenioids can survive long periods of food shortage due to their low metabolic rates. Notothenioid species have low concentrations of highly glycosylated gangliosides. These gangliosides are concentrated at the synaptic membraneous proteins like ATPase, kinase and receptors. The concentration and composition of antifreeze compounds depend on the ambient water temperature, the depth of habitat, the level of activity, and the mode of life of the specimens. Sluggish and benthic species possess higher amounts of antifreezes than they would need in their natural habitat. Antifreeze concentration of more active as well as benthopelagic or pelagic species is exactly tailored to the ambient water temperature and depth due to the metabolic cost of synthesizing these glycopeptides and peptides. High lipid contents, highly glycosylated gangliosides and freezing resistance in fish are important adaptations to life in high-Antarctic waters. However, low water temperature as such has not caused the paucity of nonnotothenioid species in the Antarctic fish fauna.

Acknowledgments

Part of this work was financed by the Deutsche Forschungsgemeinschaft (He89/55); the EPOS expedition (ANT VII/4) was funded by the European Science Foundation.

References

1. Somero GN (1990) Life at low volume change: hydrostatic pressure as a selective factor in the aquatic environment. Am Zool 30:123-135
2. Hazel JR (1989) Cold-adaptation in ectothermes: regulation of membrane function and cellular metabolism. In: Wang LCH (ed) Animal adaptation to cold. Springer, Berlin, pp 1-52
3. Cossins AR (1994) Homeoviscous adaptation of biological membranes and its functional significance. In: Cossins AR (ed) Temperature adaptation of biological membranes. Portland Press, pp 63-75
4. Becker K, Wöhrmann APA, Rahmann H (1995) Brain gangliosides and cold-adaptation in high-Antarctic fish. Biochem System Ecol 23:695-707
5. Clarke A (1983) Life in cold water: the physiological ecology of polar marine ectotherms. Oceanogr Mar Biol Ann Rev 21:341-453
6. Eastman JT (1993) Antarctic fish biology. Academic Press, San Diego
7. Hubold G (1992) Zur Ökologie der Fische im Weddellmeer. Ber Polarforsch 103:1-157
8. Ekau W (1990) Demersal fish fauna of the Weddell Sea, Antarctica. Antarct Sci 2:129-137
9. Eastman JT (1985) The evolution of neutrally buoyant notothenioid fishes: their specialisations and potential interactions in the Antarctic marine food web. In: Siegfried WR, Condy PR, Laws RM (eds) Antarctic nutrient cycles and food webs. Springer, Berlin, pp 430-436
10. Shul'man GE (1974) Life cycles of fish. Physiology and biochemistry. John Wiley and Sons, New York
11. Lund ED, Sidell BD (1992) Neutral lipid compositions of Antarctic fish tissues may reflect use of fatty acyl substrates by catabolic systems. Mar Biol 112:377-382

12. Sidell BD (1991) Physiological roles of high lipid content in tissues of Antarctic fish species. In: di Prisco G, Maresca B, Tota B (eds) Biology of Antarctic fish. Springer, Berlin, pp 220-232
13. DeVries AL, Eastman JT (1978) Lipid sacs as a buoyancy adaptation in an Antarctic fish. Nature 271:352-353
14. Eastman JT, DeVries AL (1989) Ultrastructure of the lipid sac wall in the Antarctic notothenioid fish *Pleuragramma antarcticum*. Polar Biol 9:333-335
15. Clarke A, Doherty N, DeVries AL, Eastman JT (1984) Lipid content and composition of three species of Antarctic fish in relation to buoyancy. Polar Biol 3:77-83
16. Wöhrmann APA, Hagen W, Kunzmann A (1997) Adaptations of *Pleuragramma antarcticum* (Pisces: Nototheniidae) to pelagic life in high-Antarctic waters. Mar Ecol Progr Ser 151:205-218
17. Eastman JT, Grande L (1989) Evolution of the Antarctic fish fauna with emphasis on the recent notothenioids. In: Crame JA (ed) Origins and evolution of the Antarctic biota. Special Publications of the Geological Society, No 47, London, pp 241-252
18. Rahmann H, Hilbig R (1983) Phylogenetic aspects of brain gangliosides in vertebrates. J Comp Physiol B 151:215-224
19. Hayashi K, Mühleisen M, Probst W, Rahmann H (1984) Binding of Ca^{2+} to phosphoinositols, phosphatidylserines and gangliosides. Chem Phys Lipids 34:317-322
20. Rahmann H (1992) Calcium - ganglioside interaction and modulation of neuronal functions. In: Osborne NN (ed) Current aspects of the neurosciences. Macmillan Press, London, pp 87-125
21. Wöhrmann APA (1996) Antifreeze glycopeptides and peptides in Antarctic fish species from the Weddell Sea and the Lazarev Sea. Mar Ecol Progr Ser 130:47-59
22. Lewis EL, Perkin RG (1985) The winter oceanography of McMurdo Sound, Antarctica. In: Jacobs SS (ed) Antarctic Research Series, Vol 43, Oceanology of the Antarctic continental shelf. American Geophysical Union, Washington, pp 145-165
23. DeVries AL (1988) The role of antifreeze glycopeptides in the freezing avoidance of Antarctic fishes. Comp Biochem Physiol B 90:611-621
24. Wöhrmann APA (1997) Freezing resistance in Antarctic and Arctic fishes: its relation to mode of life, ecology and evolution. Cybium 21(4): 423-442

The Oxidative Metabolism of Antarctic Fish: Some Peculiar Aspects of Cold Adaptation

Bruno Giardina, Alvaro Mordente, Bruno Zappacosta, Cinzia Callà, Luigi Colacicco, Maria L. Gozzo, and Silvio Lippa

Institute of Chemistry and Clinical Chemistry, Catholic University of Rome and CNR Center for Receptors Chemistry, L.go F. Vito 1, 00168 Rome, Italy

Introduction

Environmental factors are key determinants in the process of adaptation and evolution of living organisms. In this respect, analysis of biological systems from Antarctic fish may provide valuable insights on the molecular and cellular strategies which have been adopted in response to the very low environmental temperature which characterizes Antarctica. Hence, the temperature of the Antarctic Ocean varies from –1.86 °C at high latitudes to +5 °C towards the Antarctic Convergence. Moreover, seasonal variations in temperature are less than 0.2 °C close to the permanent ice shelf [1] and only 2.5 °C in the northern Antarctic [2]. Very likely these conditions have existed relatively unchanged for several million years [3]. Antarctic fish have therefore been forced to become highly specialized to cold conditions [4]. This extreme specialization is outlined by the fact that a water temperature of – 1.8 °C, near the freezing point of sea water, is a condition which would be lethal for virtually all other fish. Among antarctic fishes the strong challenge of this extreme environmental condition has been overcome by a number of physiological and biochemical adaptation mechanisms, the most striking of which is the synthesis of freezing-point depressing molecules (antifreeze molecules), either peptides or glycopeptides, which protect the organism from ice crystal formation in a noncolligative way [5]. In addition, hematocrit of the blood is either substantially reduced or erythrocytes and hemoglobin are absent as in Channichthyidae [6-8]. This phenomenon could find an explanation considering that the very low temperature of sea water would result in a great increase of blood viscosity; hence, the decrease in the number of erythrocytes and in the amount of hemoglobin counteract the temperature-induced viscosity increase, greatly facilitating the cardiac work. However, these key adaptations do not play any role in establishing the adaptive metabolic strategies which make it possible for Antarctic fish to survive under these extreme conditions. Thus, the work done in the last

G. di Prisco, E. Pisano, A. Clarke (Eds)
Fishes of Antarctica. A biological overview
© Springer-Verlag Italia 1998

decade has clearly shown that great evolutionary changes have occurred in several enzymatic and structural proteins [9,10] allowing Antarctic fish to sustain metabolic activities substantially comparable to those found in fish with higher cell temperature. These molecular and cellular adaptations are so strongly shaped for low temperatures that Antarctic fish die when water temperature rises from –1.9 °C to +5 °C [4]. This is the lowest temperature of heat death known for any animal and the lowest temperature at which synthesis of heat shock proteins is induced [11].

On this basis we may say that the fitness of Antarctic fishes to their environment is paid by an extreme degree of stenothermality and therefore by a significant loss of biological flexibility.

Here we recall some new and some already published results trying to give a picture of the main characteristics of the oxidative metabolism of Antarctic fish.

Material and Methods

Specimens were collected in the vicinity of Terra Nova Bay (Ross Sea) and of Palmer Station (Antarctica Peninsula). All the specimens were stored at –80 °C until analysis.

Mitochondria from fish liver (~15 g) were prepared, after homogenization of tissue with a Potter-Hevelheim homogenizer, by gradient centrifugation in H-medium (0.7 M sucrose, 0.21 M D-mannitol, 0.002 M HEPES, BSA 0.05% w/v BSA), as reported by Pedersen et al. [12].

Aliquots of mitochondria were used to determine coenzyme Q, vitamin E, NADH oxidase and NADH CoQ reductase. Coenzyme Q and vitamin E were also determined on debris fraction.

Coenzyme Q was determined according to Lippa et al. [13]. Vitamin E was measured with the method of Ericson and Soerensen [14].

The overall activity of mitochondrial respiratory chain was tested as NADH oxidase activity as reported by Mackler [15].

Furthermore the activity of the first and third complexes of respiratory chain was evaluated as they are more substrate specific and species-representative. The first complex was measured testing the activity of NADH CoQ reductase with the method of Youssef [16] and the third complex was measured testing the activity of NADH cytochrome c reductase according to Youssef and Stiggall [17].

Mitochondrial Respiration

As centres of oxidative phosphorylation, mitochondria are critical sites of adenosine triphosphate (ATP) production in aerobic tissues and therefore should represent a key system to study evolutionary adaptations. In this perspective, the increase in mitochondrial volume density observed during cold acclimation implies that mitochondrial function becomes limiting during activity at low temperatures [18].

Previous studies have shown that mitochondrial oxygen consumption is an excellent indicator of temperature adaptation [19-21]. In addition, all lipid bilayers are extremely sensitive to temperature changes and temperature-induced structural changes at the level of the membrane may seriously affect physiological function [22-24]. This is probably due to the sophisticated interactions which are established at the level of the membrane between lipids and proteins. Hence in order to study the characteristic stenothermality of a given organism an integrated system such as that represented by mitochondria is certainly more sensitive with respect to isolated components such as proteins. In this respect it is worthwhile to mention the case of the LDH (lactate dehydrogenase, A-type) of *Pagothenia borchgrevinki*, whose denaturation temperature (40 °C), although significantly lower than that of its homologues from thermophilic species, is well above the lethal temperature of the animal.

The effect of temperature changes on mitochondrial oxygen consumption exhibits a common pattern in all species: the rate of respiration increases, as a function of temperature, up to a certain temperature whereupon any further increase leads to a sharp reduction in respiration rate. This pattern is shown most clearly when data are presented in Arrhenius plots. The temperature at which the sharp change in slope is observed is termed the Arrhenius break temperature (ABT) and appears to be linked to the habitat temperature of the species. Although the biochemical mechanisms underlying the ABT phenomenon per se are not fully understood, disruption of lipid-protein interactions may be partially responsible for ABTs.

The strong stenothermic character of Antarctic fishes has been strongly outlined by recent experiments on mitochondria from the icefish *Chaenocephalus aceratus*. As evident from the data reported in Figs. 1 and 2 in the case of *C. aceratus* the Arrhenius plot obtained for the activity of NADH oxidase as well as of NADH coenzyme Q reductase is reversed up to 2 °C indicating that, within the range of temperature examined, we are to the left of the ABT of the systems under investigation. Even if on the basis of the present data we do not know the precise temperature of the break point, it seems in the proximity of 2 °C outlining the fragility of these systems with respect to temperature increases. In any case a comparison with the data obtained in the case of trout is astonishing. In addition, at 2 °C, the V_{max} values of the enzymatic systems investigated appear significantly higher in the case of the icefish with respect to those obtained for trout (V_{max} NADH oxidase$_{icefish}$ / V_{max} NADH oxidase$_{trout}$ ≅ 2 and V_{max} NADH CoQ reductase$_{icefish}$ / V_{max} NADH CoQ reductase$_{trout}$ ≅ 3.5). On the whole, these results indicate that the biological activity of mitochondria is a site of interesting functional and structural adaptations which still remains a fruitful area of investigation.

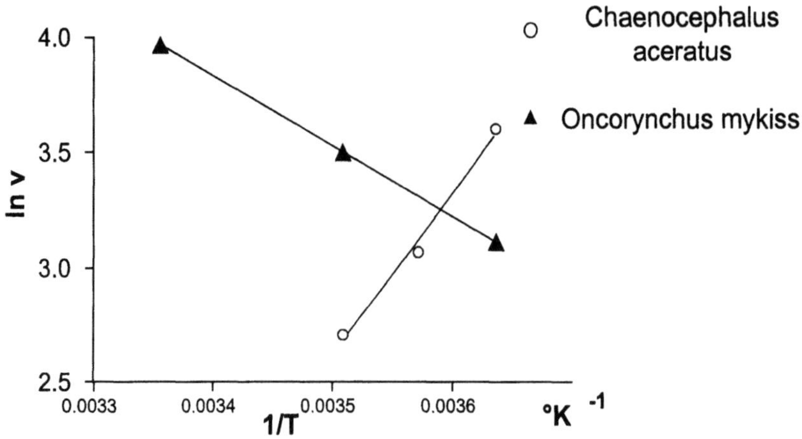

Fig. 1. Arrhenius plot of NADH oxidase activity for liver mitochondria expressed in terms of V_{max} (nmol/min/mg protein)

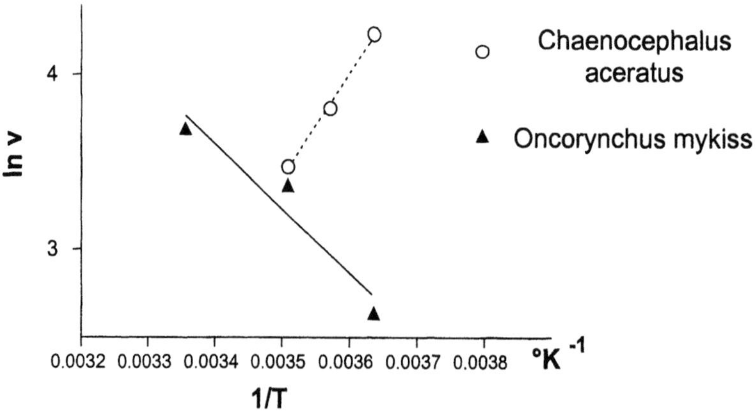

Fig. 2. Arrhenius plot of NADH coenzyme Q reductase activity for liver mitochondria expressed in terms of V_{max} (nmol/min • mg/protein^{-1})

Some Peculiar Features of the Lipid Metabolic Pathway

A peculiar feature of Antarctic fishes is their preferential utilization of the lipid metabolic pathway [25]. Hence, "Antarctic" enzymes of central pathways of aerobic energy metabolism, such as citrate synthase and cytochrome oxidase, do show maximal activities which are 1.5- to 5-fold higher with respect to their temperate zone counterparts. Along the same line, fatty acid oxidation as indicated by specific enzyme markers, palmitoyltransferase and β-hydroxyacil CoA dehydrogenase, is 1.3- to 27-fold higher in tissues from polar species while those for both aerobic and anaerobic metabolism of carbohydrates are generally lower with respect to temperate-zone fish.

All these data indicate that Antarctic fish display a significant cold adaptation of aerobic energy metabolism and suggest that this metabolism is reliant primarily upon oxidation of lipid rather than carbohydrate fuels.

It is worthwhile to mention that tissues from Antarctic fishes have high concentrations of polyunsaturated fatty acids such as eicosapentenoic acid and docosahexaenoic acid. These polyunsaturated fatty acids range from 8 to 26% of total fatty acid pools from skeletal muscle and serum and occur as result of diet. Thus many Antarctic marine invertebrates, prominent prey of Antarctic fish, are rich in these polyunsaturated fatty acids.

The very high corporeal lipid content of Antarctic fish appears therefore not only linked to buoyancy regulation [26] but also to specific requirements of energy metabolism. However, besides these two fundamental roles, other aspects, not minor, have to be considered. These are based upon the observation that some temperate-zone fish display dramatic accumulation of intracellular lipid during cold acclimation at the level of their oxidative muscles. Hence, on the basis of this observation and of the combined solubility and diffusivity characteristics of oxygen in lipids two other additional roles of intracellular lipid inclusions in the oxygen economy of oxidative tissues of Antarctic fish have been hypothesized. These may be summarized as follows:

1 The presence of lipid droplets may accelerate the rate of oxygen diffusion between capillaries and mitochondria.
2 Intracellular lipids may display a role of oxygen store, damping out otherwise large variations in oxygen supply to mitochondria during changes in muscular activities.

These peculiar characteristics, together with the major activation of the oxidative metabolism could be at the basis of an increased production of lipoperoxides and oxygen derived free radicals. Therefore, it has been hypothesized that Antarctic fish should be confronted by an increased oxidative stress which could be very harmful for some of their biological structures [27]. To counteract the effect of toxic oxygen metabolites the cells are endowed with specific scavenging systems. Oxidants that are not detoxified may attack structural proteins and enzymes, membrane lipids, and nucleic acids, thus affecting cell function.

Lipid peroxidation has been proposed to be a major mechanism of oxygen free radical toxicity. Unsaturated fatty acids are particularly susceptible to oxygen radical attack, owing to the presence of double bonds which can undergo peroxidation through a chain of oxidative reactions [28]. In the process of mitochondrial electron transport, oxygen is normally reduced to water through several steps in which hydrogen atoms act as electron donors. However, studies on isolated mitochondria have shown that oxygen can also undergo 1-electron reduction, with formation of superoxide radicals (O_2^-) and hydrogen peroxide (H_2O_2).

Although these two chemical species alone are generally accepted as being insufficiently reactive in aqueous media to initiate deleterious reactions such as lipid peroxidation, their potential danger lies in their ability to lead to the formation of more strongly oxidizing species such as the OH• radical. At least two sites in the mitochondrial respiratory chain have been identified where oxygen radicals may be generated: NADH dehydrogenase and CoQ. It is estimated that under normal conditions 1-2% of oxygen utilized by mitochondria leads to the formation of superoxide radicals.

This "physiologic" generation of oxygen radicals is normally inactivated by endogenous scavenger mechanisms present within the cells. On this basis, the presence, in various tissues from different Antarctic fishes, of those ions and molecules which are at the basis of the antioxidant defense

Table 1. Coenzyme Q and vitamin E content of plasma in Antarctic fishes. The homolog CoQ_{10} plasma level for the temperate freshwater fish trout is 0.7 ± 0.4 μg/mL, and 11.0 ± 2.1 μg/mL is the plasma level for vitamin E

	Coenzyme Q_9 (μg/mL)		Vitamin E (μg/mL)	
Champsocephalus gunnari	3.5	± 0.5	17.9	± 3.3
Notothenia coriiceps	1.9	± 0.4	142.0	± 8.3
Gobionotothen gibberifrons	1.6	± 0.4	40.0	± 6.1

Table 2. Coenzyme Q and vitamin E content of liver mitochondria and debris of Antarctic fish and a temperate freshwater fish

	Coenzyme Q (ng/mg protein)		Vitamin E (ng/mg protein)	
	Mitochondria	*Debris*	*Mitochondria*	*Debris*
C. aceratus	802[a]	164[a]	127	117
N. coriiceps	960[a]	275[a]	333	279
O. mykiss	1233[b]	448[b]	15	10

[a] = Coenzyme Q_9
[b] = Coenzyme Q_{10}

mechanisms was recently investigated [27]. Owing to the obvious evolutionary importance of these particular aspects the attention has been focused on the concentration of coenzyme Q and of vitamin E and on the presence of copper, zinc and selenium (not shown) which display an essential role respectively for the activity of superoxide dismutase, glutathione peroxidase and catalase. The obtained results seem to confirm the above outlined hypothesis showing an overall increase of the antioxidant defenses as reported in Tables 1 and 2. In fact the vitamin E content of liver mitochondria from Antarctic fishes is between 10 and 30 times higher than that observed in the case of trout, and CoQ_9 , even though present in a slightly lower amount with respect to CoQ_{10} in the trout, seems to show a higher antioxidant activity [29]. Furthermore, it is worthwhile to outline that in the blood, the content of CoQ_9 in all the species of the Antarctic fishes examined is 2 to 5 times higher than that of CoQ_{10} in trout.

The presence of the Q_9 form of coenzyme Q in substitution of the Q_{10} form normally found in cells from temperate fishes as well as from most vertebrates may well be another interesting characteristic of Antarctic fish.

Thus, although the physiological reason for such substitution is still matter of research it may be important to outline that the two different forms of coenzyme Q differ in their respective crystallization temperature which is + 0.5 °C for Q_9 and + 9.7 °C for Q_{10}. In the microenvironments, in vivo, where crystallization of the compound cannot be excluded, this could impair the right functioning of the respiratory chain with great troubles for the whole organism. In this respect it is particularly suggestive that in Antarctic fishes Q_{10} is substituted by Q_9 whose crystallization temperature is only slightly higher than sea water temperature. On the whole, the presence of Q_9 seems the result of an adaptive strategy by which the efficiency of one of the main metabolic pathways could be maintained at a reasonable level.

Conclusions

As far as the oxidative metabolism is concerned, specific adaptive features are apparent in Antarctic fishes. These features emerge from a number of observations relative to:

1. the aerobic energy metabolism that appears to be predominantly reliant upon the oxidation of fatty substrates i.e. upon a preferential utilization of the lipid metabolic pathway;
2. the significant increase of antioxidant defenses as evident by the amount of vitamin E, coenzyme Q, selenium, zinc and copper which characterize Antarctic fish tissues. This significant increase is in full agreement with point 1 and with the additional roles which have been hypothesized for the peculiar lipid accumulation observed in these organisms (see also point 4);
3. the specific form of coenzyme Q involved in the reactions of the respiratory chain. Although this finding needs to be further investigated it could be related to the different crystallization temperatures displayed by CoQ_9 and CoQ_{10} and/or by its higher antioxidant activity;
4. the peculiar lipid accumulation very likely linked to acceleration of oxygen diffusion between capillary and mitochondria and to the necessity of an oxygen store to minimize changes in oxygen supply to mitochondria at the muscular level.

However, in parallel with these fascinating biochemical adaptations, we have to consider the very narrow range of temperature in which the various complexes of the respiratory chain maintain their biological activity. The stringent stenothermality which characterizes the biological activity of mitochondria of Antarctic fish and in particular the icefish appears to be the cost that has to be paid for the impressive fitness of these organisms to their extreme environment.

References

1. Littlepage JL (1965) Oceanographic investigations in McMurdo Sound, Antarctic. In: Llano GA (ed) Biology of Antarctic seas II. Antarctic Res Ser, Am Geophys Union, Washington DC 5:1-37
2. Everson I (1970) The population dynamics and energy budget of *Notothenia neglecta* Nybelin at Signy Island, South Orkney Islands. Bull Br Antarct Surv 23:25-50

3. Kennet JP (1977) Cenozoic evolution of Antarctic glaciation, the Circum-Antarctic Ocean and their impact on global paleoceanography. J Geophys Res 82(27):3843-3860
4. Somero GN, De Vries AL (1967) Temperature tolerance of some Antarctic fishes. Sciences 156:257-258
5. De Vries AL (1988) The role of antifreeze glycopeptides and peptides in the freezing avoidance of Antarctic fishes. Comp Biochem Physiol 90B:539-545
6. Hemmingsen EA, Douglas EL (1970) Respiratory and circulatory adaptations to the absence of hemoglobin in Chaenichthyid fishes. In: Llano GA (ed) Adaptations within Antarctic ecosystems. Smithsonian Inst, Washington DC, pp 479-487
7. Hureau JC, Petit D, Fine JM, Marneux M (1977) New cytological, biochemical and physiological data on the colorless blood of the Channichthyidae (Pisces, Teleosteans, Perciformes). In: Llano GA (ed) Adaptations within Antarctic ecosystems. Smithsonian Inst, Washington DC, pp 459-477
8. Wells RMG, Ashby MD, Duncan SJ, Macdonald JA (1980) Comparative studies of the erythrocytes and haemoglobins in nototheniid fishes from Antarctica. J Fish Biol 17:517-527
9. Somero GN, Hochachka PW (1976) Biochemical adaptation to temperature. In: Newell RC (ed) Adaptation to environment: essays on the physiology of marine animals. Butterworths, London, pp 125-190
10. Detrich HW (1991) Polymerization of microtubule proteins from Antarctic fish. In: di Prisco G, Maresca B, Tota B (eds) Biology of Antarctic fish. Springer-Verlag, pp 248-262
11. Maresca B, Patriarca E, Goldenberg C, Sacco M (1988) Heat shock and cold adaptation in Antarctic fishes: a molecular approach. Comp Biochem Physiol 90B: 623-629
12. Pedersen PL, Greenwalt JW, Reynafarje B, Hullihen J, Decker GL, Soper JW, Bustamente E (1978) Preparation and characterization of mitochondria and submitochondrial particles of rat liver and liver-derived tissues. Methods Cell Biol 20:411-481
13. Lippa S, Littarru GP, Oradei A (1985) Determinazione routinaria del Coenzima Q_{10} mediante HPLC in campioni biologici. I[a] Conferenza Nazionale su La cromatografia liquida ad alta risoluzione in analitica clinica: situazione attuale e prospettive. Verona, p. 51
14. Ericson T, Soerensen B (1977) High performance liquid chromatography of vitamin E. Acta Pharmac Suec 14: 478-483
15. Mackler B (1961) Electron transport particle of yeast. Biochim Biophys Acta 50:141
16. Youssef H (1967) Preparation and properties of NADH: ubiquinone oxidoreductase (complex I) E.C.1.6.5.3. Methods Enzymol 10:13
17. Youssef H, Stiggall DL (1967) Preparation and properties of NADH cytochrome C oxidoreductase (complex I-III). Methods Enzymol 10:9
18. Egginton S, Sidell BD (1989) Thermal acclimation induces adaptive changes in subcellular structure of fish skeletal muscle. Am J Physiol 256:R1-R9

19. Van den Thillart G, De Bruin G (1981) Influence of environmental temperature on mitochondrial membranes. Biochim Biophys Acta 640: 439-447
20. Dahlhoff E, O'Brien J, Somero GN, Vetter RD (1991) Temperature effects on mitochondria from hydrothermal vent invertebrates: evidence for adaptation to elevated and variable habitat temperature. Physiol Zool 64:1490-1508
21. Van den Thillart G, Modderkolk J (1978) The effect of acclimation temperature on the activation energy of state III respiration and on the unsaturation of membrane lipids of goldfish mitochondria. Biochim Biophys Acta, 510:38-51
22. Hazel JR, Williams EE (1990) The role of alterations in membrane lipid composition in enabling physiological adaptation of organisms to their physical environment. Prog Lipid Res, 29:167-227
23. Williams EE, Hazel JR (1992) The role of docosahexaenoic acid-containing molecular species of phospholipid in the thermal acclimation of biological membranes. Proc Am Oil Chem Soc
24. Cossins AR (1983) The adaptation of membrane structure and function to changes in temperature. In: Cossins AR, Sheterline P (eds) Cellular acclimatization to environmental change. Cambridge University Press, pp 3-32
25. Sidell BD (1988) Diffusion and ultrastructural adaptive responses in ectotherms. In: Jones DP (ed) Microcompartmentation. CRC Press, Boca Raton, pp 71-92
26. Eastman JT, De Vries AL, (1982) Buoyancy studies of notothenioid fishes in McMurdo Sound, Antarctica. Copeia 1982:385-393
27. Giardina B, Gozzo ML, Zappacosta B, Colacicco L, Callà C, Mordente A, Lippa S (1998) Oxydative metabolism and adaptive strategies in Antarctic fish. Polar Biol (in press)
28. Slater TF (1984) Free radical mechanisms in tissue injury. Biochem J 222:1-15
29. Matsura T, Yamada K, Kawasaki T (1992) Difference in antioxidant activity between reduced coenzyme Q_9 and reduced coenzyme Q_{10} in the cell: studies with isolated rat and guinea pig hepatocytes treated with a water-soluble radical initiator. Biochim Biophys Acta 1123:309-315

Molecular Adaptation of Microtubules and Microtubule Motors from Antarctic Fish

H. William Detrich, III

Department of Biology, Northeastern University, Boston, MA 02115, USA

Introduction

Temperature, through its effects on the equilibria and kinetics of weak, noncovalent molecular interactions, plays an important role in governing enzyme activity and controlling macromolecular assembly reactions.

Microtubule assembly, an example of the latter, entails the formation of a cylindrical polymer composed of αβ-tubulin dimers and decorated by the heterogeneous microtubule-associated proteins (MAPs). Polymerization is entropically driven and mediated in part by the release of structured water from sites of intersubunit contact [1]. As a consequence, microtubule formation *in vitro* by the tubulins and MAPs of homeotherms is highly sensitive to thermal perturbation, with assembly favored by physiological temperatures near 37 °C and depolymerization caused by lower temperatures (<15 °C). Similarly, the microtubule motors (dyneins and kinesins) of mesophiles (e.g., mammals), which are mechanochemical enzymes that use the energy released upon ATP hydrolysis to produce directed movement along microtubules, display little activity at psychrophilic temperatures. Thus, one may ask, "How have the microtubule proteins of cold-living ectotherms evolved to function efficiently at low temperatures?"

The coastal fish of the Antarctic, which now experience body temperatures between -1.86 °C (the freezing point of seawater) and ~ +2 °C, diverged from temperate osteichthyans approximately 38 million years ago as the Southern Ocean began to cool [2]. The teleostean suborder Notothenioidei, which is largely endemic to the Antarctic, provides numerous cold-adapted species suitable to biochemical, physiological, and molecular-biological analysis. To learn about the evolutionary adaptations of microtubule systems, my laboratory has been studying the microtubule proteins of two Antarctic rockcods, *Notothenia coriiceps* and *Gobionotothen gibberifrons* and an icefish, *Chionodraco rastrospinosus*. In this review, I summarize our current understanding of microtubule cold adaptation and place our results in the context of related work on microtubules from mesophilic organisms.

G. di Prisco, E. Pisano, A. Clarke (Eds)
Fishes of Antarctica. A biological overview
© Springer-Verlag Italia 1998

Functional Adaptation of Antarctic Fish Tubulin

Microtubule Formation by Purified Brain Tubulin

Upon warming from 0 to 5 °C, a solution of pure brain tubulin from *G. gibberifrons* [3-5] forms large polymers as judged by turbidimetry (Fig. 1A) [6]. Polymerization follows a lag, during which polymer nuclei are formed, and growth continues to a steady state in polymer mass, indicated by the approach to a plateau turbidity value. Microtubules of normal morphology are the major product of the in vitro assembly reaction (Fig. 2). At higher temperatures, assembly occurs more rapidly, and comparable final turbidities (polymer masses) are attained at lower tubulin concentrations (Fig. 1B,C). Progressive cooling of such solutions causes the turbidity to decrease to a new plateau value at each temperature (data not shown), consistent with the partial disassembly of microtubules. Together, these results demonstrate that brain tubulins from Antarctic fish polymerize reversibly at temperatures near and above the physiological.

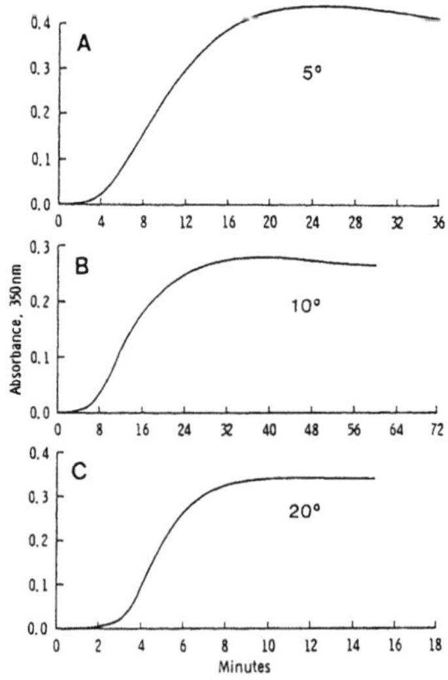

Fig. 1. Assembly of brain tubulin from an Antarctic rockcod, *G. gibberifrons*. Brain tubulin was purified by DEAE ion-exchange chromatography and one round of microtubule polymerization [3-5]. **A-C** Polymerization of brain tubulin at near- and supra-physiological temperatures. Samples of tubulin (in a polymerization buffer containing 0.1 mM GTP and a GTP-regenerating system [6]) were warmed from 0 °C to final temperatures of 5 (**A**), 10 (**B**), or 20 °C (**C**) at zero time, and microtubule assembly was monitored by turbidimetry (apparent A_{350}). Tubulin concentrations: 5 °C, 1.3 mg/ml; 10 °C, 0.96 mg/ml; 20 °C, 0.60 mg/ml. From Himes and Detrich [6] with permission. Copyright 1989 American Chemical Society

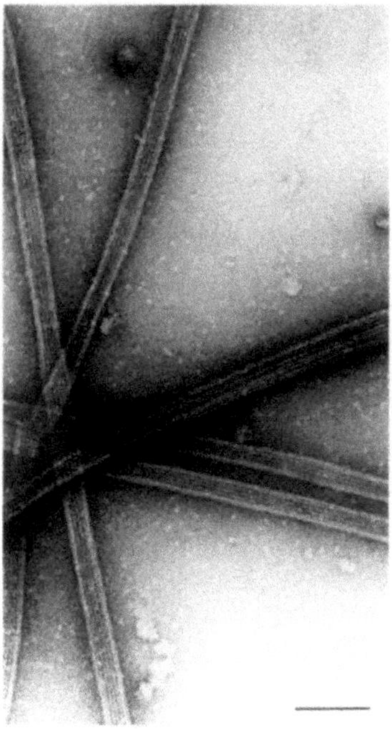

Fig. 2. Electron micrograph of microtubule polymer assembled in vitro from an Antarctic fish tubulin. A solution of *N. coriiceps* tubulin (0.64 mg/ml in a polymerization buffer containing 1 mM GTP [3]) was warmed from 0 to 20 °C at zero time, and a negatively stained specimen was prepared 30 min after the start of assembly. The protofilamentous substructure of these microtubules is readily apparent. The magnification bar represents 100 nm. Reprinted from Detrich et al. [3] with permission. Copyright 1989 American Chemical Society

Energetics of Microtubule Assembly

At a given temperature, the assembly of microtubules *in vitro* requires a minimal, or "critical", concentration of tubulin dimers [7,8]. Below the critical concentration polymerization does not occur, whereas above it the extent of microtubule assembly is a linear function of the total tubulin concentration.

Table 1 summarizes measurements of the critical concentrations of brain tubulins from *G. gibberifrons* and *N. coriiceps* at temperatures between 0 and 18 °C by quantitative sedimentation [3,9]. The critical concentrations of *G. gibberifrons* tubulin, for example, decrease from 0.88 mg/ml at 0 °C to 0.02 mg/ml at 18 °C. Under comparable solution conditions, the assembly of microtubules from pure mammalian brain tubulin at 37 °C requires protein concentrations in excess of 2 mg/ml [10]. Thus, brain tubulins from Antarctic fish form microtubules in vitro at temperatures (0 to +5 °C) and protein concentrations (~1 mg/ml) that approximate the physiological.

Table 1. Temperature dependence of critical concentrations[a]

Temperature (°C)	Critical concentration (mg/ml)	
	N. coriiceps[b]	*G. gibberifrons*[c]
0	-[d]	0.88
3	0.21 (±0.02)	-
5	-	0.26
6	-	0.16
8	0.11	-
10	-	0.12
11	0.062	-
12	-	0.086
15	0.034	0.078
18	0.025	0.020

[a]Reprinted from Detrich *et al.* [9] with permission. Copyright 1992 American Society for Biochemistry and Molecular Biology.
[b]Three tubulin preparations.
[c]Two tubulin preparations.
[d]Not determined.

By application of van't Hoff analysis, the temperature dependence of the critical concentration yields quantitative estimates of the thermodynamic parameters that govern microtubule assembly [3,8-10]. In this analysis, the apparent association constant for addition of a tubulin dimer to the end of a microtubule is assumed to equal the reciprocal of the critical concentration [8]. From the pooled data for Antarctic fish brain tubulins (Table 1), we obtain estimates for the apparent standard enthalpy and entropy changes (ΔH^{o}_{app} = +26.5 kcal/mol, ΔS^{o}_{app} = +121 eu) that are large and positive [3,9]. These values are two- to three-fold greater than those observed for tubulins from organisms with mesophilic body temperatures (e.g., temperate sea urchins, mammals) (reviewed by [3,10]). Thus, the assembly of microtubules from the brain tubulins of Antarctic fish is strongly entropy driven, and entropic control of microtubule assembly apparently increases with decreasing body temperature. The greater entropic control over, and the minimal salt perturbation of [3], the formation of microtubules by the tubulins of Antarctic fish suggest that cold adaptation in this system is based, at least in part, on an increased dependence on hydrophobic interactions, most likely located at sites of interdimer contact.

Conservation of the Critical Concentration

One important adaptive consequence of the interspecific differences in polymerization thermodynamics of tubulins is conservation of the critical concentration for microtubule assembly. Measured in vitro at physiologically relevant temperatures, critical concentrations for tubulins from cold-adapted Antarctic fish, from several temperate invertebrates (sea urchins, clams), and from mammals cluster in the range 0.4-2 mg/ml (reviewed by [3]). Thus, organisms from disparate thermal regimes are able to form microtubules efficiently within the normal limits of their body temperatures.

Structural Determinants of the Polymerization of Antarctic Fish Tubulins

The results presented above indicate that Antarctic fish have evolved brain tubulins that have an enhanced capacity to polymerize at low temperatures.

Our challenge now is to explain these functional properties in terms of molecular changes to the tubulin subunits.

Strategies to Enhance Protein Polymerization Reactions at Low Temperatures

Functional adaptation of Antarctic fish tubulins might result from two general evolutionary strategies: 1) increased reliance on polymer-stabilizing interactions; and 2) reduction in polymer-destabilizing interactions.

Constituting the first category might be primary sequence changes that increase the hydrophobicity and/or flexibility of tubulin domains involved in interdimer contact. In the latter we place reduction of electrostatic repulsion between the exceptionally acidic tubulin dimers. This most likely would be mediated by differential posttranslational polyglutamylation of the tubulin chains. We have examined these possibilities, which are not mutually exclusive, by comparative analysis of the primary sequences and posttranslational modifications of tubulins from Antarctic fish and from temperate mesophiles.

Primary Sequences and Posttranslational Modifications of Antarctic Fish Tubulins

To investigate the potential contribution of primary sequence variation to cold adaptation of microtubule assembly, my laboratory has initiated the cloning and sequence analysis of α- and β-tubulin cDNAs from *N. coriiceps*. As an example, we isolated and sequenced a 1.8-kilobase (kb) neural β-tubulin cDNA, Ncnβ1 [11]. The β tubulin encoded by Ncnβ1 is most closely related to the neural β chains (classes I-IV) of other vertebrates [12-14] and probably belongs to the type II β-isotypic class [11].

We compared its primary sequence to those of other β tubulins to identify potentially adaptive residue changes in Ncnβ1. With respect to vertebrate β chains, the Ncnβ1 polypeptide contains several unique amino acid substitutions and an unusual, carboxy-terminal residue insertion (Table 2). Three unique, but conservative, replacements are clustered near the center of the Ncnβ1 sequence (Ile for Met at position 267, Gly for Ser at 278, and Ser for Ala/Gly at 283).

Replacement of tyrosine by phenylalanine at position 200, although conservative, is unusual in vertebrate β chains, where it occurs only in the erythroid-specific class VI. Substitution of alanine for a bulky aliphatic (Val/Ile) at position 333 is unique and nonconservative, and the tyrosine inserted at position 442 is novel. Overall, most of the unique changes in this 446-residue β chain are located in its amino-terminal domain (first 300 residues), which participates directly in longitudinal bonding between dimers in a microtubule.

At neutral pH, native tubulin dimers from Antarctic fish brain possess one to two fewer net negative charges than do mammalian brain tubulins [4].

Table 2. Comparative analysis of a class-II β tubulin from *N. coriiceps*[a]

		Vertebrate β-tubulin isotypes						
Pos.	Ncnβ1	II	IVb	IVa	I	III	V	VI
18	S	A	A	A	A	A	T	A
126	G	S	S	S	S	N	H	S
200	F	Y	Y	Y	Y	Y	Y	F
267	I	M	M	M	M	M	M	M
278	G	S	S	S	S	S	S	S
283	S	A	A	A/G	A	A	A	A
333	A	V	V	V	V	I	I	I/V
442	Y	-	-	-	-	-	-	-

[a]The table shows eight sequence positions that uniquely differentiate the Ncnβ1 polypeptide from vertebrate neural β tubulins of isotypic classes I, II, and IV (the isotypes most closely related to Ncnβ1). For completeness, residue assignments (single-letter code) are shown for all six vertebrate β-tubulin classes, which are ordered from left to right by sequence homology (high to low) to Ncnβ1. With the exception of position 442, the heterogeneous carboxyl termini (i.e., the 15-20 residues beyond position 430) are not included in this comparison. Reference vertebrate β-tubulin sequences are from chicken (classes I, II, III, IVb, V, VI), mouse (I, II, IVa, IVb, VI), human (I, IVa, IVb), pig (II), rat (II), and *Xenopus laevis* (II). Reprinted from Detrich and Parker [11] with permission. Copyright 1993 Wiley-Liss, Inc

Thus, we have suggested that a decrease in the acidic residue content of the solvent-exposed carboxy-terminal tails of the α- and β-tubulin chains might contribute to microtubule formation at cold temperatures by reducing electrostatic repulsion between dimers as they approach one another [4,15]. However, the neural Ncnβ1 polypeptide [11] and six α-tubulin isotypes from *N. coriiceps* (H.W. Detrich and S. K. Parker, unpublished results) do not, relative to vertebrate tubulin isotypes, contain reduced numbers of glutamyl and/or aspartyl residues in the primary sequences of their carboxyl termini.

Furthermore, the critical concentrations of brain tubulins from an Antarctic fish and a mammal decrease similarly in response to carboxyl-group neutralization largely, but not entirely, restricted to the tubulin C termini [16]. However, carboxy-terminal charge reduction might still play an adaptive role should posttranslational polyglutamylation [17] of the C termini of Antarctic fish tubulins be reduced relative to mammalian tubulins [11]. Indeed, preliminary analyses of the polyglutamyl content of the α and β isotypes of brain tubulins from Antarctic fish (A. Frankfurter and H. W. Detrich, unpublished results) are consistent with this hypothesis.

Multiple Structural Contributions to Microtubule Cold Adaptation

The results reviewed in this section allow us to draw several interesting, if not yet definitive, conclusions regarding the structural bases of the cold adaptation of microtubule assembly in Antarctic fish. First, the hypothesis of hydrophobic remodeling of the interdimer interfaces [3,9] remains provisional. Although most of the unique residue replacements observed in Ncnβ1 tubulin (Table 2) are found in the amino-terminal domain that participates in longitudinal interdimer contacts, no clear physicochemical pattern emerges from consideration of the unique residue substitutions and insertions of this single brain isotype. A slight tendency towards increased polarity (e.g., Ser for Ala at position 18, Ser for Ala/Gly at 283, Tyr insertion at 442) is offset by reciprocal changes (Gly for Ser at 126, 278). Phe^{200}, Ile^{267}, and Ala^{333} increase, maintain, and decrease, respectively, the hydrophobicity of the Ncnβ1 chain relative to other vertebrate β isotypes. Validation or rejection of the hydrophobicity-enhancement hypothesis must await further structural analyses of other α- and β-tubulin isotypes from Antarctic fish.

Second, increased flexibility of domains involved in dimer-dimer contact very likely contributes to the favorable polymerization energetics of Antarctic fish tubulins. The tubulin dimer apparently undergoes a guanine nucleotide-dependent conformational change upon incorporation into or loss from a microtubule end [18-20]. The energy cost associated with this conformational change would be expected to decrease concomitant with a decrease in structural rigidity of the relevant domains. In this context, class-II Ncnβ1 tubulin contains two Ser→Gly substitutions, two Ala→Ser replacements, and one Val→Ala substitution (Table 2), all of which would likely increase the flexibility of its amino-terminal, interdimer contact domain [21,22].

Finally, decreased electrostatic repulsion between tubulin dimers mediated by reduced posttranslational carboxy-terminal polyglutamylation remains a viable contributor to cold adaptation of microtubule assembly. Due to shielding of carboxylate sidechains by counterions and electrostricted water molecules, its effects may be limited, yet synergistic with the two other strategies described above.

One key issue that remains is reconciliation of the three categories of proposed structural adaptations with the strongly entropic control of the polymerization reaction of Antarctic fish tubulin. The overall energetics of microtubule elongation must result from summation of the individual contributions made by all the chemical transformations that accompany addition of a tubulin dimer to a microtubule. Both the conformational-change and the electrostatic-repulsion mechanisms described above would be anticipated to *decrease* the systemic entropy change for assembly of Antarctic fish tubulin. Thus, some other structural change, be it an increase of hydrophobic interactions at dimer-dimer contact sites or another as yet unrecognized, must enhance the entropic character of the polymerization of Antarctic fish tubulin relative to the assembly of tubulins from mesotherms.

Cold Adaptation of Microtubule Motors

With respect to microtubule motor proteins of Antarctic fish, our goal is to determine the molecular adaptations that have evolved to maintain efficient mechanochemical function at low temperatures. As a first step, we have evaluated the temperature dependence of the ATPase activities of *N. coriiceps* inner and outer arm dyneins at temperatures from 0 to 37 °C [23]. Figure 3 shows that the basal, nonmicrotubule-stimulated ATPase activities of the fish dyneins, although modest in absolute terms, significantly exceed that of *Tetrahymena* outer arm dynein at temperatures ≤ 15 °C (at 0 °C, 0.069 and 0.038 μmol P_i/min/mg for inner and outer arm dyneins from *N. coriiceps* *vs.* 0.009 μmol P_i/min/mg for the *Tetrahymena* enzyme). The inner and outer arms attain their maximal activities (~0.1μmol P_i/min/mg) at 9 and 19 °C, respectively. Above these temperatures, the ATPase activities of the fish dyneins decline substantially (to 0.050 and 0 μmol P_i/min/mg for inner and outer arms, respectively, at 37 °C), whereas *Tetrahymena* dynein activity continues to increase (e.g., 0.43 μmol P_i/min/mg at 37 °C). The most

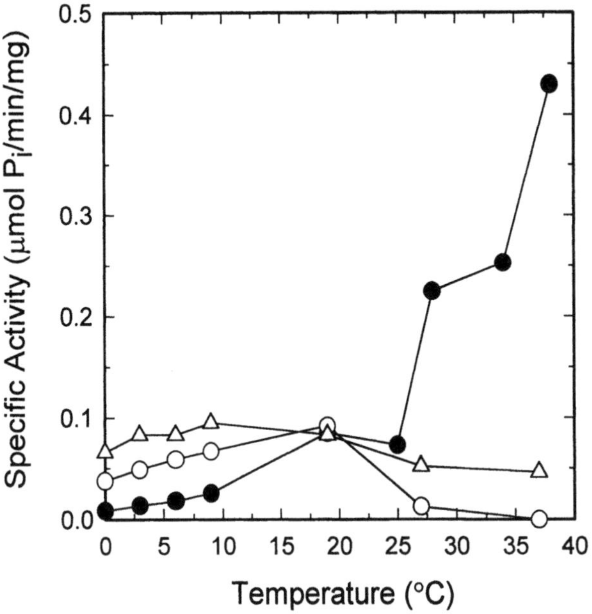

Fig. 3. Temperature dependence of inner and outer arm dynein ATPases. The specific ATPase activities of *N. coriiceps* flagellar inner arm (Δ) and outer arm (O) dyneins and of the 22 S ciliary outer arm dynein from *Tetrahymena* (●) were determined over the temperature range 0-37 °C. *N. coriiceps* samples containing predominantly inner arm dynein and outer arm dynein were obtained from sucrose density gradients [23]. Enzyme activity is expressed in units of μmol phosphate released/min/mg protein. Standard errors of the mean associated with the points (n = 3-6 measurements per datum) are smaller than the graph symbols. Reprinted from King et al. [23] with permission. Copyright 1997 American Chemical Society

plausible interpretation of these results is that the greater catalytic efficiencies of the Antarctic fish dyneins at low temperature result from evolution of greater flexibility in the active site(s) of their protein subunits, which also renders the complexes more susceptible to denaturation at elevated, nonphysiological temperature. Preliminary results from functional and structural studies of the motor domain of brain kinesin from an Antarctic icefish, *C. rastrospinosus* (A. M. Ramsey and H. W. Detrich, unpublished results), are consistent with this hypothesis.

Concluding Remarks

Both the tubulin dimers and the microtubule motors of Antarctic fish show clear evidence of cold adaptation. At temperatures near 0 °C, their functional properties approximate those of the tubulins and motors of mesotherms at their much higher body temperatures. These substantial functional adaptations result, apparently, from rather subtle changes to the primary sequences and posttranslational modifications of the tubulin and motor subunits. Thus, cold adaptation of protein function in Antarctic fish is compatible with substantial conservation of protein structure.

References

1. Correia JJ, Williams RC Jr (1983) Mechanisms of assembly and disassembly of microtubules. Annu Rev Biophys Bioeng 12:211-235.
2. DeWitt HH (1971) Coastal and deep-water benthic fishes of the Antarctic. In: Bushnell VC (ed) Antarctic map folio series, Folio 15. Am Geogr Soc, New York, pp 1-10
3. Detrich HW III, Johnson KA, Marchese-Ragona SP (1989) Polymerization of Antarctic fish tubulins at low temperatures: energetic aspects. Biochemistry 28:10085-10093
4. Detrich HW III, Overton SA (1986) Heterogeneity and structure of brain tubulins from cold-adapted Antarctic fishes: comparison to brain tubulins from a temperate fish and a mammal. J Biol Chem 261:10922-10930
5. Detrich HW III (1991) Polymerization of microtubule proteins from Antarctic fish. In: di Prisco G, Maresca B, Tota B (eds) Biology of Antarctic fish. Springer, Berlin Heidelberg New York, pp 248-262
6. Himes RH, Detrich HW III (1989) Dynamics of Antarctic fish microtubules at low temperatures. Biochemistry 28:5089-5095
7. Gaskin F, Cantor CR, Shelanski ML (1974) Turbidimetric studies of the in vitro assembly and disassembly of porcine neurotubules. J Mol Biol 89:737-758
8. Johnson KA, Borisy GG (1975) The equilibrium assembly of microtubules in vitro. In: Inoue' S, Stephens RE (eds) Molecules and cell movement. Raven, New York, pp 119-139
9. Detrich HW III, Fitzgerald TJ, Dinsmore JH, Marchese-Ragona SP (1992) Brain and egg tubulins from Antarctic fishes are functionally and structurally distinct. J Biol Chem 267:18766-18775
10. Williams RC Jr, Correia JJ, DeVries AL (1985) Formation of microtubules at low temperatures by tubulin from Antarctic fish. Biochemistry 24:2790-2798
11. Detrich HW III, Parker SP (1993) Divergent neural β tubulin from the Antarctic fish *Notothenia coriiceps neglecta*: potential sequence contributions to cold adaptation of microtubule assembly. Cell Motil Cytoskel 24:156-166
12. Sullivan KF (1988) Structure and utilization of tubulin isotypes. Ann Rev Cell Biol 4:687-716
13. Wang D, Villasante A, Lewis SA, Cowan NJ (1986) The mammalian β-tubulin repertoire: hematopoietic expression of a novel, heterologous β-tubulin isotype. J Cell Biol 103:1903-1910
14. Monteiro MJ, Cleveland DW (1988) Sequence of chicken cβ7 tubulin: analysis of a complete set of vertebrate β-tubulin isotypes. J Mol Biol 199:439-446

15. Detrich HW III, Prasad V, Ludueña RF (1987) Cold-stable microtubules from Antarctic fishes contain unique α tubulins. J Biol Chem 262:8360-8366
16. Singer WD, Parker SK, Himes RH, Detrich HW III (1994) Polymerization of Antarctic fish tubulins at low temperatures: role of carboxy-terminal domains. Biochemistry 33:15389-15396
17. Alexander JE, Hunt DF, Lee MK, Shabanowitz J, Michel H, Berlin SC, Macdonald TL, Sundberg RJ, Rebhun LI, Frankfurter A (1991) Characterization of posttranslational modifications in neuron-specific class III β-tubulin by mass spectrometry. Proc Natl Acad Sci USA 88:4685-4689
18. Carlier M-F (1983) Kinetic evidence for a conformation change of tubulin preceding microtubule assembly. J Biol Chem 258:2415-2420
19. Melki R, Carlier M-F, Pantaloni D, Timasheff SN (1989) Cold depolymerization of microtubules to double rings: geometric stabilization of assemblies. Biochemistry 28:9143-9152
20. Shearwin KE, Timasheff SN (1992) Linkage between ligand binding and control of tubulin conformation. Biochemistry 31:8080-8089
21. Feller G, Thiry M, Gerday C (1991) Nucleotide sequence of the lipase gene *lip2* from the Antarctic psychrotroph *Moraxella* TA144 and site-specific mutagenesis of the conserved serine and histidine residues. DNA Cell Biol 10:381-388
22. Fontana A (1991) How nature engineers protein (thermo) stability. In: di Prisco G (ed) Life under extreme conditions. Springer, Berlin Heidelberg New York, pp 89-113
23. King SM, Marchese-Ragona, SP, Parker, SK, Detrich, HW III (1997) Inner and outer arm axonemal dyneins from the Antarctic rockcod *Notothenia coriiceps*. Biochemistry 36:1306-1314

Metallothionein in Antarctic Fish

Vincenzo Carginale[1], Antonio Capasso[1], Clemente Capasso[1], Peter Kille[2], Elio Parisi[1], Gianluca L. Passaretti[1], Guido di Prisco[1], Marilisa Riggio[1], and Rosaria Scudiero[1]

[1]Institute of Protein Biochemistry and Enzymology, CNR, Via Marconi 12, 80125 Napoli, Italy
[2]School of Molecular and Medical Biosciences, University of Wales, Cardiff, UK

Introduction

Zinc occupies a prominent position among transition and d^{10} elements because of the role played in many biological processes [1,2], including catalysis [3], transcription and translation [4]. The functional role of zinc depends on the large variety of metallorganic complexes produced by this element as a result of its interaction with different protein families [3,5].

Although the existence of cellular mechanisms for absorption, secretion and distribution of zinc have been inferred [6,7], very little is known about regulation of zinc metabolism. Increasing evidences suggest that metallothioneins (MTs), i.e. low-molecular weight cysteine-rich metal-binding proteins, play an important role in zinc homeostasis [8,9]. In addition, these proteins are supposed to have also a potential role in attenuating the cytotoxic effects caused by heavy metals and superoxide radicals [10,11].

The understanding of the biological function of MT relies mostly on its molecular features. Mammalian MTs are single chain proteins usually made of 60 amino acids, including 20 cysteines with no histidine and aromatic residues; all the cysteines are in a reduced form and engaged in the binding of 7 equivalents of metal ions through metal-thiolate bonds [12]. As each metal ion is tetrahedrally coordinated, some cysteines are bridged with adjacent metal ions to form metal-thiolate clusters arranged in two distinct domains: the N-terminus beta domain of three-metal clusters having a cyclohexane-like structure, and the C-terminus alpha domain of four-metal clusters having a bicyclononane-like structure [13,14].

Both X-ray diffraction [13] and two dimensional NMR [14] studies revealed the dynamic structure of MT which allows a rapid exchange of metals within clusters and between MT and other proteins. These features

G. di Prisco, E. Pisano, A. Clarke (Eds)
Fishes of Antarctica. A biological overview
© Springer-Verlag Italia 1998

support a metal storage function of MT, which may acquire a particular relevance at specific stages of organism life, such as embryonic development and growth [15]. This does not rule out the protective role played by MT against toxic heavy metals and free radicals. A direct evidence in support of such a role is given by the observation that divalent metals like cadmium and zinc are known to induce MT synthesis by increasing the level of MT gene expression [16,17]. In addition, several other organic compounds, oxidative stress and inflammatory conditions have been shown to induce MT synthesis [18-20]. Recently, a number of evidences have assessed the importance of MT by using mice with null alleles of MT genes [21]. These results confirm the role of MT in the homeostasis of heavy metals and its participation in the control of heavy metal and free radical toxicity.

In this review, we present the results of our studies on MT in Antarctic fishes. As these organisms have remained isolated south of the Antarctic Convergence for more than 20 million years, they may represent a useful tool to study the strategies evolved to adapt themselves to particularly severe environmental conditions.

The Zinc Status in Antarctic Fish

We have investigated the status of hepatic zinc in different species of Nothothenioidei by determining the intracellular levels of MT. We found a large difference in MT content between the hemoglobinless Channichthyidae and the red-blooded Nothothenidae. The results in Table 1 show the levels of MT (expressed in μg/mg protein) in acetone powder preparations obtained from hepatic tissues of *Chionodraco hamatus*, *Chionodraco rastrospinosus*, *Chaenocephalus aceratus* and *Trematomus bernacchii*. In the three icefish, the MT levels varied from a minimum of 0.002 μg MT/mg protein in *C. aceratus* to a maximum of 0.2 μgMT/mg protein in *C. hamatus*, whereas in the red-blooded *T. bernacchii*, the MT content was significantly higher.

Despite these marked differences in MT content, all the species investigated showed significant levels of protein-associated zinc, of which part was recovered in a high-molecular weight and part in a low-molecular weight protein fraction, as shown by the chromatographic profiles reported in Fig. 1.

As most of the metal-binding proteins possibly involved in zinc homeostasis are low-molecular weight proteins, we focused our attention on the protein fraction included into the resin. Indeed, the low-molecular weight zinc-containing fractions eluted from the G-75 columns were purified by a combination of anion-exchange chromatography and HPLC.

Table 1. MT content in Antarctic fish livers. MT content was determined by silver saturation assay. Quantification of MT was performed on the basis of standard curves obtained with rabbit MT in the silver assay

	μg MT/ mg protein
	Silver assay
C. hamatus	0.2 ± 0.01
C. rastrospinosus	0.004 ± 0.002
C. aceratus	0.0025 ± 0.002
T. bernacchii	1.3 ± 0.2

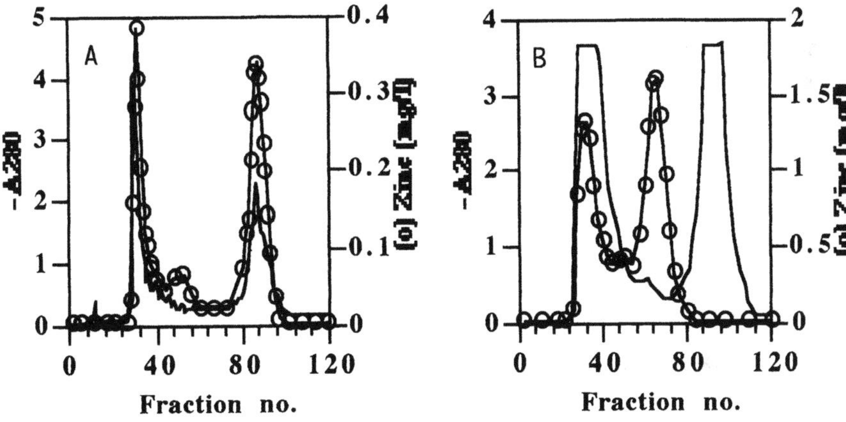

Fig. 1. A Sephadex G-75 elution profile of an extract prepared from icefish liver. **B** Sephadex G-75 elution profile of an extract prepared from *T. bernacchii* liver

When analyzed by SDS-PAGE, the *C. hamatus* protein showed a single band of 11 kDa, whereas the *T. bernacchii* protein displayed a molecular mass of 6 kDa.

In Table 2 are reported the amino acid compositions of the icefish and red-blooded Zn-binding proteins.

From the results of this analysis, the differences between the two proteins are clear: the icefish protein displays a cysteine content which is only one-third that of *T. bernacchii*, and is particularly rich in glutamate and aspartate. Other distinctive features are the presence of histidine and aromatic residues in the icefish protein, but not in the *T. bernacchii* protein; the latter, on the other hand, has a cysteine content (30%) which is typical of MT. Hence, red-blooded fish differ substantially from icefish

Table 2. Amino acid composition of zinc-binding proteins from *T. bernacchii* (MT) and *C. hamatus* (Zn-bp)

Amino acid	Zn-bp *C. hamatus* (mol%)	MT *T. bernacchii* (mol %)
Cys[a]	4.75	30.1
Asp	10.92	6.8
Thr	5.43	11.2
Ser	5.98	14.3
Glu	17.57	4.3
Pro	8.00	5.7
Gly	11.99	11.9
Ala	6.94	2.1
Val	5.68	1.5
Met	1.63	0
Ile	4.12	0
Leu	5.10	0
Tyr	1.98	0
Phe	2.05	0
His	1.94	0
Lys	3.88	12.1
Arg	1.89	0

[a]Determined as cysteic acid

as far as concerns the zinc status: in fact, very little, if any, zinc-thionein is present in the icefish liver, most of zinc being associated with a non-MT low-molecular weight protein. No information is available at the moment on the role played by such a protein: apparently, it resembles a family of still ill-characterized metal-binding proteins described in a number of vertebrate and invertebrate species. The situation is entirely different in the red-blooded fish where the predominant species is given by MT.

Sequence and Structure of *T. bernacchii* MT

Fish MTs have a blocked amino terminus. Hence, in order to obtain information on MT sequence and structure, MT cDNA was generated by reverse-transcriptase polymerase-chain-reaction (RT-PCR) using as primers oligo-dT and an oligonucleotide designed on the amino terminal sequence of piscine MT [22]. Electrophoresis of the PCR reaction showed a band of about 350 bp. This fragment was eluted from the gel, ligated into a pGEM-T plasmid and cloned in *E. coli*. The plasmid DNA obtained from cloned colonies was sequenced on both strands by the dideoxy method using the T7 sequencing kit (Pharmacia Biotech.). The translated amino acid sequence reported in Fig. 2 shows a protein made of 60 residues with 20 cysteines arranged in a fashion typical of vertebrate MTs; the homology to rat MT-2 is also shown in Fig. 2.

```
TberMT   MDP-CQCSKS GTCNCGGSCT CTNCSCTSCK KSCCPCCPSG CTKCASGCVC KGKTCDTSCC Q
RatMT2   MDPNCSCATD GSCSCAGSCK CKQCKCTSCK KSCCSCCPVG CAKCSQGCIC KEASDKCSCC A
         *** * *.   *.*.*.***  *  * ***** ****.*** * * **. **.* *  .   ***
```

Fig. 2. Primary amino acid sequences of *T. bernacchii* and rat MTs.

As MT lacks well-defined secondary structure elements such as α-helices and β-strands, it is not well-suited for comparative model building. However, since the domains that bind heavy metals in MT are likely to be structurally conserved, a model building based on the replacement of the substituted amino acids in a suitable template could be allowed. Modeling of the *T. bernacchii* MT was performed by Swiss PDB Viewer using ProMod Tools [23] and rat MT-2 structure as template. The three-dimensional structure of *T. bernacchii* MT is shown in Fig. 3 together with the experimentally determined structure of rat MT-2.

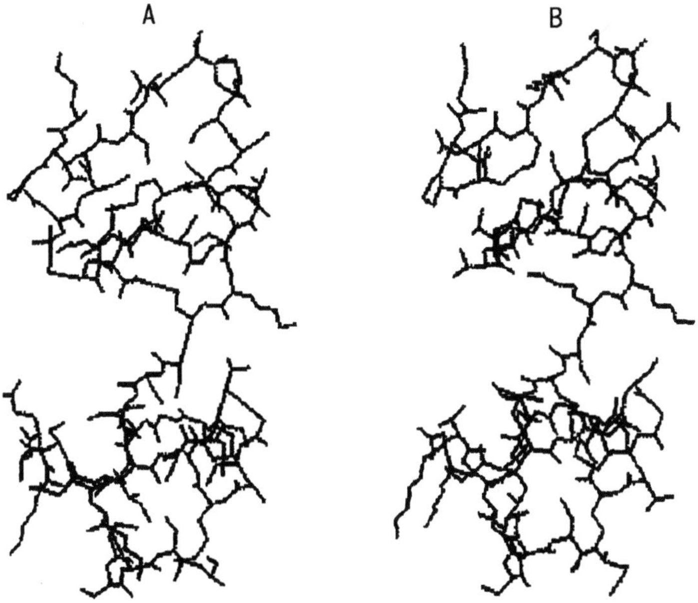

Fig. 3. Comparison between the experimental three-dimensional structure of rat MT (**A**) and the homology modeled *T. bernacchii* MT (**B**)

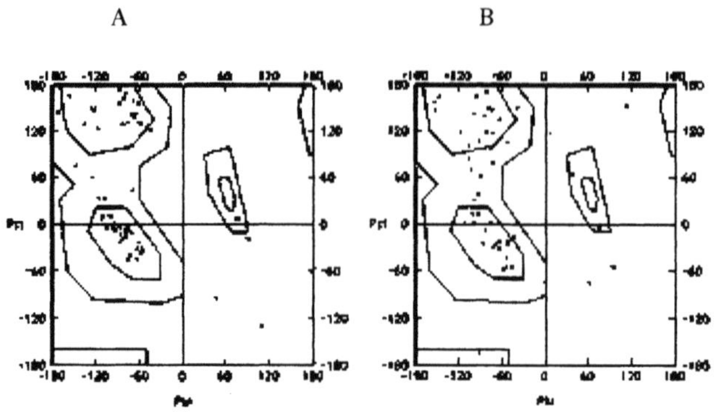

Fig. 4. Ramachandran plots of rat (**A**) and fish (**B**) MTs

A comparison between the Ramachandran plots performed on rat and fish MT is reported in Fig. 4. In the case of mammalian MT, the residues falling out of the allowed regions are gly 11, ala 16, gly 47, ser 54, and asp 55; for fish MT these residues are gly 16, cys 20, ser 38, gly 46 and thr 53.

The fact that the modeled protein shows only few residues falling out of the allowed regions indicates that the new sequence fits well into the three-dimensional structure of rat MT-2.

Structure of the MT Transcripts and Transcriptional Status of MT Genes in Red-Blooded and Hemoglobinless Fish

In order to ascertain whether the low level of MT in icefish is a consequence of a lack of MT transcripts, RT-PCR reactions were performed as described above, using RNA extracted from the livers of various kinds of icefish. Surprisingly, the reaction products analyzed by electrophoresis on 1.4% agarose gels stained with ethidium bromide showed the presence of the same 350 bp band found in *T. bernacchii*. Hence, despite the very low level of MT protein detected, a certain amount of MT transcript is present in the population of mRNAs in icefish.

The fragments obtained by RT-PCR were recovered from the gel and ligated into a pGEM-T vector; the resulting constructs were cloned in *E. coli* and sequenced. The nucleotide sequences of these cDNAs are reported in Fig. 5 aligned with the sequence of *T. bernacchii* MT cDNA.

While very few differences exist at the level of the coding regions, significant modifications can be observed in the 3'-UTRs, just after the stop codon TGA (position 184) and in a region spanning from position 270 to 291, upstream of the polyadenilation signal AATAAA.

```
hamatusMT   ATGGACCCCT GCGACTGCTC CAAAAGTGGG ACCTGCAACT GCGGAGGATC CTGCACTTGC
rastroMT    ATGGATCCCT GTGAGTGCTC CAAAAGTGGG ACCTGCAACT GCGGAGGATC CTGCACTTGC
bernMT      ATGGATCCCT GCGAGTGCTC CAAAAGTGGG ACCTGCAACT GCGGAGGATC CTGCACTTGC
acerMT      ATGGATCCCT GCGAGTGCAC CAAAAGTGGG ACCTGCAACT GCGGAGGATC CTGCACTTGC
            ***** **** * ** *** * ********** ********** ********** **********

ACAAACTGCT CCTGCACCAG CTGCAAGAAG AGCTGCTGCC CATGCTGCCC ATCCGGCTGC ACCAAATGCG
ACAAACTGCT CCTGCACCAG CTGCAAGAAG AGCTGCTGCC CATGCTGCCC ATCCGGCTGC ACCAAATGCG
ACAAACTGCT CTTGCACCAG TTGCAAGAAG AGCTGCTGCC CATGCTGCCC ATCCGGCTGC ACCAAATGCG
ACAAACTGCT CCTGCACCAG CTGCAAGAAG AGCTGCTGCC CATGCTGCCC ATCCGGCTGC ACCAAATGCG
********** * ********  ********* ********** ********** ********** **********

CCTCTGGCTG CGTGTGCAAA GGGAAGACTT GTGACACAAG CTGCTGTCAG TGAAG..... GACCTCAGCC
CCTCTGGCTG CGTGTGCAAA GGGAAGACTT GTGACACAAG CTGCTGTCAG TGAA...... GACCTCAG.C
CCTCTGGCTG TGTGTGCAAA GGGAAGACTT GTGACACAAG TTGCTGTCAG TGAAGAGCCT GACCTCCAGT
CCTCTGGCTG CGTGTGCAAA GGGAAGACTT GTGACACAAG CTGCTGTCAG TGAAGAAC.. .....TCAGC
**********  ********* ********** **********  ********* ****

CGCTTCTGCT CTTGGAATGG AGCCTTTGTG AACTACTTTG  ACTACATTCC TAGTTGCAAA TGTCTACAGA
CGCTTCTGCT CTTGGAATGG AGCCTTTGTG AACTACTTTG  ACTACATTCC T.GTTGCAAA TGTCTACAGA
CGCTTCTGCT CTTGGAATGG AGCCTTTGTG AACTACTTTA  ACTACATTCC T.GTTGCCAA TGTCTACAGA
CGCTTCTGCT CTTGGAATGG AGCCTTTGTG AACTACTTTG  AATACATTCC TGCTTCTGCT CTTGGAATGG
********** ********** ********** *********   * ******** *  **          *  *  *

GAATGGTG.. ....AA.TTG TGTACTTGTT TACG.ATGTT GAAATAAATG CACTTCCTTG
.......G.. ....AA.TTG TGTACTTGTT TACG.ATGTT GAAATAAATG CAGTTACTTG
GAATGGTG.. ....AATTTT TGTACTTGTT TATGAATGTT GAAATAAATA CAGTTCCTTG
AGCCTTTGTG AACTACTTTG AATACATTCC TTACTATGTT GAAATAAATG CAGTTCCTTG
       *        * **    *** *     *    ***** *********  ** ** ****
```

Fig. 5. Aligned nucleotide sequences of MTs from *C. aceratus*, *C. hamatus*, *C. rastrospinosus* and *T. bernacchii*

In order to obtain information on the 5'-UTRs, a 5'-RACE protocol was applied using a 5'-Marathon cDNA amplification kit and the fish MT-specific primer (5'-GCAGGAGCAGTTTGTGCAAGTGCA-3'). The amplified fragments were cloned as usual in the pGEM-T vector and sequenced. As shown in Fig. 6, the leader sequence of the *T. bernacchii* MT mRNA lacks the initial segment of 21 nucleotides present in the icefish MT mRNA; in addition, a mismatching segment is found 42 nucleotides upstream of the AUG codon.

```
      1
hamat GACAACACAACACGCUCUCAUCUGAGACCAGAUACACUGCACCUACUCACGAGGACGAGAGACAUCACCUG
                           || |||||  | ||||||||||||||||||||||| ||||||||||||||
bern                       CUCAGACCGAAGACACUGCACCUACUCACGAGGACAAGAGACAUCACCUG
                           1

                           94
      AGAACAUGGACCCCUGCGACUGC
      ||||||||||||| ||| | |||
      AGAACAUGGACCCUUGCCAGUGC
                           73
```

Fig. 6. Sequences of the 5' noncoding regions (5'-UTR) of MT mRNAs from *C. hamatus* and *T. bernacchii*

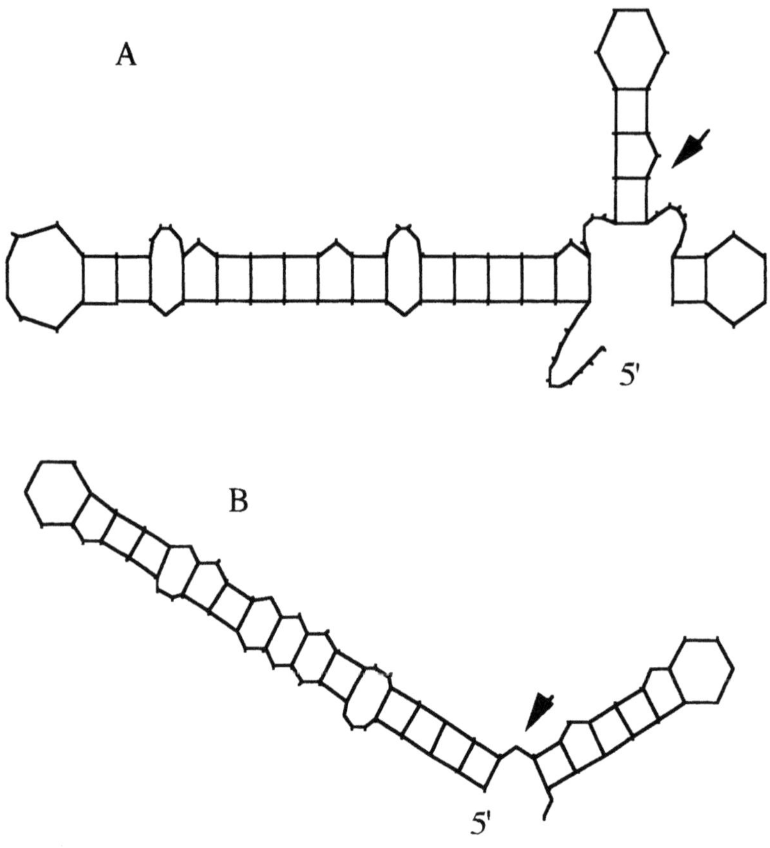

Fig. 7. Predicted secondary structures of the 5'-leader sequences of the MT mRNAs from icefish (**A**) and *T. bernacchii* (**B**). The arrow indicates the position of the start codon AUG

Possible secondary structures of the 5'-leader sequences from *C. hamatus* and *T. bernacchii* mRNAs are depicted in Fig. 7. These structures were determined by using the program MFOLD designed by the University of Wisconsin Genetics Computer Group. As shown in Fig. 7, the 5'-leader sequence of the *C. hamatus* mRNA (A) displays an additional stem loop structure which is absent in the 5' UTR of the *T. bernacchii* mRNA (B).

The expression of MT transcripts was assessed by Northern blotting of total liver RNA using *C. rastrospinosus* MT cDNA as probe. The results depicted in Fig. 8 show that the amounts of mRNA in icefish are markedly higher than those in the red-blooded fish.

1 2 3

Fig. 8. Northern blots of total RNA from liver tissues of *C. rastrospinosus* (1), *T. bernacchii* (2) and *C. aceratus* (3)

Conclusions

The results reported here show that Antarctic icefish contain low MT levels compared to red-blooded fish. The only low-molecular weight zinc-binding protein detected in icefish livers is a 11 kDa protein with a low cysteine content and rich in acidic amino acid residues [24], whereas in the red-blooded *T. bernacchii* the predominant metal-binding protein is an MT with molecular characteristics similar to rat MT [25].

Despite the failure to detect a normal amount of MT in icefish livers, a cDNA encoding MT was produced from total RNA by reverse-transcriptase PCR. Full information on 5'-leader sequences of the MT mRNAs was obtained by 5'-RACE. A comparison of the MT mRNAs from icefish and *T. bernacchii* reveals major differences at the level of 5'- and 3'-UTR; computer analysis suggests that these differences are responsible for the formation of an additional hairpin loop upstream of the starting codon of the icefish MT mRNA. Since such an additional stem loop structure is absent in the 5'-UTR of *T. bernacchii* MT mRNA, we suggest that the translation of icefish MT mRNA is under the control of specific elements present in the 5'- and 3'-UTR. Such an hypothesis is also corroborated by the results of Northern blot analysis of total RNA showing the presence of large amounts of untranslated MT mRNA in icefish livers.

Acknowledgments

This research is part of the Italian National Programme for Antarctic Research (PNRA).

References

1. Danks DM (1986) Of mice and men, metals and mutations. J Med Genet 23:99-106
2. Prasad AS (1991) Discovery of human zinc deficiency and studies in an experimental human model. Am J Clin Nutr 53:403-412
3. O'Dell BL (1992) Zinc plays both structural and catalytic roles in metalloproteins. Nutr Rev 50:48-50
4. Giedroc DP, Keating KM, Martin CT, Williams KR, Coleman JE (1986) Zinc metalloproteins involved in replication and transcription. J Inorg Biochem 28:155-169
5. Vallee BL, Auld DS (1995) Zinc metallochemistry in biochemistry. Exs 73:259-277
6. Coto JA, Hadden EM, Sauro M, Zorn N, Hadden JW (1992) Interleukin 1 regulates secretion of zinc-thymulin by human thymic epithelial cells and its action on T-lymphocyte proliferation and nuclear protein kinase C. Proc Natl Acad Sci U S A 89:7752-7756
7. Saris NE, Niva K (1994) Is Zn^{2+} transported by the mitochondrial calcium uniporter? Febs Lett 356:195-198
8. Kille P, Hemmings A, Lunney EA (1994) Memories of metallothionein. Biochim Biophys Acta 1205:151-161
9. Vallee BL (1995) The function of metallothionein. Neurochem Int 27:23-33
10. De SK, Dey SK, Andrews GK (1990) Cadmium teratogenicity and its relationship with metallothionein gene expression in midgestation mouse embryos. Toxicology 64:89-104
11. Camhi SL, Lee P, Choi AM (1995) The oxidative stress response. New Horiz 3:170-182
12. Kagi JHR, Nordberg M (1979) Metallothionein. Birkhaser Verlag. Basel
13. Robbins AH, Stout CD (1991) X-ray structure of metallothionein. Methods Enzymol 205:485-502
14. Narula SS, Brouwer M, Hua Y, Armitage IM (1995) Three-dimensional solution structure of *Callinectes sapidus* metallothionein-1 determined by homonuclear and heteronuclear magnetic resonance spectroscopy. Biochemistry 34:620-631
15. Andersen RD, Piletz JE, Birren BW, Herschman HR (1983) Levels of metallothionein messenger RNA in foetal, neonatal and maternal rat liver. Eur J Biochem 131:497-500
16. Bonham K, Gedamu L (1984) Induction of metallothionein and metallothionein mRNA in rainbow-trout liver following cadmium treatment. Biosci Rep 4:633-642
17. George SG, Todd K, Wright J (1996) Regulation of metallothionein in teleosts: induction of MT mRNA and protein by cadmium in hepatic and extrahepatic tissues of a marine flatfish, the turbot (*Scophthalmus maximus*). Comp Biochem Physiol C Pharmacol Toxicol Endocrinol 113:109-115

18. Schwarz MA, Lazo JS, Yalowich JC, Allen WP, Whitmore M, Bergonia HA, Tzeng E, Billiar TR, Robbins PD, Lancaster J, Jr (1995) Metallothionein protects against the cytotoxic and DNA-damaging effects of nitric oxide. Proc Natl Acad Sci U S A 92:4452-4456
19. Dalton T, Palmiter RD, Andrews GK (1994) Transcriptional induction of the mouse metallothionein-I gene in hydrogen peroxide-treated Hepa cells involves a composite major late transcription factor/antioxidant response element and metal response promoter elements. Nucleic Acids Res 22:5016-5023
20. Xie T, Belinsky M, Xu Y, Jaiswal AK (1995) ARE- and TRE-mediated regulation of gene expression. Response to xenobiotics and antioxidants. J Biol Chem 270:6894-6900
21. Lazo JS, Kondo Y, Dellapiazza D, Michalska AE, Choo KH, Pitt BR (1995) Enhanced sensitivity to oxidative stress in cultured embryonic cells from transgenic mice deficient in metallothionein I and II genes. J Biol Chem 270:5506-5510
22. Kille P, Stephens PE, Kay J (1991) Elucidation of cDNA sequences for metallothioneins from rainbow trout, stone loach and pike liver using the polymerase chain reaction. Biochim Biophys Acta 1089:407-410
23. Peitsch MC (1996) Promod and Swiss model, internet-bond tools for automated comparative protein modelling. Biochem Soc Trans 24:274-279
24. Scudiero R, De Prisco PP, Camardella L, D'Avino R, di Prisco G, Parisi E (1992) Apparent deficiency of metallothionein in the liver of the Antarctic icefish *Chionodraco hamatus*. Identification and isolation of a zinc-containing protein unlike metallothionein. Comp Biochem Physiol 103B:201-207
25. Scudiero R, Carginale V, Riggio M, Capasso C, Capasso A, Kille P, di Prisco G, Parisi E (1997) Difference in hepatic metallothionein content in Antarctic red-blooded and haemoglobinless fish: undetectable metallothionein levels in haemoglobinless fish is accompained by accumulation of untranslated metallothionein mRNA. Biochem J 322:207-211

Respiration and Activity of Arctic and Antarctic Fish with Different Modes of Life: A Multivariate Analysis of Experimental Data

Christopher Zimmermann[1] and Gerd Hubold[1,2]

[1]Institut für Polarökologie der Universität Kiel, Wischhofstr. 1-3, Geb. 12, 24148 Kiel, Germany,
[2]Institut für Seefischerei, Bundesforschungsanstalt für Fischerei, Palmaille 9, 22767 Hamburg, Germany

Introduction

Traditionally, polar fishes have been considered to be restricted in their physiological performance by low water temperatures [1]. This restriction could even limit the types of life style in polar seas [2]. However, detailed ecological investigations based on in situ and long term aquarium observations show a surprisingly wide scope of life types and behavioral patterns for polar ocean fishes [3-8]. In today's fish fauna of both polar seas all ecotypes are present, except for scombrid-like forms [9]. Uncertainties persist about the metabolic performance potential of polar fishes. Following the expected respiration trend of temperature, ectotherm organisms should all display an extremely low metabolism at polar temperatures [10], which ought to result in limited activity and physiological performance. Early results of Scholander et al. [11] and Wohlschlag [12] found elevated respiration rates in polar fish species, which was in accordance with Krogh's [10] suggestions, and developed the concept of a "Metabolic Cold Adaptation" (MCA). More sophisticated experimental conditions (i.e. the consideration of handling stress [1] and spontaneous activity [13]) rendered lower oxygen consumption values which led to a rejection of the MCA concept [1, 14-15]. However, oxygen consumption rates even within one species yielded contradictory results, which can be attributed to the various experimental set-ups used (see [16] for a review). In most cases, no account was made for the ecological profile of the respective species nor for its spontaneous activity. In order to resolve the persistent controversy, it is essential to investigate a wide spectrum of fish ecotypes with a standardized method. Furthermore, means to account for spontaneous activity have to be developed to exclude

G. di Prisco, E. Pisano, A. Clarke (Eds)
Fishes of Antarctica. A biological overview
© Springer-Verlag Italia 1998

activity effects on the standard metabolism. Standard metabolism is defined as the metabolism without any external or internal work, i.e. without spontaneous activity. Ideally, the complete range of fish species from polar, temperate and tropical environments should be investigated to separate the effect of temperature from the ecologically determined ("mode of life") effect on the metabolism. A quantitative criterium for activity and life type should be developed which surpasses the simple subjective consideration of a fish's "mode of life" as just being "sluggish" or "active."

The aim of the present study was to link behavioral activity (as expressed e.g. by movements) and metabolic activity (in terms of oxygen consumption) of different fish species from both polar seas. The quantitative effects of a fish's activity on its metabolism are presented by means of multivariate statistics. The consequences for the three competing concepts on fish metabolism are discussed. These concepts are:

- The standard oxygen consumption (SOC) is strictly related to the thermodynamic principle given by the environmental temperature. This should hold true, if a proper account for the exclusion of spontaneous activity, stress etc. is made in the experimental set-up.
- Polar fishes deviate from the thermodynamic principle by an elevated standard metabolism: the metabolic cold adaptation concept [10-12].
- The "mode of life" is the dominant factor determining the SOC of fishes; the "mode of life" can be expressed by the level of spontaneous activity.

Material and Methods

Experiments were conducted at two field stations (European Large Scale Facility Ny-Ålesund, Spitsbergen: June-August 1994, US McMurdo Station, Ross Sea/Antarctica: January-February 1995) and at the home laboratory in Kiel (January-June 1996). At field stations fish were caught in traps and fyke nets; the species measured at the Kiel lab were obtained from bottom trawls, conducted from RV "Polarsten" during various cruises to the Antarctic (for more detailed information see [9]). Seven fish species were chosen to cover a wide range of different ecotypes: the cryopelagic hunters *Boreogadus saida* (polar cod, Gadidae) from the Arctic, and *Pagothenia borchgrevinki* (Nototheniidae) from the Antarctic; three demersal species, *Anarhichas minor* (Anarhichadidae), *Trematomus bernacchii* (Nototheniidae) and *Gymnodraco acuticeps* (Bathydraconidae); and two strictly benthic forms, the Arctic sculpin *Myoxocephalus scorpius* (Cottidae) and the Antarctic genus *Pogonophryne* (Artedidraconidae). The sculpin is the only investigated species not exclusively found in polar waters.

To accommodate the animals to the experimental conditions, fish were kept in tanks for at least 10 days prior to the experiment at in situ temperature and salinity (for the Kiel lab experiments more than 15 months). For a synchronous recording of oxygen consumption, spontaneous activity and basic parameters (e.g. temperature) during the experiment, a modified intermittent flow respirometer [17-19] was used (Fig. 1): single fish were kept in respiration chambers of various sizes allowing for spontaneous movements. Water was continuously subsampled, the oxygen content automatically determined and the water recirculated to the chamber (closed measurement cycle). When the O_2 saturation dropped below 89%, the water was exchanged with oyxgenated seawater from a separate tank (open cycle). The activity of the fish was monitored with an infrared (IR) video camera under IR illumination and recorded with a time lapse recorder. The recordings were evaluated manually to achieve maximum precision. An exact runtime could be assigned to every movement of the fish [9]. Subsequently, the counts of individual movements were summarized for 2 min intervals. In addition, mean and maximum movement rates (projected to 1 h) and elapsed time between two single movements were determined.

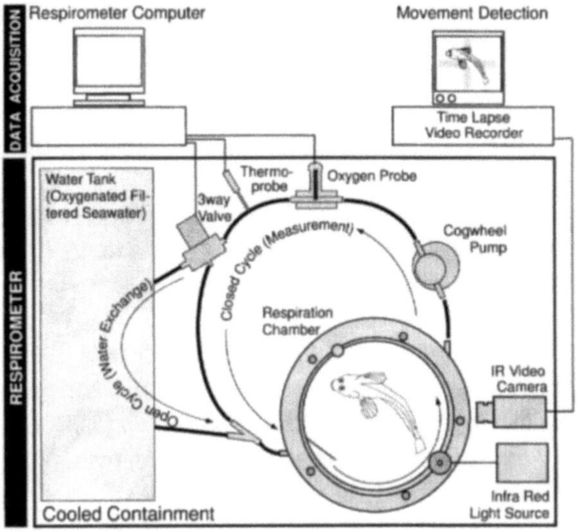

Fig. 1. Intermittent flow respirometer (schematic view from above) with the set-up for movement detection. IR, infrared

After a synchronization of the parameters, a linear regression between activity (in terms of movements per unit time, usually 2 min) and respiration (for the identical interval) was computed. Standard oxygen consumption rate was determined by extrapolation to zero movements, following a procedure used earlier this century [20-22] with different set-ups. With this approach, determination of SOC was possible even when the fish never displayed phases of inactivity. Standardization to SOC is a prerequisite to compare the metabolism of active and inactive fish, although the SOC value has no direct relevance for an active species like the polar cod.

A number of different variables were recorded in each experiment to quantify fish activity and respiration as well as experimental conditions. An intricate pattern of intercorrelations exists between these variables. It can be assumed that complex factors (such as the "mode of life"), which cannot be measured directly, are underlying these variables. A factor analysis was conducted to identify these hypothetical parameters, based on the correlation matrix of the variables [23-24]. To avoid collinearities, variables with correlation coefficients >0.75 were excluded. The 9 variables chosen for the factor analysis were:

- parameterizing activity:

- mean movement rate ("move mean")
- quotient of maximum and mean movement rate obtained during one experiment ("move quot")
- mean duration of phases between two movements ("pause mean")

- parameterizing respiration:

- calculated standard oxygen consumption ("SOC")
- factorial scope for activity, the ratio between respiration during maximum activity and SOC, which gives a measure for the metabolic capacity of an animal [25]
- slope of the regression curve between activity and respiration as a raw measure for the energy requirements of a single move ("slope act/resp")

- parameterizing experimental conditions:

- mean temperature ("temp mean")
- maximum temperature range during one experiment ("temp diff") and
- relation between fish volume and respiration chamber volume, as a measure for the space available for the specimen ("space").

All values were adequately transformed (reciprocals of "move" and "pause" and logarithms of all other variables) prior to correlation computations. Principal components were extracted from the correlation matrix, orthogonally rotated and orthotran/varimax transformed. Finally, the component scores were calculated.

Results and Discussion

Activity and Respiration

Twenty respiration experiments, lasting for a total of 2300 h (up to 380 h per experiment) were used for the factor analysis (Table 1); 1500 h of video recordings were evaluated and more than 82 000 movements detected.

Differences between active and sluggish representatives were clearly visible, especially in the movement rate. While the most active cryopelagic individual (*Boreogadus*) showed a mean movement rate of up to 600 per hour (max.: 2000 moves), the most sluggish specimen (a *Pogonophryne*) did not move more than 2.7 times an hour (max.: 360). The quotient (*ratio)* of max. and mean movement rate is high for sluggish species compared to active ones (Table 1b). The most sluggish Antarctic animals exhibited periods of up to 22 h without any significant movement (*Pogonophryne* sp.). Mean intervals between moves were 22 min for this species. The most torpid Arctic specimen investigated was a *Myoxocephalus scorpius* with maximum inactivity intervals of 1.1 h and mean periods of inactivity of 3.5 min.

The calculated individual SOC ranged from 8.0 (± 0.04) mg O_2 h^{-1} kg WM (wet mass) $^{-1}$ for an Antarctic *Pogonophryne* sp. to 68 (± 2) mg O_2 h^{-1} kg WM^{-1} for a large and very active Arctic *Myoxocephalus scorpius*. The individual variability was generally high. Routine oxygen consumption (including phases with spontaneous activity) was up to 6 times and respiration during phases with maximum spontaneous activity up to 8 times higher than the calculated SOC.

Factor Analysis

The first three factors separated by factor analysis had Eigen values of 3.31, 1.96 and 1.54. They explained almost three-quartes of the total variance among the original variables. The communalities for the 9 variables analyzed ranged between 71% and 91%, which means that all variables were well explained by the extracted factors.

The factor loadings of the 9 variables are visualized by the orientation and length of vectors, which are projected in the planes between factor 1 and 2 as well as factor 1 and 3 (Fig. 2).

The first factor explains almost 35% of the variance in the standardized variable values. It is related mainly to the mean movement rate and the standard oxygen consumption rate of the investigated fishes. The second factor represents temperature-related variables and is responsible for almost 22% of the variation. The scope for activity has the strongest relation to the third factor, which still has a variance share of about 17%.

Table 1. Experiments used for the factor analysis. **a.** Values of the parameters describing the experimental conditions. **b.** Values of the parameters describing respiration and activity, which were used for the factor analysis. Given is (from left to right) the species name, experiment code (a short form of the species name and a letter), wet mass of the fish, duration of the experiment, duration of the video observation, mean temperature, maximum temperature range, space available for the animal (ratio of the volumes of the respiration chamber and the fish), calculated standard oxygen consumption, slope of the linear regression between activity and respiration, factorial scope for spontaneous activity, mean movement rate, ratio of maximum and mean movement rate, and mean duration of phases without activity.

Species	Exp.	a: WM (g)	Duration (hh:mm)	Video Observ. (hh:mm)	Temp mean (°C)	Temp range (°C)	rel. Space	b: cSOC (mgO_2 h^{-1} $kgWM^{-1}$)	Slope a/r (mgO_2 $kgWM^{-1}$ $movement^{-1}$)	Scope for Activity	Move mean (h^{-1})	Move Quot.	Pause mean mm:ss
Anarhichas minor	Anarh A	27.7	49:45	49:15	2.72	0.72	81	35.3	0.08	1.7	82.4	10.9	0:39
Boreogadussaida	Boreo Aa-c	110.6	143:45	128:00	2.02	1.15	42						
	Boreo Aa							17.2	0.16	8.4	589.1	2.6	0:06
	Boreo Ab							44.6	0.13	2.8	356.0	5.1	0:10
	Boreo Ac							47.4	0.10	2.7	610.7	3.2	0:05
Myoxocephalus scorpius	Myoxo A	175.4	22:30	22:30	2.01	0.25	26	68.2	0.14	1.9	299.5	2.2	0:12
Myoxocephalus scorpius	Myoxo B	33.3	191:30	100:30	2.82	0.60	67	16.3	0.17	2.9	51.6	12.8	1:09
Myoxocephalus scorpius	Myoxo C	56.9	54:30	53:15	2.73	0.31	39	25.0	0.10	1.2	17.4	17.2	3:25
Myoxocephalus scorpius	Myoxo D	56.0	59:00	52:15	2.75	0.19	40	25.4	0.10	1.4	21.1	35.5	2:50
Trematomus bernacchii	Bernac A	55.2	106:00	48:00	0.16	1.65	40	12.5	0.17	2.2	17.3	22.5	3:26
Gymnodraco acuticeps	Gymno A	87.2	47:30	14:00	-0.62	0.59	54	34.4	0.17	1.9	20.4	20.6	2:55
Pagothenia borchgrevinki	Borch A	66.7	98:00	56:00	0.09	1.06	71	52.7	0.25	2.3	116.0	8.3	0:30
Pagothenia borchgrevinki	Borch C	78.9	98:00	60:00	-0.32	1.11	60	33.4	0.12	1.8	43.0	16.8	1:23
Pagothenia borchgrevinki	Borch D	135.0	216:30	127:30	0.55	0.90	34	35.7	0.12	1.9	57.5	13.6	1:02
Pagothenia borchgrevinki	Borch E	45.3	46:00	46:00	0.72	1.18	49	36.7	0.44	5.0	140.1	4.9	0:25
Pagothenia borchgrevinki	Borch F	65.5	185:15	145:00	0.48	1.00	34	46.4	0.21	2.7	70.6	10.2	1:13
Pagothenia borchgrevinki	Borch G	48.0	135:30	78:30	0.31	1.45	46	25.9	0.16	4.4	136.3	7.7	0:26
Pogonophryne sp. 1	Pogo A	13.6	160:00	152:00	1.04	0.86	45	14.6	0.05	1.3	2.7	134.8	21:32
Pogonophryne sp. 2	Pogo B	63.6	385:15	186:30	1.25	0.54	35	8.5	0.28	2.4	2.8	87.0	6:13
Pogonophryne sp. 2	Pogo C	63.6	162:30	141:00	0.94	0.39	35	8.1	0.20	3.3	9.6	31.3	12:42
Pogonophryne sp. 3	Pogo D	124.4	141:00	107:30	0.95	0.38	17	13.5	0.12	1.9	4.6	65.8	22:24

Furthermore, correlations between variables can be deduced from the plot of the factor loadings in the factorial space: highly correlated variables have a rather narrow angle to each other, while uncorrelated ones are perpendicular to each other. It is evident in Fig. 2a that standard oxygen consumption (SOC) and mean movement rate were highly correlated in our experiments (the opposite direction of both vectors is caused by the reciprocal transformation of the latter variable). In contrast, temperature had an insignificant effect on both activity and SOC: the vectors have an angle of almost 90°. Even in view of the rather narrow in situ temperature range in our experiments (about 3.4 °C) this result is surprising. According to the Q_{10}-rule, one should expect a significantly higher standard rate at higher temperatures. The scope for activity was independent both from activity and from temperature.

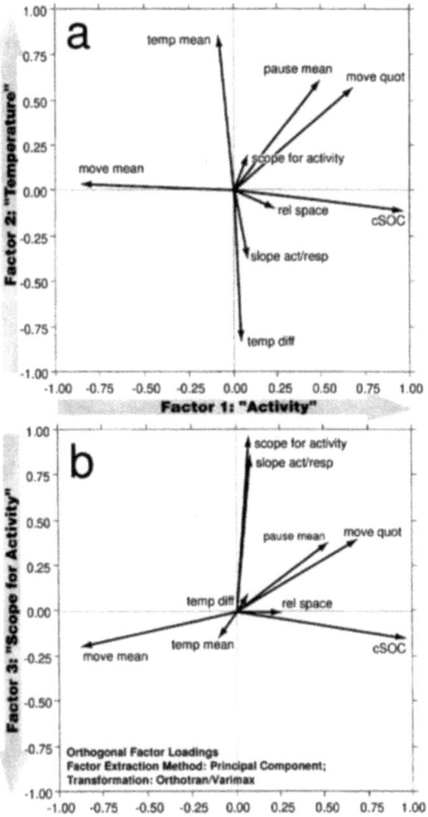

Fig. 2. Factor analysis: plot of the variables' factor loadings in factorial planes 1-2 (**a**) and 1-3 (**b**). Factors were named according to the characterizing variables. Abbreviations are the same as in Table1

A different approach to the data analysis is shown in Fig. 3: instead of factor loadings, the component scores are now projected into the factor planes. Component scores are the coordinates of the 20 single experiments (or fish individuals) in the factorial space. The axes correspond to those in Fig. 2. The ordination of the single experiments permits the ecological interpretation of the data.

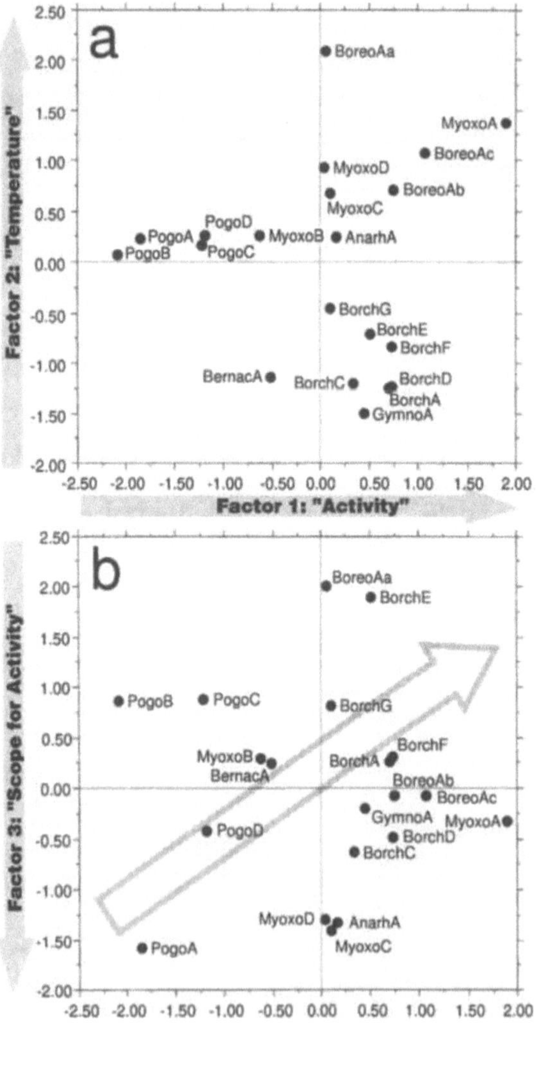

Fig. 3. Factor analysis: plot of the experiment's component scores in factorial planes 1-2 (**a**) and 1-3 (**b**). The grey outlined arrow represents a suggested "mode of life" gradient. Factors were named according to the characterizing variables. Abbreviations of are the same as in Table 1

In factorial plane 1-2, experiments are clearly separated by activity and temperature. The measurements of the extremely sluggish *Pogonophryne*, conducted at temperatures of about 1 °C, have negative scores on factor 1, whereas the more active species are characterized by positive scores. Of these experiments, the Arctic fish (2.0 to 2.8 °C) had positive scores on factor 2, and the Antarctic fish (–0.6 to +0.7 °C) had negative scores.

The picture is less clear for scores in the factorial plane 1-3 (Fig. 3b). From the distribution pattern, a "mode of life" gradient can be discerned between sluggish, strictly benthic animals in the lower left corner and active, cryopelagic hunters in the upper right corner. This gradient is symbolized by the grey outlined arrow. The direction of this arrow is similar to that for the factor loading of the variable "quotient of maximum and mean movement rate" (Fig. 2b). This similarity suggests to use the latter variable for the quantification of the weak term "mode of life".

The qualitative observations on life style and feeding strategy of the investigated species (e.g. [26-28]) were compared to the quantitative value for the move ratio (Fig. 4).

The strong correspondence between the variable move ratio and the ecological species profiles is obvious, except for the catfish *Anarhichas*.

The one juvenile *Anarhichas* specimen investigated may not be

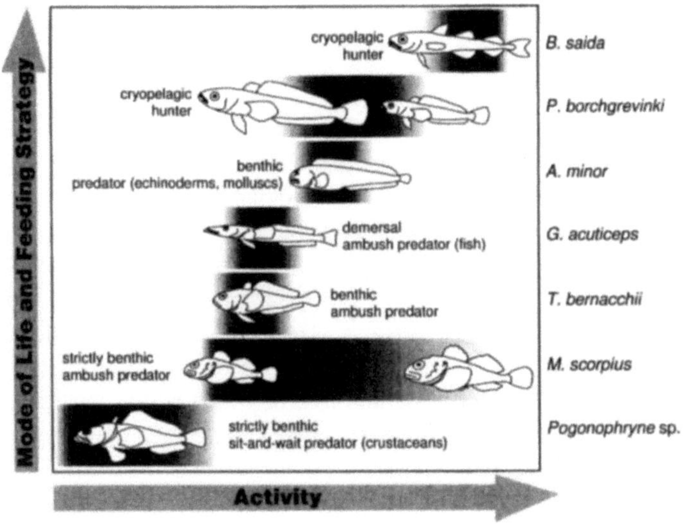

Fig. 4. Comparison of literature values on the life style of the investigated fish species and the values for the ratio of maximum and mean movement rate (1/move quot), which is proposed to be a parametric measure for "mode of life." The species are sorted from up to down after increasing affinity to the sea floor, i.e. from cryopelagic to benthic. The shaded box shows the range of values; the darkest part represents the mean

representative for the benthic life style of the adult specimens [26]. In addition to the general ordination, intraspecific effects of body mass or age can be derived from Fig. 4. For *Myoxocephalus* and *Pagothenia* there are indications for a correlation between activity and size - surprisingly in opposite directions: younger *Pagothenia* are more active than older ones, the opposite is true for *Myoxocephalus*. A reduction of activity with increasing age is also known for *Boreogadus* [29], *Pagothenia's* ecological equivalent in the Arctic [30], and for a number of species from lower Antarctic latitudes [13]. This ontogenetic activity shift should also be considered for the comparison of metabolic rates between species.

Conclusions

The results of this study show that within the limited range of polar ocean temperatures, the standard oxygen consumption is mostly dependent on the fish's potential activity (as a measure for its "mode of life"). In contrast, differences are much less related to the direct effects of the environmental temperature. Temperature effects on fish metabolism can only be determined when the predominant influence of the mode of life, including differences in ontogenetic activity, is identified and eliminated in the experimental set-up, or if a wide temperature range is used. The still unresolved question of different metabolic performance of tropical, temperate and polar fishes can be only answered when truly comparable fishes of the same ecological profile are compared in carefully designed, low-stress experiments.

Therefore, the knowledge of a fish's ecotype and activity pattern is essential for a reasonable comparison of metabolic rates. This was not always taken into account in previous investigations, resulting in contradictions about the adaptations of polar fish.

The determination of a fish's life style is hardly possible without a careful recording of its spontaneous activity. On the other hand, the evaluation of a fish's activity might even provide us with a parametric measure for "mode of life," which was so far not quantifiable. For this purpose we propose to use the quotient of maximum and mean movement rate, which is easy to determine and avoids - as a *ratio* - difficulties due to different set-ups used for a movement detection.

Acknowledgments

The critical comments of D. Piepenburg, C. Krembs, W Hagen and an anonymous referee improved the manuscript significantly. Thanks are due to A. L. DeVries, who made the stay of C.Z. at the US McMurdo Station possible. Part of this work was supported by the Deutsche Forschungsgemeinschaft (DFG Hu 461/2-1 to G.H.) and the National Science Foundation (S-005 to A.L. DeVries).

References

1. Holeton GF (1974) Metabolic cold adaptation of polar fish: fact or artefact? Physiol Zool 47 (3): 137-152
2. Clarke A, Johnston IA (1996) Evolution and adaptive radiation of Antarctic fishes. Trends Ecol Evol 11 (5): 212-220
3. Moreno CA (1980) Observations on food and reproduction in *Trematomus bernacchii* from the Palmer Archipelago. Copeia 1: 171-173
4. Daniels RA (1982) Feeding ecology of some fishes of the Antarctic Peninsula. Fish Bull 80 (3): 575-588
5. Ekau W, Gutt J (1991) Notothenioid fishes from the Weddell Sea and their habitat, observed by underwater photography and television. Proc NIPR Symp Polar Biol 4: 36-49
6. Janssen J, Sideleva V, Montgomery J (1991) Under-ice observations of fish behavior at McMurdo Sound. Antarct J US 26: 174-175
7. Janssen J, Montgomery JC, Tien R (1992) Social rotational feeding in *Pagothenia borchgrevinki* (Pisces: Nototheniidae). Copeia 2: 559-562
8. North AW (1996) Locomotory activity and behaviour of the Antarctic teleost *Notothenia coriiceps*. Mar Biol 126 (1): 125-132
9. Zimmermann C (1997) Zur Ökologie arktischer und antarktischer Fische: Aktivität, Sinnesleistungen und Verhalten. Ber Polarforsch 231: 1-137 [in German]
10. Krogh A (1914) The quantitative relation between temperature and standard metabolism in animals. Internat Zeitschr Physikal-Chem Biol 1: 491-508
11. Scholander PF, Flagg W, Walters V, Irving L (1953) Climatic adaptations in arctic and tropical poikilotherms. Physiol Zool 26: 67-92
12. Wohlschlag DE (1964) Respiratory metabolism and ecological characteristics of some fishes in McMurdo Sound, Antarctica. In: Lee MO (ed) Biology of the Antarctic seas 1. American Geophysical Union, Washington DC, pp 33-62
13. Morris DJ, North AW (1984) Oxygen consumption of five species of fish from South Georgia. J Exp Mar Biol Ecol 78: 75-86
14. Clarke A (1991) What is cold adaptation and how should we measure it? Am Zool 31 (1): 81-92
15. Steffensen JF, Bushnell PG, Schurmann H (1994) Oxygen consumption in four species of teleosts from Greenland: no evidence of metabolic cold adaptation. Polar Biol 14 (1): 49-54
16. Wells RMG (1987) Respiration of Antarctic fishes from McMurdo Sound. Comp Biochem Physiol 88A: 417-424
17. Forstner H (1983) An automated multiple-chamber intermittent-flow respirometer. In: Gnaiger E, Forstner H (eds) Polarographic oxygen sensors. Springer, Berlin, pp 111-126.
18. Steffensen JF, Johansen K, Bushnell PG (1984) An automated swimming respirometer. Comp Biochem Physiol 79A: 437-440
19. Steffensen JF (1989) Some errors in respirometry of aquatic breathers: how to avoid and correct for them. Fish Physiol Biochem 6 (1): 49-59

20. Spoor WA (1946) A quantitative study of the relationship between the activity and oxygen consumption of the goldfish, and its application to the measurement of respiratory metabolism in fishes. Biol Bull (Woods Hole) 91: 312-325
21. Fry FEJ (1947) Effects of the environment on animal activity. Univ Toronto Stud, Biol Ser 55: 1-62
22. Beamish FWH, Mookherjii PS (1964) Respiration of fishes with special emphasis on standard oxygen consumption. Can J Zool 42: 161-175
23. Flury B, Ridwyl H (1983) Angewandte multivariate Statistik. Fischer, Stuttgart [in German]
24. Legendre L, Legendre P (1983) Numerical ecology. Developments in environmental modelling 3. Elsevier, Amsterdam
25. Fry FEJ (1971) The effect of environmental factors on the physiology of fish. In: Hoar WS, Randall DJ (eds) Fish physiology, Vol. 6. Academic Press, New York, pp 1-98
26. Nikolskii GW (1957) Spezielle Fischkunde. VEB Verlag der Wissenschaften, Berlin [in German]
27. Pethon P (1989) Aschehougs store fiskebok. Aschehaug & Co A/S, Oslo [in Norwegian]
28. Gon O, Heemstra PC (eds) (1990) Fishes of the Southern Ocean. JLB Smith Institute of Ichthyology, Grahamstown
29. Lønne OJ, Gulliksen B (1989) Occurrence and ecological importance of sympagic fauna in the Fram Strait, Svalbard area, and western Barents Sea. Rapp P-v Réun Cons Int Explor Mer 188: 170
30. Eastman JT (1997) Comparison of the Antarctic and Arctic fish faunas. Cybium, 21(4): 335-352

Antarctic Fish Immunoglobulins: Preliminary Data on Structure and Antibody Specificity

Maria Rosaria Coscia and Umberto Oreste

Institute of Protein Biochemistry and Enzymology, CNR, Via Marconi 12, 80125 Naples, Italy

Fish immunoglobulins resemble mammalian IgM in terms of size (five Ig domains), carbohydrate composition of the heavy chain (12-16%), and polymeric assembly of the basic unit H_2L_2 (tetrameric in bony fish and pentameric in cartilaginous fish). However, the overall features of the mammalian secondary immune response (rise in titre, shift to high affinity antibodies, isotypic switch) are not found in fish [1].

For an adequate understanding of fish immunity, many groups have focused their research efforts on the Ig gene structure: in particular, regarding teleosts, the organisation of H chain genes is shown to be of mammalian type, whereas that of L chain genes is of multicluster type [2]. Little is known about the VH diversity. Several teleost species are known to have multiple Ig VH families [3-5].

We think that studying the Antarctic fish Ig is a suitable approach to learn more about the functional adaptation of Ig to particularly extreme environmental conditions. In fact it is known that cold temperatures tend to depress life functions including the immune response.

We have dealt with the isolation and characterization of Ig from a number of Antarctic fish species. We purified Ig from plasma of *Trematomus bernacchii*, *Notothenia coriiceps*, *Chaenocephalus aceratus*, *Chionodraco hamatus*, from eggs of *C. hamatus*, from bile of *T. bernacchii* and from cutaneous mucus of *Gobionotothen gibberifrons* [6,7]. The purification procedure used was a combination of thiophilic adsorption chromatography on T-gel and size exclusion fast performance liquid chromatography (FPLC) on Superose 6 column. We found the T-gel chromatography as an efficient alternative to other Ig purification methods, such as the ammonium sulfate precipitation, anion exchange chromatography or Protein A Sepharose CL-4B affinity chromatography, which yielded lower amounts of purified fish Ig.

The Ig/total protein ratio was found lower for the bile compared with the other samples analyzed [6]. The plasma or bile fractions bound by T-gel were analyzed in FPLC (Fig. 1A). The relative molecular mass of the major peak (780 kDa) of plasma Ig was interpreted to correspond to the tetrameric assembly (Fig. 1B). Indeed, the FPLC elution profile of bile Ig

G. di Prisco, E. Pisano, A. Clarke (Eds)
Fishes of Antarctica. A biological overview
© Springer-Verlag Italia 1998

disclosed a major peak, (190 kDa) corresponding to the monomer [6]. Both plasma and bile FPLC purified peaks were then resolved, by SDS-PAGE under reducing conditions, into two separated bands, corresponding to H and L chains (Fig. 1C). In addition, bile Ig gave at least two distinct L-chain bands, of very similar molecular mass (about 25-27 kDa), in agreement with data reported by Lobb et al. for *Ichtalurus punctatus* serum Ig [8]; it is not clear if these L chain variants represent distinct L chain classes. We did not find any difference in relative molecular mass between plasma and bile Ig H chain, as seen in *Archosargus probatocephalus* [9].

The pI spectrum of purified plasma Ig, ranging between 4.0 and 6.0, was found lower than that of higher vertebrates (Fig. 1D). This result suggests the prevalence of acidic residues in the carbohydrate and/or amino acid moieties. In fact, the sialic acid linked α(2-3) to galactose was selectively recognized as terminal residue by *Maackia amurensis* agglutinin (Coscia et al., in preparation).

Antisera specific for either H or L chains were obtained by immunizing rabbits with the H-chain or L-chain band, electroeluted from an SDS-PAGE slab gel of purified *T. bernacchii* Ig, run under reducing conditions. The anti-L chain antiserum reacted with human IgG, whereas *T. bernacchii* Ig was recognized, in double immunodiffusion, by commercial antihuman L chain (λ) antiserum and antimouse H chain (γ) antiserum; nonimmune rabbit sera were used as controls [7]. The antigenic crossreactivity observed between *T. bernacchii* Ig and mammalian Ig suggests the conservation of structurally important epitopes during evolution.

To better understand how the Antarctic fish Ig can work at very low temperatures we looked for the structural characteristics accounting for this property. For this purpose, and to overcome the difficulties related to the amino acid sequencing, we constructed a cDNA library in ZAP Express vector.

Total RNA was extracted from the spleen of a *T. bernacchii* specimen by the guanidinium thiocyanate method. Poly(A)$^+$ mRNA was purified from total RNA through an oligo(dT)-cellulose column and was used for cDNA synthesis. Rabbit anti-*T. bernacchii* Ig H chain antibodies, purified by MabTrap G affinity chromatography, were used to immunoscreen the cDNA library. Fourteen clones were found positive; five of them were partially sequenced in both directions by the Sanger dideoxynucleotide chain termination method with T7 DNA polymerase, using T3 or universal primers. Due to its length (1.9 kb), the clone 2C2.3 was expected to represent the most length of the Ig H chain transcript.

When the partial nucleotide sequences were used to search the databases, significant homologies were obtained with the VH region of a primitive teleost (*Elops saurus*), rainbow trout (*Oncorhynchus mykiss*), *Xenopus laevis*, human, mouse, and rabbit. On the basis of this homology, the clone

2C2.3, although lacking the main part of the leader, was shown to contain the VH domain and the poly-A tail; in addition there might be a 3' untranslated region. Sequencing of the entire C region is in progress.

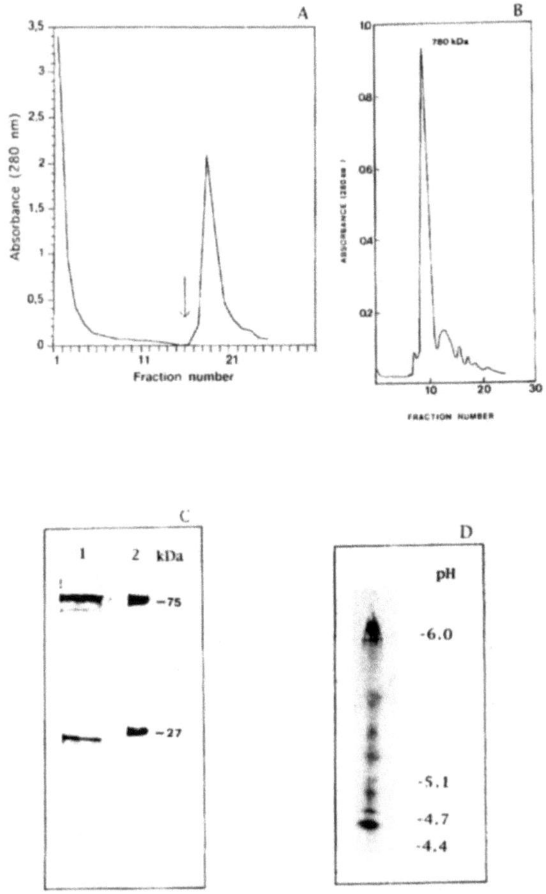

Fig. 1. Purification of *T. bernacchii* Ig. **A**: thiophilic adsorption chromatography on T-gel. Arrow indicates the recovery of bound fraction. **B**: FPLC elution profile of T-gel bound fraction. **C**: SDS-PAGE, under reducing conditions, of bile Ig (lane 1) and plasma Ig (lane 2). **D**: isoelectricfocusing (IEF) of plasma Ig

The molecular cloning approach will help extend our analysis of Antarctic fish immunoglobulin structure by verifying: the VH region diversity and the relationship to VH families from temperate teleost species; the conservation of the domain structure in heavy chain; the distribution of the cysteine residues involved in interchain and intersubunit linkages; the conservation of the carbohydrate acceptor sites.

Once obtained, the complete nucleotide and deduced amino acid sequences will help understand the structural characteristics accounting for the adaptation of the Antarctic fish Ig to low temperatures.

Humoral Immune Response of Antarctic Fish Against Nematode Parasites

The nematode species most frequently found infesting fish are *Anisakis simplex* [10], *Contracaecum osculatum* [11] and *Pseudoterranova decipiens* [12], all belonging to Anisakidae family. Fish are intermediate hosts of the third stage nematode larvae and marine mammals are the final hosts of the adult parasites; the liver is the main site of infestation [13]. The nematode proteins involved in fish immune response certainly include secreted/excreted protein and the components of the cuticle (collagens, cuticlins and surface-associated proteins)[14].

Fish immune response against parasites has been poorly investigated: the presence of *Anguilla anguilla* antibodies, specific for the nematode *Anguillicola crassus* [15], the monogenean *Pseudodactylogyrus bini* [16] and the trematode *Telogaster opisthorchis* [17] has been demonstrated.

We addressed our interests to the molecular bases of the parasitism of Antarctic fish by nematodes: our first aim was to verify the presence of Antarctic fish antibodies specific for nematode antigens; a further goal was the purification of both antibodies and nematode antigens.

Nematodes were collected from the liver of specimens of different teleostean species: those recovered from *N. coriiceps*, *C. aceratus*, *T. bernacchii*, *Chionodraco rastrospinosus*, *Pseudochaenichthys georgianus* and *Cryodraco antarcticus* were identified as third stage larvae of *Pseudoterranova decipiens*, whereas those from *C. hamatus* as third stage larvae of *Contracaecum osculatum* complex. Both the nematode species belong to the superfamily of Ascaridoidea, family Anisakidae, subfamily Anisakinae.

Some *P. decipiens* nematodes were individually maintained alive in vessels containing phosphate buffer saline (PBS) and the proteins released within 8 days were collected: this preparation, containing the secreted/excreted proteins, was referred to as *Pd*RP. Dead larvae were incubated, at 37 °C, in PBS containing a cocktail of protease inhibitors; the extract obtained (*Pd*AP) contained, presumably, the cuticular components corresponding to surface-associated proteins. Finally, the larvae were boiled for 10 min in the lysis buffer containing 2% SDS and 10% β-mercaptoethanol and incubated in the same buffer, at 37 °C overnight. This preparation (*Pd*SP) presumably contained collagens and cuticlins in addition to solubilized cellular components. The extract of *C. osculatum* nematodes, available in smaller amount, was prepared only under reducing conditions (*Co*SP) [18].

A high molecular heterogeneity was found when the nematode protein preparations were analyzed by different methods: in SDS-PAGE, *Pd*SP showed about 26 protein bands whereas *Pd*AP and *Pd*RP showed a lower number of components (12 and 6 respectively). The main band (40 kDa) of *Pd*SP was revealed also in *Pd*AP and *Pd*RP, besides components specific for each preparation (Fig. 2). *Pd*SP was found very heterogeneous also in isoelectric point as demonstrated by two-dimensional electrophoresis (Coscia et al. in preparation). *Pd*RP was found less heterogeneous also when analyzed by reverse-phase HPLC as only six peaks were obtained [6].

The nematode extracts obtained from the two species by the same method (*Pd*SP and *Co*SP) shared at least three components (Fig. 2).

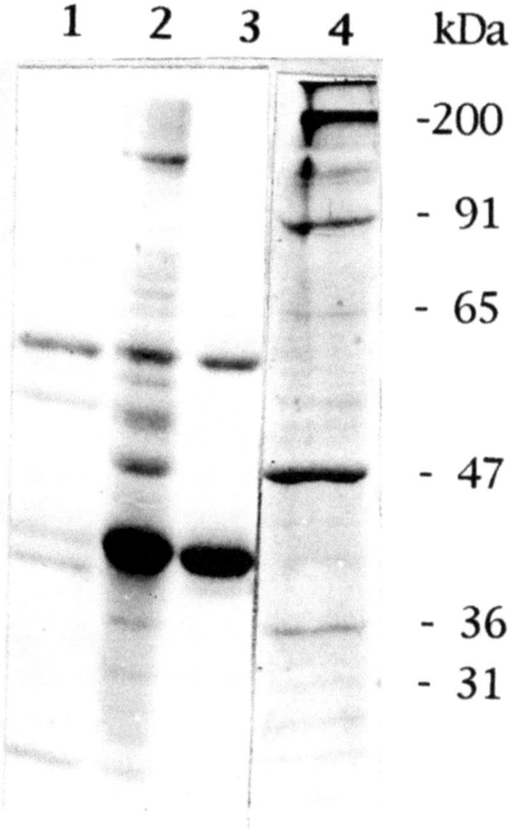

Fig. 2. SDS-PAGE, under reducing conditions, of the nematode preparations *Pd*RP (lane 1), *Pd*SP (lane 2), *Pd*AP (lane 3), *Co*SP (lane 4), Coomassie brilliant blue (CBB) stained

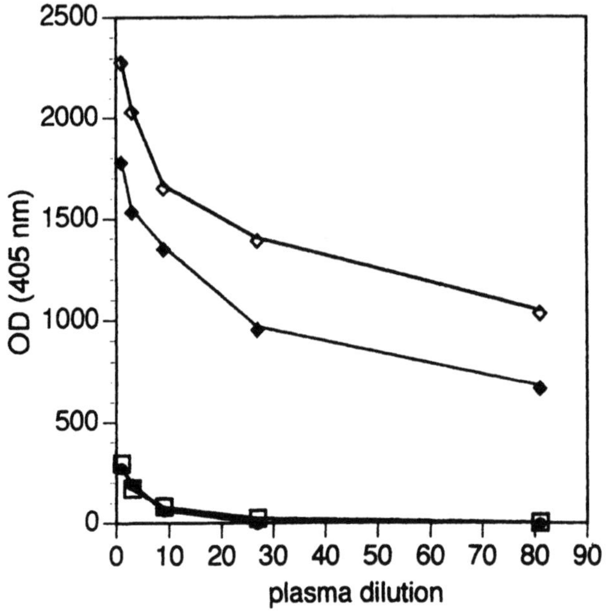

Fig. 3. ELISA analysis of the binding of *T. bernacchii* plasma Ig to *Pd*AP (-◊-◊-) and *Pd*SP (-◆-◆-). *Dicentrarchus labrax* plasma Ig was used as negative control of the binding to *Pd*AP (-□-□-) and *Pd*SP (-•-•-)

The ability of the nematode proteins to recognize Antarctic fish Ig was tested by different methods. By ELISA, the binding of *T. bernacchii* Ig to *Pd*AP was found higher than that to *Pd*SP (Fig. 3). Consistent results were obtained by affinity chromatography when *Pd*AP and *Pd*SP were eluted through an *N. coriiceps* Ig coupled column: the fraction of *Pd*AP retained by fish antibodies was more abundant than that of *Pd*SP, indicating the presence in *Pd*AP of antigens at higher concentration or with higher affinity.

To investigate the *Pd*SP binding activity of antibodies from plasma and secretions, the same amounts of *T. bernacchii* plasma, bile Ig, and of *G. gibberifrons* cutaneous mucus Ig were tested by dot-blot immunoassay (results not shown): in autoradiography, the spot incubated with bile Ig gave a signal clearly higher than that with plasma Ig; in addition, the cutaneous mucus Ig gave a detectable signal [6]. Our results indicate the presence in Antarctic fish of a mucosal immune response against parasites.

Preliminary investigations were carried out on the prevalence of the immune response to nematode parasites in different species of Antarctic fish: plasma of individual fish of four species was tested, in double immunodiffusion assay, for the presence of anti-*C. osculatum* antibodies.

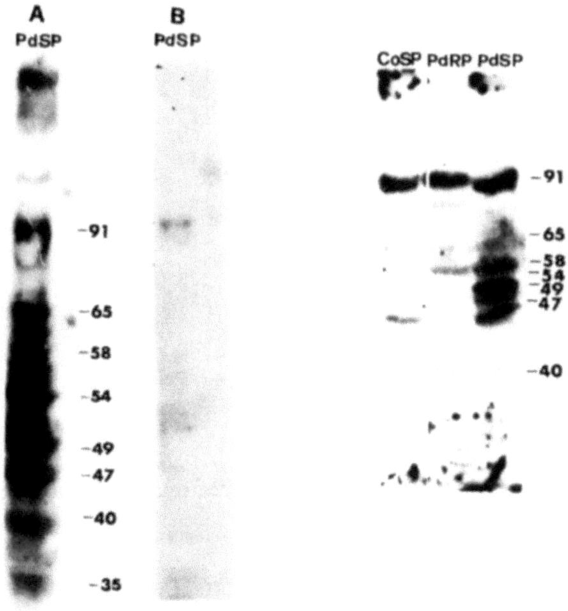

Fig. 4. Immunoblotting analysis of the nematode preparations *Pd*RP, *Pd*SP, *Co*SP. On the left, *Pd*SP was incubated with *T. bernacchii* bile Ig (A) and, as negative control, with *D. labrax* bile Ig (B). On the right, *Pd*RP, *Pd*SP antigens were revealed by *T. bernacchii* plasma Ig, while *Co*SP antigens were revealed by *C. hamatus* plasma Ig

Immunoprecipitates were observed with 2 out of 10 *T. bernacchii* plasma, 2 out of 4 *G. gibberifrons* plasma, 3 out of 3 *N. coriiceps* plasma and none out of 2 *P. georgianus* plasma.

However, the paucity of the data we collected so far does not allow us to evaluate the susceptibility to the parasitic infestation among these species.

To identify the antigenic components, nematode proteins were separated by SDS-PAGE, blotted onto nitrocellulose sheet and then incubated with *T. bernacchii* plasma Ig; bound antibodies were revealed by ^{125}I-rabbit IgG (anti-*T. bernacchii* Ig H chain). Five out of 6 *Pd*RP components and only 7 out of 26 *Pd*SP components were identified as antigens by autoradiography indicating that *Pd*RP is more antigenic (Fig. 4). A higher number (13) in *Pd*SP bands was found positive when *T. bernacchii* bile Ig was used as first antibody, whereas no bands were detected when *Dicentrarchus labrax* bile Ig was used as negative control (Fig. 4). These results confirm the presence of a mucosal immune response and suggest differences in specificity between plasma and mucosal Ig.

Immunoblotting analysis of *Co*SP, using *C. hamatus* Ig, revealed at least three antigens, two of them in common with *Pd*SP (Fig. 4).

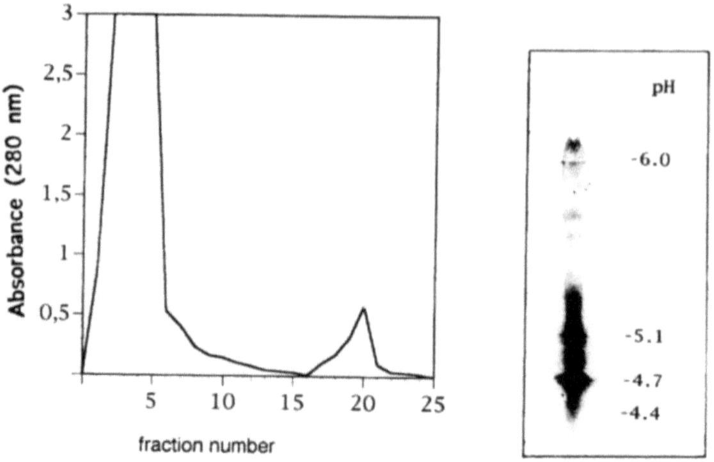

Fig. 5. Affinity purification of anti-*Pd*SP antibodies from *N. coriiceps* plasma and IEF analysis of purified specific antibodies

To find out the biological role of the major protein component (40 kDa) of *Pd*SP, its N-terminal amino acid sequence was determined, after blotting onto an Immobilon sheet.

The database search for identities of the sequence of the first 20 animo acid residues gave a high percent of identity with the *P. decipiens* extracellular globin. This is an oxygen-transporter molecule associated to the cuticle external surface [19]. The subsequent step in our study on the antigenic molecules was the purification of the *Pd*SP 54 kDa antigen by FPLC using a Superdex 200 column [6].

Moreover, *Pd*SP specific antibodies were purified from *N. coriiceps* plasma by *Pd*SP-bound Sepharose 4B affinity chromatography (Fig. 5); the retained fraction, run in SDS-PAGE under reducing conditions, gave the typical Ig pattern as that shown in Fig. 1C. The isoelectrofocusing pattern of the purified specific antibodies consisted of at least 13 focused bands, giving a measure of the antibody diversity (Fig. 5).

To investigate the antigen/antibody interaction and its temperature dependence, ELISA was performed at -2 °C, +4 °C and +20 °C, using *D. labrax* plasma as negative control (Fig. 6).

The antibody recognition, though lower, was detected also at -2 °C, suggesting that *T. bernacchii* Ig can carry out its function at the temperature of the aquatic environment.

In conclusion, our results demonstrate the presence, in plasma and secretions of infested fish, of antibodies specific for nematode antigens; clearly, bile antibodies showed a relatively high affinity and heterogeneity, compared with plasma antibodies.

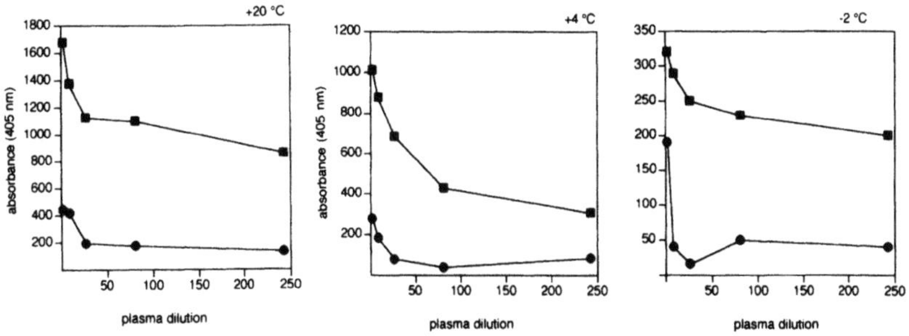

Fig. 6. ELISA analysis: effects of temperature on the reaction of *T. bernacchii* plasma Ig (-■-■-) with *Pd*SP antigens. *D. labrax* plasma Ig (-●-●-) was used as negative control

The antigenic protein components of different nematode extracts were compared, the *Pd*SP 40 kDa antigen was identified as extracellular globin and the *Pd*SP 54 kDa antigen was purified. Finally, *Pd*SP specific *N. coriiceps* antibodies were isolated and found heterogeneous in isoelectric point.

Acknowledgments

This research was supported by a Grant from the Italian National Programme for Antarctic Research. We thank Prof. G. di Prisco, Institute of Protein Biochemistry and Enzymology, CNR, Naples, for providing us with nematodes and fish plasma and secretions and Dr. S. Mattiucci, Institute of Parasitology, University "La Sapienza," Rome, for the identification of nematode larvae.

References

1. Wilson MR, Warr GW (1992) Fish immunoglobulins and the genes that encode them. Annual Rev of Fish Diseases 201-221
2. Warr GW (1995) The immunoglobulin genes of fish. Dev Comp Immunol 19: 1-12
3. Warr GW, Middleton DL, Miller NW, Clem LW, Wilson MR (1991) An additional family of VH sequences in the channel catfish. Eur J Immunogen 18: 393-397
4. Bengtén E, Strömberg S, Pilström L (1994) Immunoglobulin VH regions in Atlantic cod (*Gadus morhua* L.): their diversity and relationship to VH families from other species. Dev Comp Annual 18: 109-122
5. Roman T, Charlemagne J (1994) The immunoglobulin repertoire of the rainbow trout (*Oncorhynchus mykiss*): definition of nine IgH-V families. Immunogenetics 40: 210-216
6. Coscia MR, Oreste U (1997) Immunoglobulins from Antarctic fish: structure and antibody specificity. Proceedings of the third meeting on "Antarctic Biology". Camerino University Press, pp 61-67

7. Coscia MR, Oreste U (1997) Structure and antibody specificity of immunoglobulins from plasma of Antarctic fishes. Antarctic Communities. Cambridge University Press, Cambridge. pp 273-276
8. Lobb CJ, Olson MOJ, Clem WL (1984) Immunoglobulin light chain classes in a teleost fish. J Immunol 132: 1917-1923
9. Lobb CJ, Clem LW (1981) Phylogeny of immunoglobulin structure and function. XII. Secretory immunoglobulins in the bile of the marine teleost *Archosargus probatocephalus*. Mol Immunol 18: 615-619
10. Nascetti G, Paggi L, Orecchia P, Smith JW, Mattiucci, Bullini L (1986) Electrophoretic studies on the *Anisakis simplex* complex (Ascaridida: Anisakidae) from the Mediterranean and North-east Atlantic. Int J Parasitol 16: 633-640
11. Koie M, Fagerholm HP (1995) The life cycle of *Contracaecum osculatum* (Rudolphi, 1802) *sensu stricto* (Nematoda, Ascaridoidea, Anisakidae) in view of experimental infections. Parasitol Res 81:481-489
12. Palm H, Andersen K, Klöser, Plötz J (1994) Occurrence of *Pseudoterranova decipiens* (Nematoda) in fish from the southeastern Weddell Sea (Antartic). Polar Biol 14: 539-544
13. Kock KH (1992) Antarctic fish and fisheries. Cambridge University Press, Cambridge. pp 153-159
14. Politz SM, Philipp M (1992) *Caenorhabditis elegans* as a model for parasitic nematodes: a focus on the cuticle. Parasitology Today 8: 6-12
15. Buchmann K, Pedersen L, Glamann J (1991) Humoral immune response of European eel *Anguilla anguilla* to a major antigen in *Anguillicola crassus* (Nematoda). Dis aquat Org 12: 55-57
16. Buchmann K (1993) A note on the humoral immune response of infected *Anguilla anguilla* against the gill monogenean *Pseudodactylogyrus bini*. Fish & Shellfish Immunol 3: 397-399
17. McArthur CP (1978) Humoral antibody production by New Zealand eels, against the intestinal trematode *Telogaster opisthorchis* (Macfarlane 1945). J Fish Dis 1: 377-387
18. Coscia MR, Oreste U. Humoral immune response of Antarctic fish to *Contracaecum osculatum*, a nematode parasite. Submitted to Fish & Shellfish Immunol
19. Dixon B, Walker B, Kimmins W, Pohajdak B (1991) Isolation and sequencing of a cDNA for an unsual hemoglobin from the parasitic nematode *Pseudoterranova decipiens*. Proc Natl Acad Sci USA 88:5655-5658

Vascular Adaptations for a Low Pressure / High Flow Blood Supply to Locomotory Muscles of Antarctic Icefish

Stuart Egginton[1] and J.Cliff Rankin[2]

[1]Department of Physiology, University of Birmingham, UK
[2]Institute of Biology, Odense University, Denmark

Introduction

The extremely stable Antarctic environment may be expected to have produced a number of cardiovascular adaptations to offset increased blood viscosity and/or decreased oxygen diffusion rates in the cold [1,2]. The ichthyofauna is dominated by Notothenioids which have a modest haematocrit and a reduced blood haemoglobin concentration [3], a situation taken to the extreme in the icefish (channichthyids) that possess only a few vestigial erythrocytes and no respiratory pigment. Low blood pressure and large venous return reported for icefish [4] would best be accommodated by a relatively low afterload on the heart [5]. Evidence for the origin of the assumed low peripheral resistance in icefish was sought in the morphology of the blood vascular system; preliminary results from this study were reported as abstracts [6,7].

Material and Methods

Fish used for vascular casting were caught by bottom or benthopelagic trawling at 360 – 600 m depth from the RV "Polarstern" in the Weddell Sea (71-75°S, 13-28°W) on the ANTARKTIS VII/4 (EPOS) expedition in 1989. Seventeen individuals of 6 species of the family Channichthyidae (icefish) were used for vascular casting with Acrifix (A) plus 27 individuals of 8 species with Microfil (M) as follows: *Chaenocephalus aceratus* Lönnberg (A, M); *Cryodraco antarcticus* Dollo (M); *Chionodraco myersi* DeWitt and Tyler (A, M); *Chionodraco hamatus* Lönnberg (A, M); *Chionodraco rastrospinosus* DeWitt and Hureau (A, M); *Champsocephalus gunnari* Lönnberg (A, M); *Pseudochaenichthys georgianus* Norman (A, M); *Chionobathyscus dewitti* Andriashev and Neelov; *Neopagetopsis ionah* Nybelin (M). In addition, Microfil casts

G. di Prisco, E. Pisano, A. Clarke (Eds)
Fishes of Antarctica. A biological overview
© Springer-Verlag Italia 1998

were made in one specimen each of *Gobionotothen gibberifrons* Lönnberg and *Trematomus eulepidotus* Regan. Specimens of *C. aceratus* for histological investigation were caught around the British Antarctic Survey's station on Signy Island (60°43'S, 45°36'W) and held in temperature controlled tanks until sampling (T_{amb} = 0 °C). For purposes of comparison, rainbow trout (*Oncorhynchus mykiss* Walbaum) were sampled at a trout farm during autumn (T_{amb}=11 °C).

Vascular Casting

After capture specimens were injected with heparinised Ringer followed by either Microfil (Canton Bio-Medical Products, Boulder, Colorado) or Acrifix (Röhm, Darmstadt) via the bulbus arteriosus, or in some cases the hypobranchial artery, and left overnight at ambient temperature. Acrifix injected fish were kept frozen until the tissues were dissolved with concentrated KOH. These casts were then washed with, and stored in, distilled water. Microfil injected fish were dissected, examined and photographed or drawn on the ship, and then preserved in formalin, followed by dehydration in graded alcohols. Details of the techniques were varied as it was not possible to judge which procedure gave the best results until after the cruise. The best Acrifix filling of the pectoral musculature was achieved in a 660 g specimen of *C. hamatus*. Six hours after capture it was given an intraperitoneal injection of the vasodilator papaverine (Sigma: 0.5 ml of $5x10^{-3}$ M in Ringer adjusted to pH 4 with HCl). The heart continued beating strongly and to determine the direction of blood flow in the hypobranchial artery (H) it was ligated before cannulation. It swelled anterior to the ligature (preliminary experiments in which Evans Blue was injected into anaesthetized fish had already shown that flow in the hypobranchial artery was from the gills to the pectoral muscles; this connectivity was confirmed in other specimens by injection of Microfil). Blue-pigmented Acrifix (7 parts resin/3 parts hardener/0.3 parts pigment) was then injected into the cannulated H to fill the pectoral muscle vasculature; 0.3 ml of heparinized Ringer was then injected through the cannulated bulbus arteriosus, followed by 8.5 ml of red Acrifix in ca. 7 min. This also entered the dorsal aorta and its branches, most of the blue Acrifix in the pectoral musculature being displaced by the red, which retrogradely penetrated into H.

Rainbow trout, *O. mykiss,* were anaesthetized in MS222 and injected with 500 U heparin. Following cannulation of the dorsal aorta near complete perfusion of the trunk musculature was possible. The bulbus arteriosus was severed to allow drainage, and heparinized saline perfused via a Watson-Marlow rotary pump at a rate of 10 ml min^{-1}, passing through a 0.8 μm in-line filter. After a clear exudate was observed (5-10 min, depending on the degree of vasodilatation), the perfusate was changed to 10% neutral formalin while the acrylic casting compound was being prepared. Batson's No. 17 anatomical corrosion compound is a suitable casting agent for vascular networks of homeotherms, although it is quite viscous and particularly unsuitable at low temperatures if complete filling of the smallest vessels (capillaries) is to be achieved. This was therefore diluted by a monomeric methyl methacrylate (Sevriton) which significantly reduced its viscosity. Mixing on ice to retard polymerization and setting time also improved the extent of filling as judged by methyl salicylate-cleared material. The final recipe was: 10.0 ml Batson's (A), 3.0 ml Catalyst (B), 0.2 ml Promotor (C), 0.1 ml Colorant, 6.0 ml Sevriton.

Microscopy

Muscle samples were fixed by immersion in buffered paraformaldehyde/ glutaraldehyde at resting length [8]. Transverse sections of resin-embedded material were double stained (lead citrate/uranyl acetate) and electronmicrographs taken with a JEOL 100 cx II operating at 60 kV. Casts were mounted on stubs and gold or platinum coated at low power in a Polaron E5000 sputter coated to a thickness of 10-20 nm before examination in a Hitachi S-520 Scanning Electron Microscope (SEM) at 10 kV.

Results

In both *C. hamatus* and *C. aceratus* a complete set of arterial arches from the first three gill arches joined to form a vessel which ran posteriorly to a position ventral to the heart, and then divided into two vessels which passed laterally and posteriorly to the pectoral muscles, anastomosing with the subclavian arteries coming from the dorsal aorta (Fig. 1). The ventral efferents from the 4th gill arches were smaller and divided to supply local tissues. All icefish species examined showed a similar pattern, which is illustrated by a drawing of a freshly injected specimen of *C. hamatus* (Fig. 2). This functional hypobranchial network [6,9] thus serves to provide an anastomosing dual arterial blood supply to the main locomotory (pectoral) muscles. Perfusion resistance will be lowered by some indeterminate amount, but given the similar diameters of both dorsal aorta and hypobranchial afferents this is likely to be around a factor of 2.

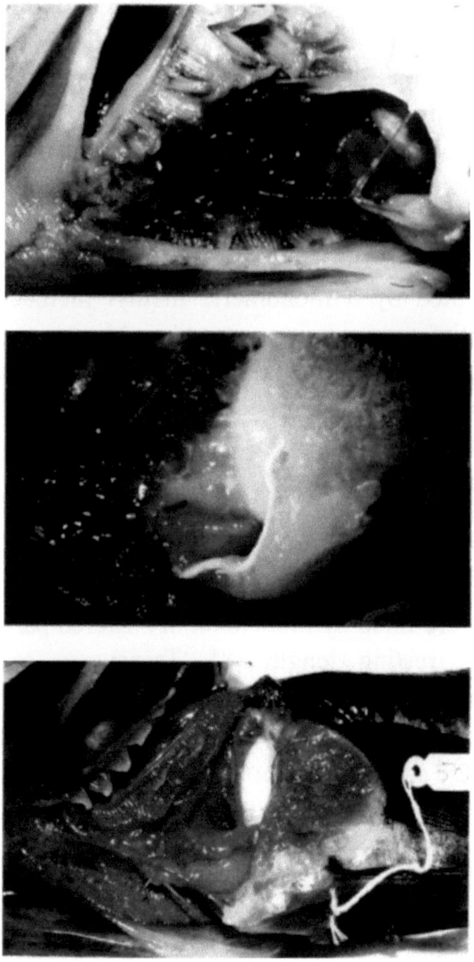

Fig. 1. Photomicrographs of injected specimens, showing filling via the hypobranchial system of both gills and musculature. Top panel: an anaesthetized specimen of *Chionobathyscus dewitti* dissected to show the heart and hypobranchial system; 0.25 ml of 2% Evans blue dye was injected into the bulbus arteriosus to reveal the hypobranchial artery, ventral to the heart (which was still beating). Note the bifurcation at the level of the ventricle (asterisk), giving rise to major feed arteries for both sets of pectoral muscles (arrows). G gills. Middle panel: *Chionodraco myersi*. White Microfil injected retrogradely to fill the pectoral muscles via the hypobranchial system, which bifurcates (arrowhead) into feed branches (arrow). Note the intensity of pectoral muscle vasculature (asterisk). Subsequent injection of red Microfil via the bulbus arteriosus filled gills (G) and trunk vessels. Bottom panel: *Cryodraco antarcticus*. Red Microfil injected into the bulbus arteriosus fills both the gill (G) and pectoral muscle vasculature (V), via the dorsal aorta and the hypobranchial artery. Note the amount of Microfil visible in the skinned pectoral musculature

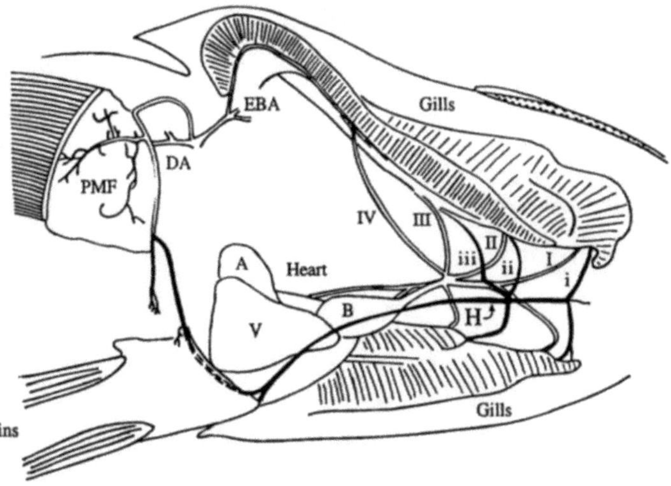

Fig. 2. The anterior arterial (hypobranchial) system of *Chionodraco hamatus*: dissected Microfil-injected specimen. The oesophagus has been removed, and the heart and hypobranchial system reflexed to reveal the dorsal aorta. This runs in the midline of the body but in this oblique ventral view appears to be on one side. The operculae have been removed to show the gills. The hypobranchial artery was attached to the ventral wall of the pericardium.
Abbreviations: A atrium; V ventricle; B bulbus arteriosus; I-IV afferent branchial arteries; EBA efferent branchial artery; DA dorsal aorta; PMF pectoral musculature; i-iii hypobranchial arches; H hypobranchial artery

Nototheniid casts showed evidence for a hypobranchial system, attached to the pericardium in *T. eulepidotus*, while in *G. gibberifrons* and *N. coriiceps* a very small vessel ran ventral to the pericardium dividing into two minute vessels below the pectoral girdle. None of the channichthyid or nototheniid casts showed any evidence for coronary arteries derived from the hypobranchial system. The arterial supply to the lateral (trunk) locomotory muscle of trout was of the standard teleostean form, as described previously [10]. This consists of segmental arterial branches from the dorsal aorta dividing in the midmyoseptal region to form anterior and posterior-directed microvascular beds (Egginton, unpublished). Comparison of vascular casts from the *C. hamatus* and *O. mykiss* musculus pectoralis under the SEM showed that paired feed arteries to icefish pectoral muscles were ca. 130 μm, with draining veins ca. 250 μm in diameter, which was of a similar magnitude to the segmental arteries of trout (ca. 170 μm). All microvessels were around threefold wider: arterioles 18±16 (18) and 54±22 (18), capillaries 3±0.5 (21) and 11±2

(21), venules 20±9 (3) and 54±24 μm (21) for trout and icefish, respectively (mean±SD; number of observations in parentheses). In addition to differences in capillary size (Fig. 3), those of the icefish show an unusual architecture.

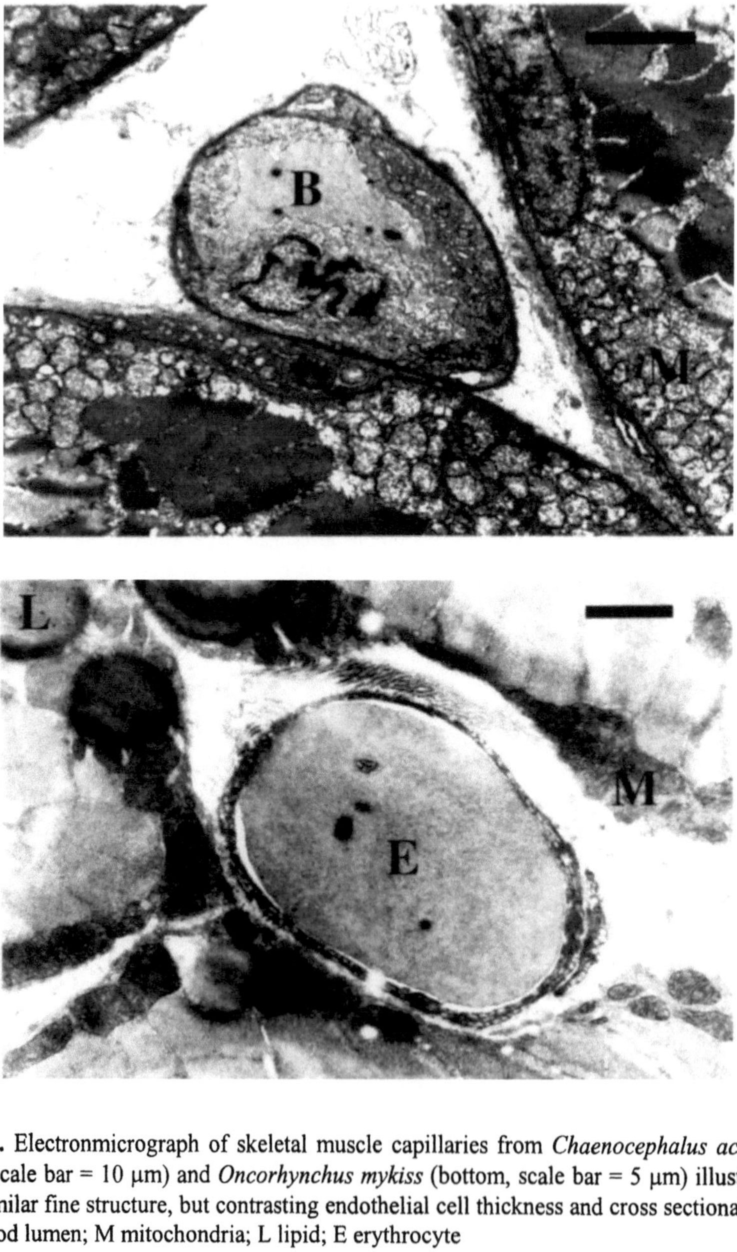

Fig. 3. Electronmicrograph of skeletal muscle capillaries from *Chaenocephalus aceratus* (top, scale bar = 10 μm) and *Oncorhynchus mykiss* (bottom, scale bar = 5 μm) illustrating the similar fine structure, but contrasting endothelial cell thickness and cross sectional area. B Blood lumen; M mitochondria; L lipid; E erythrocyte

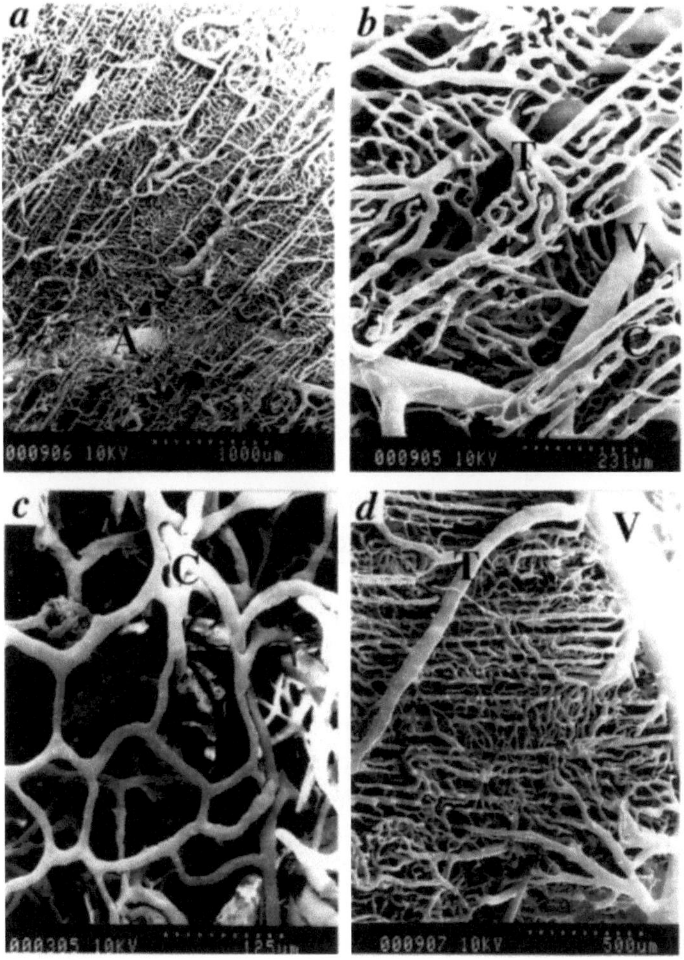

Fig. 4. Scanning electronmicrographs of vascular casts from *Chionodraco hamatus* pectoral muscle. Scale bars included for each picture. **a** Feed arteriole (A) lying under a mat of capillaries, some of which branch direct from the parent vessel. **b** The capillary bed covers muscle fibers in an evenly spaced network, orientated with the long axis. T, terminal arteriole; C capillary; V collecting venule. **c** Loops formed by convergence and divergence of capillaries (C) provide maximal coverage of the fibre surface for diffusive exchange. **d** Large drainage venules (V) have joining capillaries as well as terminal venules (T). Muscle fibre axis lies in the horizontal direction

Capillaries from trout muscle are orientated along the longitudinal axis of muscle fibres (unpublished data) with some tortuosity but little branching (anisotropic orientation) adding only ca. 5% to capillary length over that of a truly straight capillary network, as in other teleosts [8].

In contrast, capillaries from icefish show a large degree of isotropy, due largely to the presence of numerous capillary loops (Fig. 4).

These represent true divergent and convergent capillary branches, and not the arcade microvascular network observed in vertebrate mesentary. The capillary loops thus form a loose tube around each muscle fibre, thereby maximizing the contact surface area for a low capillary density.

Discussion

The oxygen content of arterial blood in icefish (*Chaenocephalus aceratus*) is only one-twentieth that of a sympatric red-blooded species (*Notothenia coriiceps*), reflecting the lack of respiratory pigment [11]. In order to support adequate tissue metabolism, peripheral oxygen transport relying on physical solubility must therefore utilize a high rate of tissue perfusion. Given the observed low blood pressure in icefish this may best be achieved by an appropriate reduction in peripheral resistance to blood flow. Despite the abnormally large heart of icefish, involving chamber dilation to accommodate the large venous return, the maximal in vitro power output of the myocardium is unremarkable [12]. This suggests there is indeed a low afterload on the in situ working heart. We have therefore examined vascular adaptations that may underlie this condition.

The dual arterial supply to the locomotory (pectoral) muscles is made possible by retention of the primitive condition of a hypobranchial artery [6,9]. A reduced hypobranchial system, giving rise to the coronary arteries where these are present, is found in other teleosts [13]. Based on measurements of external diameters, Eastman [1] estimates flow capacity of the hypobranchial system to be some 30% of cardiac output in icefish, while among the red-blooded nototheniids it was of smaller calibre and was formed by union of fewer efferent branchial arteries, suggesting its flow capacity (and hence functional relevance) is greatly reduced. This is consistent with little obvious filling of the hypobranchial system with casting compound in specimens of *N. coriiceps*, with both main dorsal aortic branches filling the pectoral muscle mass and adjacent body musculature up to the medial division (Egginton, unpublished). It is unclear whether this structure is related to labriform locomotion, where a high blood flow (i.e. low resistance vascular bed) to the pectoral muscles is required to support aerobic swimming, rather than an endemic specialization of Antarctic fishes.

Total peripheral resistance (TPR) is given as mean arterial blood pressure (MABP; mmHg)/cardiac output (CO; ml/sec). Using published values of 10/1.98 for *Chaenocephalus aceratus* [4] and 23.6/0.50 for

Oncorhynchus mykiss [14] gives estimates of TPR of 67.3×10^3 and 629×10^3 dynes^{-1} sec^{-1} cm^{-5}, respectively. Using dorsal aortic blood pressure means that we actually calculate postbranchial, rather than total peripheral resistance in fishes. The systemic resistance of icefish is therefore <10% that of rainbow trout at their preferred (optimal) temperatures of 0 and 11 °C, respectively. The reduced skeletal muscle mass in icefish compared with trout (40% vs. 60% body weight, respectively) suggests the difference in total vascular resistance located in muscle is somewhat less. Given that both the conduit arterial system and high resistance vessels show adaptations to maximize muscle perfusion, we may also try to estimate the influence of capillary size on the potential microvascular blood flow (Q). Comparative values can be obtained from solving the Hagan-Poiseuille equation:

$$Q = (\pi \cdot r^4 \cdot \text{MABP}) / (8 \cdot L \cdot \eta)$$

where r is vessel radius, L is vessel length and η is blood viscosity. As there are two unknowns (L and η) we can only calculate the likely ranges. Fletcher and Haedrich [15] have measured the viscosity of rainbow trout blood to be 9 cP at low shear rate (22.5 sec^{-1}). Studies that have directly compared blood viscosity of an icefish with a red-blooded species estimated it to be 4 (*C. antarcticus*) vs. 11 cP (*Pagothenia = Trematomus bernacchii*) at -1.8 °C [16], and 3.4 (*C. aceratus*) vs. 5.5 cP (*N. coriiceps*) at -1.0 °C [17]. Trout blood, using the same equipment and temperature, was around 5.4 cP (Egginton, unpublished). While the technique used in the former study, cone-plate viscometry at low shear rate, is likely to slightly overestimate values of η, these data are similar enough for our purposes. Substituting mean values, and assuming the same capillary length among muscle from different species, the predicted maximal capillary blood flow in icefish is some two orders of magnitude greater than that of trout (ca. x120). As yet there are no quantitative data on capillary lengths, but even if the observed tortuosity increased L by a factor of two, which is likely, the estimated Q would only be halved and hence the predicted large differential in maximal blood flow would be maintained. The modest difference between TPR and muscle vascular resistance may reflect the much higher capillary density of trout muscle, thereby offsetting to some extent the effect of larger capillaries in icefish.

The highly tortuous capillaries in icefish maximize the contact area with muscle fibres, although the density of microvessels is much less than found in other teleosts [18], such that the ratio of capillary surface area to muscle volume (surface density or index of supply to demand) is therefore compromised [19]. What appears to be conserved, however, is capillary volume density. The instantaneous blood volume of the microcirculation in locomotory muscles of icefish is quite similar to that found for other teleosts, which argues strongly for a perfusion-limited system [20]. Similar arguments may be advanced for the vasculature of the gills and skin [21,22].

Acknowledgments

The authors wish to thank the officers and crew of the PPS "Polarstern" and the British Antarctic Survey for provision of specimens, and the Natural Environment Research Council for support. Participation by J.C.R. in the EPOS 3 cruise was supported by the European Science Foundation, the Alfred-Wegener-Institut and NERC. J.-C. Hureau assisted with identification of specimens. Scanning electron microscopy was carried out in the School of Biological Sciences, University of Wales, Bangor with skilled technical assistance by A. Davies. Acrifix was kindly donated by A. Kunzmann.

References

1. Eastman JT (1993) Antarctic fish biology. Academic Press London
2. Egginton S, Cordiner S (1994) Effect of fixation protocols on muscle preservation and in situ diffusion distances. J Fish Biol 47:59-69
3. Macdonald JA, Montgomery JC, Wells RMG (1987) Comparative physiology of Antarctic fishes. Adv Mar Biol 24:321-388
4. Hemmingsen EA, Douglas EL (1972) Respiratory and circulatory responses in a haemoglobin-free fish, *Chaenocephalus aceratus*, to changes in temperature and oxygen tension. Comp Biochem Physiol A 43:1031-1043
5. Holeton GF (1970) Oxygen uptake and circulation by a hemoglobinless Antarctic fish (*Chaenocephalus aceratus* Lönnberg) compared with three red-blooded Antarctic fish. Comp Biochem Physiol 34:457-471
6. Rankin JC, Harrison B, Airey AC (1987) Branchial vascular anatomy in the icefish *Chaenocephalus aceratus* Lönnberg. Proc V Congr Europ Ichthyol, Stockholm 1985, p 453
7. Egginton S, Rankin JC (1991) The vascular supply to skeletal muscle in fishes with and without respiratory pigments. Int J Microcirc Clin Exp 10:396
8. Egginton S, Cordiner S (1997) Cold-induced angiogenesis in seasonally acclimatized rainbow trout (*Oncorhynchus mykiss*). J Exp Biol 200:2263-2268
9. Stewart JB, Douglas EL (1973) Circulatory anatomy in the icefish family Chaenichthyidae. Antarct J US 8:204-205
10. Satchell GH (1991) Physiology and form of fish circulation. Cambridge University Press, Cambridge.

11. Egginton S (1994) Stress response in two Antarctic teleosts (*Chaenocephalus aceratus* Lönnberg and *Notothenia neglecta* Nybelin) following capture and surgery. J Comp Physiol B 164:482-91
12. Tota B, Acierno R, Agnisola C (1991) Mechanical performance of the isolated and perfused heart of the haemoglobinless Antarctic icefish *Chionodraco hamatus* (Lönnberg): effects of loading conditions and temperature. Phil Trans Soc Lond, Ser B 332:191-198
13. Laurent P (1984) Gill internal morphology. In: Hoar W.S., Randall D.J. (eds) Fish physiology, Vol. 10A, Academic Press, London, pp 73-183
14. Wilson RW, Egginton S (1994) Assessment of maximum sustainable swimming performance in rainbow trout (*Oncorhynchus mykiss).* J Exp Biol 192:299-305
15. Fletcher GL, Haedrich RT (1987) Rheological properties of rainbow trout blood. Can J Zool 65:879-883
16. Wells RMG, Macdonald JA, di Prisco G (1990) Thin-blooded Antarctic fishes: a rheological comparison of the haemoglobin-free icefishes *Chionodraco katleenae* and *Cryodraco antarcticus* with a red blooded nototheniid, *Pagothenia bernacchii.* J Fish Biol 36:595-609
17. Egginton S (1996) Blood rheology of Antarctic fishes: viscosity adaptations at very low temperatures. J Fish Biol 48:513-521
18. Fitch NA, Johnston IA, Wood RE (1984) Skeletal muscle capillary supply in a fish that lacks respiratory pigments. Resp Physiol 57:201-211
19. Egginton S (1990) Morphometric analysis of tissue capillary supply. In: Boutilier RG (ed.) Vertebrate gas exchange from environment to cell. Adv Comp Environ Physiol 6:73-141
20. Egginton S (1992) Adaptability of the anatomical capillary supply to skeletal muscle of fishes. J Zool, Lond 226:691-698
21. Vogel W, Kock KH (1981) Morphometry of gill vessels in icefish. Archiv für Fischereiwissenschaft 31:139-150
22. Jakubowski M (1982) Dimensions of respiratory surfaces of the gills and skin in the Antarctic white-blooded fish, *Chaenocephalus aceratus* Lönnberg. Zeit für Mikrosk Anat Forsch 96:145-156

Cardiac and Locomotory Muscle Mass in Antarctic Fishes

Russell F. Robertson[1], Nia M. Whiteley[1], and Stuart Egginton[2]

[1]School of Biological Sciences
[2]Department of Physiology, University of Birmingham, Edgbaston, Birmingham B15 2TT, UK

Introduction

The fish fauna around Antarctica is dominated by a single perciform suborder, the Notothenioidei, whose phylogeny has been well defined by traditional and molecular techniques. They have undergone a remarkable adaptive radiation (into a variety of forms from small, sculpin-like to large, hake-like species) that is possibly unique in the marine environment, and which molecular evidence suggests is a relatively recent event [1]. The last few decades have seen the development of extensive scientific and commercial fishing in the Southern Ocean that has provided much data on the growth and reproduction of a wide range of species. In contrast, financial constraints and logistical difficulties have meant that studies into the physiological mechanisms that allow these fishes to successfully exploit this extremely cold and highly seasonal environment have mainly concentrated on a few species, predominantly from Nototheniidae and Channichthyidae [2,3]. Antarctic pelagic fishes are diverse in their origins, coming from bathypelagic, mesopelagic and, of most interest for our purpose, from coastal fish families that have become secondarily adapted to midwater life. However, most species (about 75% of those so far recorded) inhabit benthic or demersal niches, which is thought to reflect the benthic lifestyle of the parent taxa, and helps explain the lack of a functional swimbladder [2]. Those species that are now able to exploit the water column have had to evolve means of adjusting buoyancy, including reduced skeletal ossification and extensive lipid deposits [3]. Other adaptations may also exist, e.g. the Antarctic silverfish, *Pleuragramma antarcticum,* is streamlined and shows a reliance on subcarangiform locomotion. In contrast, the notothenioids have large pectoral fins which are utilised in labriform locomotion, providing a sculling movement during swimming while the trunk acts mainly as a rudder. This reflects the benthic origins of this group and the maintenance of relatively low activity

G. di Prisco, E. Pisano, A. Clarke (Eds)
Fishes of Antarctica. A biological overview
© Springer-Verlag Italia 1998

levels, even in species living in the water column. Consequently pelagic Antarctic fishes show differences in the relative utilisation of the major skeletal muscle masses, suggesting morphormetric adaptations distinguishing sluggish from more active species [4].

Some striking adaptations are found among Antarctic fishes, including the presence of antifreeze glycopeptides in the blood (see Cheng, this Vol.) together with the development of aglomerular kidneys, and, in the icefishes, enlarged hearts and absence of respiratory pigments [3]. In general, rates of physiological systems are lower than found in temperate species, and maximal power output of fast muscle in Antarctic fish at 0 °C is only a third that of tropical species at 25 °C. While the factorial scope (the difference between resting and maximal activity) is similar for Antarctic and tropical species, the absolute metabolic flux is significantly less in the former [1]. This severely limits the power available for locomotion and may explain the relative paucity of truly pelagic species in Antarctica, although it remains unclear to what degree species from different niches are restricted in their swimming capacity.

Despite the expanding literature on the strategies that have evolved to cope so successfully with life in thc cold, many questions remain unresolved. In particular, much of the physiological data derives from inshore, rather sluggish species so that at least some of the findings may be biased by the type of species available for study, and may not be representative of the broader Antarctic fish fauna. For example, an unusual and possibly unique vertebrate stress response has been documented in sluggish inshore species, where a rise in circulating catecholamines (the classic "flight or fight" hormones) is apparently absent [5]. The aim of this study was to examine a range of Antarctic fish species from different niches for signs of morphometric adaptations for active locomotion. This was achieved by scaling ventricle and locomotory muscle masses (total skeletal and lateral and medial pectoral muscles) in a red-blooded pelagic nototheniid as indices of the effort required to maintain an active lifestyle. Values were compared to those obtained from benthic species, and those obtained from fishes characterised by the absence of respiratory pigments (channichthyids), to see if this had any effect on the scaling relationships.

Material and Methods

Various sized specimens of different species were collected from around King George Island (62°S 58°W) and Elephant Island (61°S 55°W) on the recent (1996-1997) RV "Polarstern" Ant XIV/2 cruise. Fish were either caught in 30 min bottom ground trawls (grundschleppnetz) or 1 h benthopelagic trawls. Although the species list for the sampling region numbered about 50, based on the outcome of the first 5 trawls it was decided due to logistical constraints to concentrate sampling on only those species that were frequently caught and had a good size range. Five key species were selected from different niches: *Champsocephalus gunnari* (active, pelagic channichthyid), *Chaenocephalus aceratus* (sluggish, epibenthic channichthyid), *Dissostichus mawsoni* (active, pelagic nototheniid), *Gobionotothen gibberifrons* (sluggish, benthic nototheniid), and *Notothenia rossii* (relatively active, benthopelagic nototheniid).

Blunt dissection on fresh fish was performed to remove the ventricle, spleen, and the lateral and medial pectoral muscles. These were stored frozen (-30 °C) and accurately weighed on return to the UK. Hearts were cut open and residual blood removed by blotting. Samples and carcasses were transported frozen (-30 °C) back to the Alfred-Wegener Institute at Bremerhaven, where they were weighed and further samples taken for analysis. Muscle mass relative to total or lean (gonad free) body mass, to avoid complication due to different stages of sexual maturity, was examined in an allometric analysis to allow comparison with data from inshore epibenthic or benthic Antarctic and temperate marine fish species.

Results and Discussion

The relationships between body mass and mass of total skeletal muscle, ventricle, lateral and medial pectoral muscles are plotted in Figs. 1 to 4. The relative masses of the ventricle, total skeletal muscle and pectoral muscles in all five species are displayed in Table 1, along with values for the relative mass of the spleen as a comparative non-muscle organ.

The scaling of ventricular mass showed a clear separation between channichthyids and nototheniids that reflects the substantial reduction in oxygen carrying capacity of the former, and the need to generate a much higher cardiac output [6,7]. More surprising is the apparent lack of a separation according to ecotype, i.e. *C. gunnari* was similar to *C. aceratus* while *D. mawsoni* was similar to *G. gibberifrons* and *N. rossii* (Fig. 1). Interestingly, all species showed a strikingly similar allometric relationship between total skeletal muscle mass and body mass (Fig. 2).

Other organ systems showed no consistent difference among species, e.g. spleen mass (data not shown). As with many other perciform families, notothenioids rely mainly on labriform locomotion (involving pectoral muscles) using sub-carangiform locomotion (involving trunk muscle) only during burst swimming.

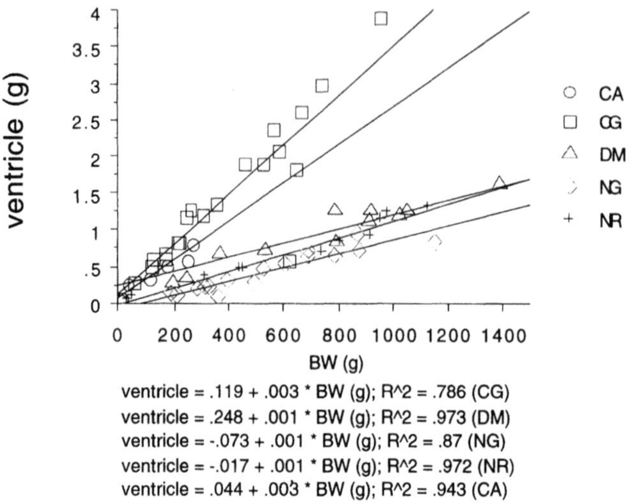

Fig. 1. Scaling of ventricle mass with body mass. CA, *Chaenocephalus aceratus*; CG, *Champsocephalus gunnari*; DM, *Dissostichus mawsoni*; NG, *Gobionotothen* (*Notothenia*) *gibberifrons*; NR, *Notothenia rossii*; BW, *body weight*

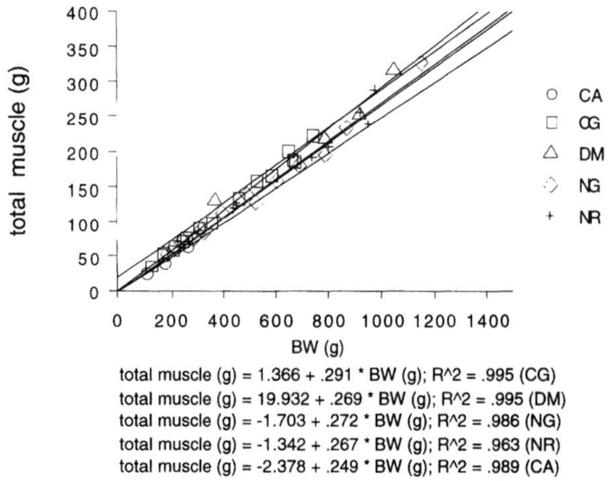

Fig. 2. Relative skeletal muscle mass. CA, *Chaenocephalus aceratus*; CG, *Champsocephalus gunnari*; DM, *Dissostichus mawsoni*; NG, *Gobionotothen* (*Notothenia*) *gibberifrons*; NR, *Notothenia rossii;* BW, *body weight*

Scaling of individual lateral (Fig. 3) and medial (Fig. 4) pectoral muscles also showed little evidence for species specificity in scaling, with different slopes reflecting the relative mass of individual muscles. Medial muscles,

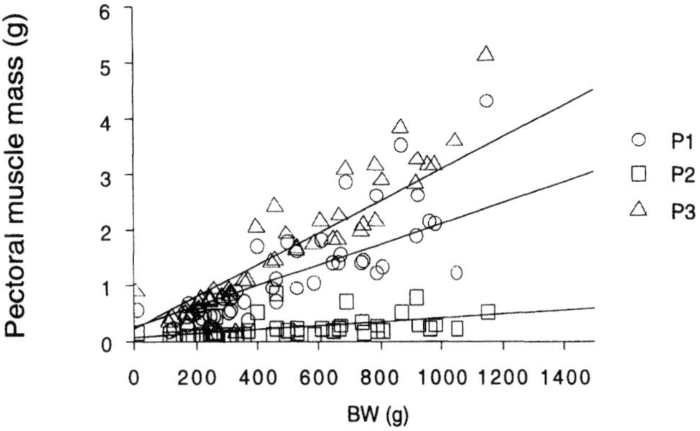

P1 = .263 + .002 * BW (g); R^2 = .732
P2 = .075 + 3.332E-4 * BW (g); R^2 = .501
P3 = .222 + .003 * BW (g); R^2 = .894

Fig. 3. Lateral pectoral muscle masses. P1, abductor superficialis; P2, erector ventralis; P3, abductor profundis; BW, body weight

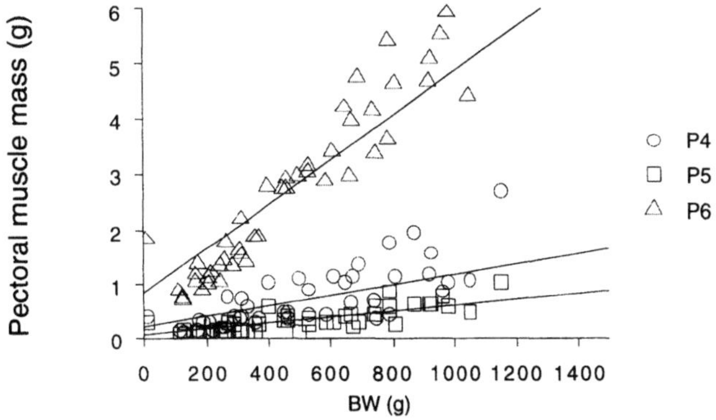

P6 = .819 + .004 * BW (g); R^2 = .874
P4 = .212 + .001 * BW (g); R^2 = .616
P5 = .045 + .001 * BW (g); R^2 = .825

Fig. 4. Medial pectoral muscle masses. P4, adductor superficialis; P5, erector dorsalis; P6, adductor profundis; BW, body weight

those used for the power stroke in labriform locomotion (adduction of the pectoral fins), showed a greater absolute mass, but scaled similarly with those used for the recovery fin stroke (abduction; lateral pectoral muscles).

Table 1. Relative mass of spleen, ventricle, total skeletal muscle and pectoral muscles (p.m.) in Antarctic fishes with and without haemoglobin

	% spleen	% ventricle	% muscle	% p. m.
CHANNICHTHYIDAE				
Champsocephalus gunnari (17) active, pelagic	0.07±0.009	0.40±0.013	59.2±0.73	2.6±0.05
Chaenocephalus aceratus (6) sluggish, epibenthic	0.19±0.008*	0.30±0.005*	47.1±1.31*	3.0±0.15
NOTOTHENIIDAE				
Dissostichus mawsoni (7) active, pelagic	0.31±0.047	0.13±0.011	59.4±2.15	2.2±0.09
Gobionotothen gibberifrons (13) sluggish, benthic	0.16±0.017*	0.08±0.004*	53.9±0.88*	3.2±0.18*
Notothenia rossii (8) relatively active, benthopelagic	0.30±0.018	0.11±0.004	53.2±1.12*	2.7±0.15*

Values are organ mass/body mass x100; mean ± SEM (number of observations).
* $p<0.05$ *vs* active species.

These data suggest that the discrepancy between cardiac and skeletal muscle scaling reflects the haemoglobinless condition of the Channichthyids, but no gross allometric adaptations for a pelagic lifestyle were evident.

In contrast, the relative muscle masses (as a percentage of body mass) did reflect the more active lifestyle of both *C. gunnari* and *D. mawsoni*, where both cardiac and total skeletal muscle mass were greater than in the other species (Table 1). The spleen, however, showed no consistent differences among species suggesting that, as expected, non-muscle organ systems were unaffected by activity levels and forms of locomotion. Interestingly, the difference in total skeletal muscle mass among ecotypes was less evident in relative pectoral muscle mass indicating differences in trunk musculature (Table 1). This suggests that the mode of locomotion may vary, with the more pelagic species utilising more of the trunk muscle during propulsion. Indeed, *D. mawsoni* had significantly lower relative pectoral muscle mass than did either *G. gibberifrons* or *N. rossii*. These data are consistent with those presented in Eastman [3] where relative heart mass varied between 0.07 and 0.21% and relative pectoral mass between 1.6 and 2.9% in ecologically diverse notothenioid species from McMurdo Sound. In this study the highest relative heart mass was found in the cryopelagic *Pagothenia borchgrevinki*, while *D. mawsoni* (0.12%) was actually a little lower than the benthic *Trematomus bernacchii* and *T. hansoni* (0.13%) [3].

In summary, various sized specimens from different ecological niches were examined for morphometric adaptations for active locomotion, including allometric scaling of ventricle and locomotory muscle masses. The channichthyids had a significantly greater slope for ventricle mass vs body mass than the nototheniids, whereas scaling of skeletal muscle mass was similar among all species studied. It was not possible to distinguish between pelagic and epi-benthic representatives of Channichthyidae and Nototheniidae based on these criteria. However, the relative mass of both cardiac and skeletal muscle was greatest in pelagic species. Thus, the origin of these differences lies in an initial offset of the growth curve, suggesting that ecotype specialisation occurs in the early life history stages. As all notothenioid larvae start off with a pelagic phase but only some juveniles persist in a pelagic lifestyle by utilising a modified morphology (see White, this volume), it is probably at the early juvenile stage where the first evidence of specialisation may be found. Further work will determine water and lipid content of muscles, as well as metabolic enzyme activity, to assess whether more subtle adaptations are found among these ecotypes.

Acknowledgments

The authors wish to thank the Alfred-Wegener Institute (AWI) for the opportunity to participate in this cruise; the officers and crew of the RV "Polarstern"; Fahrtleiter G. Kattner; H.O. Pörtner and I. Hardewig for provision of facilities at Bremerhaven; and the following people for their expertise in fish identification: P.A. 'Butch' Hulley, K.-H. Kock, V. Siegel and M. White. Financial support was provided by The Fisheries Society of the British Isles.

References

1. Clarke A, Johnston IA (1996) Evolution and adaptive radiation of Antarctic fishes. Trends Ecol Evol 11:212-218
2. Kock K-H (1992) Antarctic fish and fisheries. Cambridge University Press. pp 359
3. Eastman JT (1993) Antarctic fish biology: evolution in a unique environment. Academic Press, London. pp 322
4. Montgomery JC, Macdonald JA (1984) Performance of motor systems in Antarctic fishes. J Comp Physiol 154A:241-248
5. Egginton S, Davison W (1997) Effects of environmental and experimental stress on Antarctic fishes. In: Playle R C, Pörtner HO (eds) Cold ocean physiology. Soc Exp Biol Seminar Series 66, Cambridge University Press, Cambridge, pp 299-326
6. Hemmingsen EA, Douglas EL (1972) Respiratory and circulatory responses in a haemoglobin-free fish, *Chaenocephalus aceratus*, to changes in temperature and oxygen tension. Comp Biochem Physiol 43A:1031-1043
7. Egginton S (1994) Stress response in two Antarctic teleosts (*Chaenocephalus aceratus* Lönnberg and *Notothenia coriiceps* Richardson) following capture and surgery. J Comp Physiol B 164:482-491

Natriuretic Peptides in Antarctic Teleosts: Cardiac Receptors in *Chionodraco hamatus* and *Trematomus bernacchii*

Maria C. Cerra[1,2], Rosa Mazza[1], Daniela Pellegrino[1], and Bruno Tota[1,3]

[1]Department of Cell Biology
[2]Department of Pharmaco-Biology, University of Calabria, 87030 Arcavacata di Rende, Italy
[3]Zoological Station "A. Dohrn," Villa Comunale, 80121 Napoli, Italy

Introduction

The myocardium of teleosts, like the atrial myocardium of homeotherm vertebrates, is a major site of synthesis and release of various structuraly related hormones commonly referred to as natriuretic peptides (NPs). This family of vasorelaxant, diuretic, and natriuretic substances is involved in both cardiovascular and body fluid homeostasis of fish [1-4].

In teleosts an increase in plasma volume is a major stimulus for NPs' release by myocardiocytes [1], serving as an intrinsic feedback inhibitor that protects the heart from volume overload [4]. NPs induce peripheral vasodilation [5], associated with an in vivo reduction of cardiac output and a decrease of the stroke volume via a direct negative inotropic action and/or a decrease in venous return [6]. In several species of fish, NPs have potent relaxant properties on some large vessels in an endothelium-dependent manner [7].

Two functional aspects of the Antarctic notothenioid teleosts call for a hormonal control potentially involving NPs. The first aspect concerns the plasma osmolality of these fishes that is higher (about 500 mosmol kg^{-1}) than that of other temperate marine teleosts. This is attributed to increased NaCl concentrations [8,9] dependent on the modulation of the renal and branchial Na^+/K^+ ATPases [10]. The second aspect is related to the family Channichthyidae, or icefish, which are characterized by the lack of hemoglobin in their blood [11]. The blood volume in icefish (*Chaenocephalus aceratus*: 64-110 ml kg^{-1} [12,13]; *Chionodraco hamatus*: 127-145 ml kg^{-1} [14]) is up to ninefold higher than in other

G. di Prisco, E. Pisano, A. Clarke (Eds)
Fishes of Antarctica. A biological overview
© Springer-Verlag Italia 1998

teleosts (13-71 ml kg^{-1}[5]). This hypervolemic condition is associated with cardiac enlargment that makes the icefish heart size comparable to that of small mammals. This heart acts as a volume pump since it is able to displace large stroke volumes (*C. hamatus*: 4.02 ml kg^{-1}) against low ventral aorta pressure (*C. hamatus*: 3.0/2.0 kPa) [15].

This chapter will provide information on the presence of NP receptors in the heart of two nototheniids, *Trematomus bernacchii* and the channichthyid *Chionodraco hamatus*. These two species, which differ for the presence and absence of hemoglobin in their blood, show similar ecotype being endemic and sluggish bottom dwellers.

NPs' Structure and Receptors

In several species of teleosts at least three types of NPs have been isolated and the sequence determined. Two of them, A-type NP (ANP), and C-type NP (CNP), are structurally very similar to the NPs identified in other vertebrates, namely ANP, B-type NP (BNP) and CNP [16]. The third type of NP found in fish is mainly synthesized by the ventricle and is called ventricular NP (VNP). It is longer than the other natriuretic hormones but it shares a high sequence homology with all the other vertebrates' NPs [3]. It remains unclear whether this peptide is an additional type of NP or it belongs to the BNP family. In mammals, BNP is synthesized by the ventricle and this synthesis overrides the atrial one in some cardiovascular adjustments or pathologies, such as hypertension and/or cardiac hypertrophy [17].

As in other vertebrates [18], the NP system of teleost fish exhibits heterogeneity not only at the ligand level but also at the receptor level as demonstrated by the presence of multiple types of NP receptors (NPRs) [16]. The similarities with the mammalian NPRs and the binding potencies shown by NPRs described in fish suggested a classification into NPR-B and NPR-C [19,20]. NPR-B, coupled to a guanylate-cyclase, has low affinity for ANP and higher affinity for the other homologous and heterologous NPs. NPR-C binds with similar affinity to all NP types and is a guanylate-cyclase free receptor [19-21]. The identification of these types of NPRs in organs such as the gills, kidney, and intestine provided a molecular target for the osmoregulatory functions of NPs in fish [19,22,23]. Two NPRs have been recently described in the cardiac chambers of the eel [23,24]. Using homologous ligands, an NPR-B-like receptor was observed in the bulbus arteriosus of the eel, while an NPR-C-like receptor was found in the atrial tissue and the venticular myocardium; in addition, a receptor type with a high affinity for eel VNP was observed in the ventricular endocardium of this fish, suggesting the presence of a putative novel type of NPR [24]. The linkage of these NPRs with both homologous and/or heterologous NPs could account for some of the functional effects exerted by these hormones on the fish cardiovascular system [3,4,6].

Cardiac NP Receptor Heterogeneity in Antarctic Teleosts

NPs have not yet been isolated from Antarctic teleosts; however, the possibility of a hormonal modulation mediated by NPs in Antarctic notothenioids may be indicated by the presence of NP-like material in heart homogenates and plasma of *C. hamatus* and *T. bernacchii* [25].

Since no information is available on the presence and the nature of a signal transduction system (i.e. specific receptors) whereby these hormones exert their modulatory actions on the cardiovascular function of these fishes, the presence of NPRs in the cardiac regions of the two endemic Antarctic notothenioids, the red-blooded *T. bernacchii* and the hemoglobinless *C. hamatus* has been assessed. In vitro quantitative autoradiography was applied to localize and quantify the receptors for $[^{125}I]$-rat ANP-(1-28) ($[^{125}I]$-rANP) that, in the eel cardiac tissue, specifically binds to the NP sites [24]. Transverse cardiac sections of *T. bernacchii* and *C. hamatus* were processed as previously described [24] with some modifications. In both teleosts the analysis of the autoradiograms incubated in the presence of 100 pM of $[^{125}I]$-rANP indicated that ventricular endocardium and myocardium gave greater labeling than the other regions. Interestingly, an inverse binding pattern was observed in the bulbus arteriosus of the two fishes. Indeed, while in *C. hamatus* the inner (intimal) layer was more labeled than the outer (middle and adventitial) layer, in *T. bernacchii* the outer bulbar layer gave more dense binding levels than the inner layer. This observation is in contrast with the densest NP binding usually found in the inner layer of other teleost species [24,26]. A search of the literature failed to provide data that could explain the significance of this zonation. The extent to which this is related to the presence of NPs in the pericardial fluid of these fishes, as reported in mammals [27], requires a further investigation.

A saturation binding study was carried on in the atrium, ventricular myocardium and endocardium and inner and outer layer of the bulbus arteriosus of the two Antarctic teleosts (Fig. 1). In *T. bernacchii* the

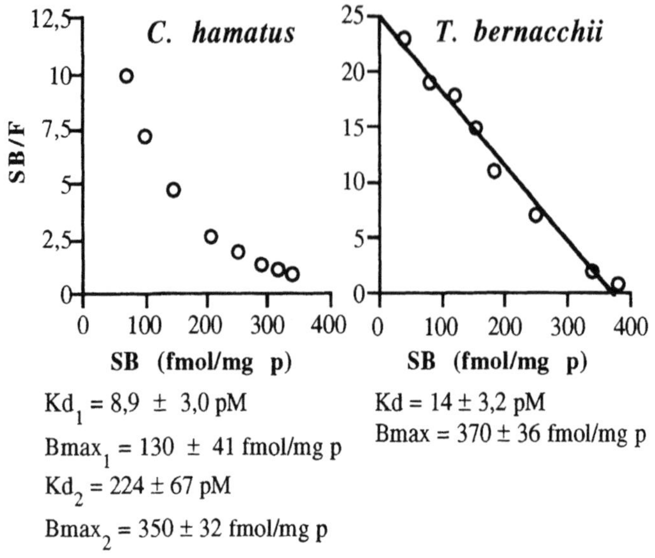

Fig. 1. Scatchard analysis of saturation curves of [^{125}I]-rANP in the atrium of *C. hamatus* and *T. bernacchii* (n=3). The sections were incubated in the presence of increasing concentrations (10^{-12} to 10^{-9} M) of [^{125}I]-rat ANP-(1-28) and of 10^{-6} M cold rANP; F, free; SB, specific binding; Kd, dissociation constant; Bmax, maximal binding capacity; p, protein

various cardiac regions were characterized by different binding patterns. As shown in Fig. 2B one class of rANP sites was observed in the atrium, ventricular endocardium and outer bulbar layer. In the atrium, an elevated maximal binding capacity (Bmax = 370±36 fmol/mg protein) and a dissociation constant (Kd) value of 14±3.2 pM were observed for rANP sites. Comparable Bmax and Kd values were also obtained in the ventricular endocardium and in the outer bulbar layer. In contrast, the ventricular myocardium and the inner bulbar layer revealed the presence of two classes of rANP binding sites. Very low Kd values (1.8±0.6 pM) were calculated for the high affinity site of the ventricular myocardium while higher Kd values (14±6.1 pM) were observed in the inner bulbar layer. For site 2, the highest dissociation constant value (209±66 pM) was calculated for the inner bulbar layer. This heterogeneous binding pattern resembles that found in the cardiac regions of the temperate *Anguilla anguilla* in which one class of NPRs was observed in the atrium, while two classes of sites were found in the inner layer of the bulbus arteriosus [24].

With respect to the heart of *T. bernacchii* all the cardiac regions of *C. hamatus* were characterized by two classes of rANP receptors (Fig. 2A).

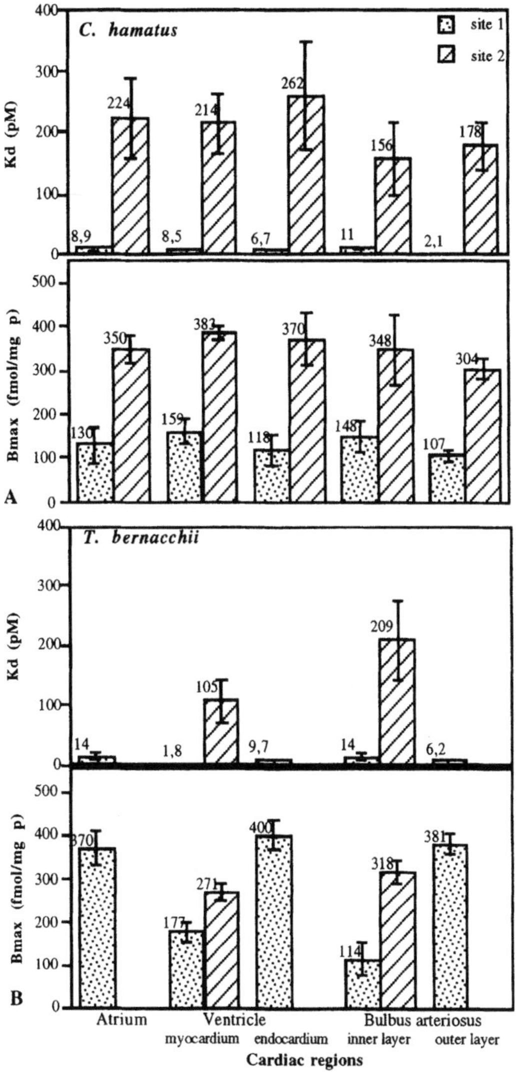

Fig. 2. Bar representation of Kd and Bmax values (indicated on each bar) ± SEM obtained from the Scatchard analysys of saturation curves of [^{125}I]-rANP in the cardiac regions of *C. hamatus* and *T. bernacchii* (n=3)

The atrium, the ventricular myocardium and endocardium contained rANP receptors with very similar Kd values for both the high and the low affinities (6.7±1.2/8.9±3 pM and 214±51/262±90 pM, respectively). Also the maximal binding capacity was found to be in the same range (118±36/159±29 fmol/mg p). In the case of the inner and outer layers of the bulbus arteriosus, site 2 exhibited slightly lower Kd values (156±61 and 178±41 pM, respectively) than the other cardiac chambers, while the maximum number of binding sites was similar to those of the other regions. These observations, together with the results of previous studies on temperate teleosts [24,26], indicate that only the inner bulbar layer constantly demonstrates the presence of two classes of binding sites. Our work does not specifically address the functional significance of such a constant receptor pattern of the inner bulbar layer, but it suggests that it is independent of either environmental influences, and/or of cardiac adjustments since it occurs equally in both the enlarged heart of *C. hamatus* and in the heart of temperate fish species.

The NPRs found in the various cardiac regions of the two Antarctic nothotenioids were further characterized by a displacement study. Given the importance of BNP as a major ventricular hormone in the mammalian normal and failing heart, the ability of porcine BNP (pBNP) to compete with rANP for the NPRs was compared in the atrium, ventricle, and bulbus arteriosus of *T. bernacchii* and *C. hamatus*. rANP and pBNP were differentially able to bind the NPRs in the various cardiac chambers of the two fishes. In *C. hamatus,* the NPRs of all cardiac regions demonstrated a higher affinity for rANP than for cold pBNP. On the contrary, in *T. bernacchii,* a higher affinity for rANP was observed in the ventricular myocardium and endocardium and in the two bulbar layers, while the NPRs demonstrated in the atrium revealed a higher affinity for cold pBNP than for cold rANP (Fig. 3). The presence

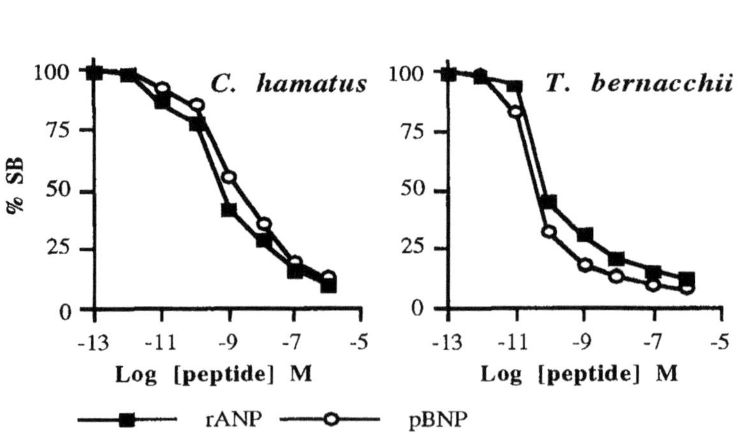

Fig. 3. Competitive curves of [^{125}I]-rANP specific binding (SB) obtained in the atrium of *C. hamatus* and *T. bernacchii* in the presence of increasing concentrations (10^{-13} to 10^{-6} M) of cold rANP and pBNP

of binding sites specific for both ANP and BNP, and interacting with BNP at different affinities was observed in the mammalian heart [28,29]. Recently, pBNP was shown to specifically bind two types of NPRs in the rectal gland of the elasmobranch *Squalus acanthias* [30]. Although BNP has not yet been identified in fish, its high sequence homology with respect to the other piscine NPs [31] suggests that in the various cardiac regions of both notothenioids, and particularly in the atrium of *T. bernacchii,* NP sites able to bind a BNP-like peptide exist.

Since homologous NPs from both *T. bernacchii* and *C. hamatus* are not available, caution must be exercised in analyzing the results using heterologous ligands. However, the present data are of interest in relation to the presence of a heterogeneous NP system in Antarctic teleosts whose potential physiological significance needs to be further investigated.

Acknowledgments

This research was supported by the Italian National Programme for Antarctic Research (PNRA) (B. Tota).

References

1. Evans DH (1990) An emerging role for a cardiac peptide hormone in fish osmoregulation. Ann. Rev. Physiol. 52: 43-60
2. Olson KR (1992) Blood and extracellular fluid volume regulation: role of the renin-angiotensin system, kallykrein-kinin system, and atrial natriuretic peptides. In: Fish physiology, Hoar WS, Randall DJ eds Academic Press Inc Orlando, Florida, Vol XIIB pp 135-254
3. Takei Y, Balment RJ (1993) Biochemistry and physiology of a family of eel natriuretic peptides. Fish Physiol Biochem 11:183-188
4. Acierno R, Axelsson A, Tota B, Agnisola C, Nilsson S (1991) Hypotensive effect of atrial natriuretic factor (ANF) in the Atlantic cod, *Gadus morhua*. Comp. Biochem Physiol 99C:11-14
5. Olson KR, Duff DW (1992) Cardiovascular and renal effects of eel and rat atrial natriuretic peptide in rainbow trout, *Salmo gairdneri*. J Comp Physiol B, 162: 408-415
6. Oudit GY, Butler DG (1995) Cardiovascular effects of arginine vasotocin, atrial natriuretic peptide, and epinephrine in freshwater eels. Am J Physiol 268: R1273-R1280
7. Sverdrup A, Helle KB (1994) Different patterns of relaxation by atrial natriuretic peptide in major blood vessels of two distantly related teleosts. Reg Peptides, 53:89-101

8. O'Grady SM, DeVries AL (1982) Osmotic and ionic regulation in polar fishes. J Exp Mar Biol Ecol 57: 219-228
9. Dobbs G, DeVries AL, (1975) Renal function in Antarctic teleost fishes: serum and urine composition. Mar Biol 29:59-70
10. Gonzales-Cabrera P, Dowd F, Pedibhotla VK, Rosario R, Stanley-Samuelson D, Petzel D (1995) Enhanced hypo-osmoregulation induced by warm-acclimation in Antarctic fish is mediated by increased gill and kidney Na^{+}/K^{+}-ATPase activities. J Exp Biol 198:2279-2291
11. Eastman JT (1993) Antarctic fish biology. Evolution in a unique environment. Academic Press, New York
12. Hemmingsen EA, Douglas EL (1970) Respiratory characteristics of the hemoglobin-free fish *Chaenocephalus aceratus*. Comp Biochem Physiol 33:733-744
13. Twelves EL (1972) Blood volume of two Antarctic fishes. Br Antarct Surv Bull 31:85-92
14. Acierno R, Macdonald JA, Tota B (1995) Blood volume in the hemoglobinless Antarctic teleost *Chionodraco hamatus* (Lönnberg). J Exp Zool 272:407-409
15. Tota B, Acierno R, Agnisola C (1991) Mechanical performance of the isolated and perfused heart of the haemoglobinless Antarctic icefish *Chionodraco hamatus*: effects of loading conditions and temperature. Phil Trans Royal Soc London, 332:191-198
16. Takei Y (1994) Structure and function of natriuretic peptides in vertebrates. Persp Comp Endocrinol 149-154
17. Lang CC, Choy AMJ, Struthers AD (1992) Atrial and brain natriuretic peptides: a dual natriuretic peptide system potentially involved in circulatory homeostasis. Clin Sci 83: 519-527
18. Anand-Srivastava MB, Trachte GJ (1993) Atrial natriuretic factor receptors and signal transduction mechanisms. Pharmacol Rev 45(4):455-497
19. Sakaguchi H, Katafuchi T, Hagiwara H, Takei Y, Hirose S (1993) High-density localization of ANP receptors in chondrocytes of the eel gill cartilage. Am J Physiol 265: R474-R479
20. Katafuchi T, Takashima A, Kashiwagi M, Hagiwara H, Takei Y, Hirose S (1994) Cloning and expression of eel natriuretic-peptide receptor B and comparison with its mammalian counterpart. Eur J Biochem 222:835-842
21. Donald JA, Toop T, Evans DH (1994) Localization and analysis of natriuretic peptide receptors in the gills of the toadfish, *Opsanus beta* (teleostei). Am J Physiol 267:R1437-R1444
22. Duff DW, Olson KR (1992) Atrial natriuretic peptide clearance receptors in trout: effects of receptor inhibition in vivo. J Exp Zool 262:343-346
23. Takashima A, Katafuchi T, Shibasaki M, Kashiwagi M, Hagiwara H, Takei Y, Hirose S (1995) Eel natriuretic peptide receptor NPR-C. Cloning, properties, site-directed mutagenesis analysis of the subunit structure,

tissue distribution, and regulation of expression. Eur J Biochem 277:673-680

24. Cerra MC, Canonaco M, Takei Y, Tota B (1996) Characterization of natriuretic peptide binding sites in the heart of the eel, *Anguilla anguilla*. J Exp Zool 275:27-35
25. Uva BM, Masini MA, Napoli L, Devecchi M (1993) Immunoreactive atrial natriuretic peptide in Antarctic teleosts. Comp Biochem Physiol A, 104(2): 291-297
26. Cerra MC, Canonaco M, Tota B (1992) A quantitative autoradiographic study of ^{125}I atrial natriuretic factor in the heart of a teleost fish (*Conger conger*). J Exp Zool 263:215-219
27. Klemola R, Laine M, Weckström M, Vuolteenhao O, Ruskoaho H, Huttunen P, Leppäluoto J (1995) High concentrations of atrial natriuretic peptide and brain natriuretic peptide in rat pericardial fluid and their reduction by reserpine in vivo. Arch Pharmacol 352:331-336
28. Oehlenschlager WF, Baron DA, Shomer H, Currie MG (1989) Atrial and brain natriuretic peptides share binding sites in the kidney and heart. Eur J Pharmacol 161: 159-164
29. Wilcox JN, Augustine A, Goeddel DV, Lowe DG (1991) Differential regional expression of three natriuretic peptide receptor genes within primate tissues. Mol Cell Biol 11: 3454-3462
30. Gunning M, Cuero C, Solomon R, Silva P (1993) C-type natriuretic peptide receptors and signaling in rectal gland of *Squalus acanthias*. Am J Physiol 264: F300-F305
31. Takei Y, Takano M, Itahara Y, Watanabe TX, Nakajima K, Conklin DJ, Duff DW, Olson KR (1994) Rainbow trout ventricular natriuretic peptide: isolation, sequencing, and determination of biological activity. Gen Comp Endocrinol 96:420-426

Atrial Natriuretic Peptides in Antarctic Fish

Maria Angela Masini[1], Maddalena Sturla[1], Eva Pisano[1], Philippe Koubbi[2], Rosella Coradeghini[1], and Bianca Maria Uva[1]

[1]Istituto di Anatomia Comparata, Università di Genova, 5 viale Benedetto XV, 16132 Genova, Italia; [2]Université du Littorel, Station Marine de Wimereux, France

Introduction

Atrial natriuretic factors (ANFs) belong to a family of peptides originally described in mammalian heart. Beside ANPs, two other different types of natriuretic peptides with high sequence similarity have been identified, B-type natriuretic peptide (BNP) in the heart of chicken and mammals, and C-type natriuretic peptide (CNP) in the brain [1]. Their physiological and pharmacological effects have been extensively studied, their structure is known [2,3], their primary effects in mammals are natriuretic, diuretic and vasorelaxant [4,5]. ANFs are also present in nonmammalian vertebrates including birds [6], reptiles [7,8], amphibians [1,9,10], jawed [1,7,11,12] and jawless fishes [13]. Moreover, there are specific binding sites for ANP in several vertebrates' tissues. Extracts from fish heart stimulate ion excretion in the trout [14] and are vasodepressor in the quail [15]. In addition, rat ANP stimulates chloride secretion from the killifish opercular membrane in vitro [16]. Together these data indicate that ANFs regulate body water, electrolyte balance and cardiovascular dynamics in vertebrates. To inhabit deep cold water, Antarctic and subAntarctic fish have developed remarkable cardiovascular adaptations like reduction of the number of red blood cells: this adaptation is extreme in Channichthyidae which have erythrocyte-free blood. To compensate for the lack of hemic pigments, the icefish possess a hypertrophic heart [17,18] that generates a high cardiac output [19]; the blood volume is also very large [20]. ANF-like peptides are likely to be involved in the regulation of blood volume, cardiovascular activity, and ion-fluid balance in Antarctic teleosts.

The present paper addresses in sequence the general characteristics of atrial natriuretic peptides in terms of their molecular structures, localization of ANP-like immunoreactivity and distribution of ANP binding sites in adults, and finally immunoreactivity in a larval stage.

Characterization of ANP Molecules in Teleosts

Immunocytochemical studies using antisera against rat ANP (rANP), human ANP (hANP) or porcine ANP have revealed the presence of ir-

G. di Prisco, E. Pisano, A. Clarke (Eds)
Fishes of Antarctica. A biological overview
© Springer-Verlag Italia 1998

ANP-like materials in the cardiocytes of several species of fish, and the immunohistochemical controls used support the specificity of the immunostaining. It would thus appear that the fish heart produces substances quite similar to mammalian ANP.

An eel atrial peptide with 27 amino acid residues has been purified and sequenced; this ANP, when injected into rats, was natriuretic, diuretic and vasorelaxant [21]. The eel peptide is structurally similar to the mammalian ANPs; it has an intramolecular ring structure with 17 amino acid residues formed by a disulfide bond. The identical sequences conserved in all ANPs so far identified are also conserved in this peptide except for four amino acid substitutions (Fig. 1).

129 (105)	130 (106)	131 (107)	132 (108)	133 (109)	134 (110)	135 (111)	136 (112)	137 (113)	138 (114)	139 (115)	140 (116)
Cys-	Phe-	Gly-	Gly-	Arg-	Ile-	Asp-	Arg-	Ile-	Gly-	Ala-	Gln-
-	-	-	-	-	Met-	-	-	-	-	-	-
-	-	-	-	Lys	Leu	-	-	-	-	Ser	Tyr
130	131	132	133	134	135	136	137	138	139	140	141

141 (117)	142 (118)	143 (119)	144 (120)	145 (121)	
Ser-	Gly-	Leu-	Gly-	Cys	a
-	-	-	-	-	b
-	-	-	-	-	c
142	143	144	145	146	

Fig. 1. Amino acid sequences of: rat α-ANP (a), human ANP (b) and eel ANP (c).

The immunodetection of ir-ANP-like substances with mammalian ANP antibodies may be therefore performed in teleostean fish, but obviously the physiological responses, after administration of a heterologous ANP, are open to questions. The molecular structure of Antarctic teleost ANP probably differs from that of mammals and perhaps is more closely related to eel ANP. Antarctic fish ANP has not yet been sequenced, but if it is presumed similar to the eel ANP there is a strong homology with the α-mammalian ANP.

ANP in Heart Extracts and Plasma of Antarctic Teleosts

ANP radioimmunoassay of heart homogenates and plasma were performed in *Chionodraco hamatus* and *Trematomus bernacchii*, using standard protocols (RIA, Peninsula Lab. Belmomt, CA) and antibodies to α-rANP (99-126) [22]. The levels of ANP-like peptides in plasma and heart of *Chionodraco* were twice the corresponding values in *Trematomus* (Fig. 2).

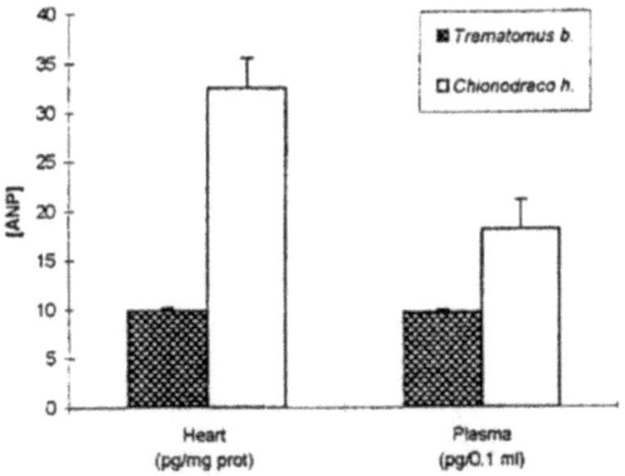

Fig. 2. ANP-like immunoreactivity in heart homogenates and plasma from *C. hamatus* and *T. bernacchii*. Values are expressed as pg/mg protein in heart homogenates and pg/0.1 ml in plasma (means ± SD). The difference between the two species is statistically significant (Student's t-test, $p<0.05$)

It is possible that the synthesis is more active in the icefish than in other fish. In the hemoglobin-free fish, blood volume and cardiac output are several fold greater than in the red-blooded species, to compensate for the reduced oxygen-carrying capacity of the blood [23]. SDS-PAGE of heart homogenates from *Champsocephalus gunnari* [24] revealed a band whose electrophoretic mobility resembled that of rANP. Western immunoblotting, using the rabbit antisera to rANP, showed that the band at that position was conjugated with the corresponding antibody (Fig. 3).

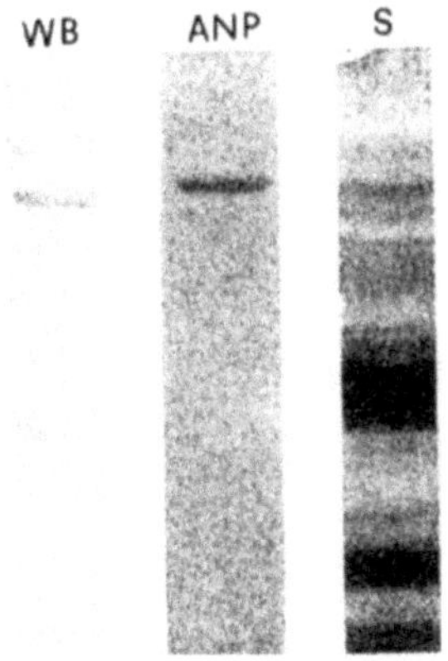

Fig. 3. SDS-PAGE and Western immunoblotting of crude extract from the heart of *C. gunnari*. Lane WB: Western blot with antibodies to ANP showing a band at about 3.0 kDa; Lane ANP: SDS-Page of standard ANP stained with Silver-staining; Lane S: separation of heart proteins by 15%T gradient SDS-PAGE, stained with Silver-staining

ANP-Like Immunoreactivity in Cardiac Structures

In teleosts the heart has four chambers, the sinus venosus, atrium, ventricle and a short, sometimes completely absent, conus arteriosus arranged in sequence. The sinus venosus and atrium will be considered together. Strong rANP-like immunoreactivity was detected immunohistochemically in the atrial myocytes of *C. hamatus*, *Cryodraco antarcticus*, *Notothenia neglecta* and *C. gunnari*. Numerous cardiocytes, in the subepicardial region and in the trabeculae, were intensely immunostained, especially in their perinuclear zone (Fig. 4). The atrium of *T. bernacchii* gave weaker immunoreactivity, the cardiocytes were less numerous, but the localization was similar in the two species. The atrial cardiocytes showed stronger ANP immunoreactivity than the ventricular cardiocytes. In the ventricle of *C. gunnari* and *N. neglecta*, a number of cardiocytes, localized in the trabeculae, were immunolabeled (Figs. 5,6). There was no immunostaining in the ventricles of *Chionodraco* and *Trematomus*.

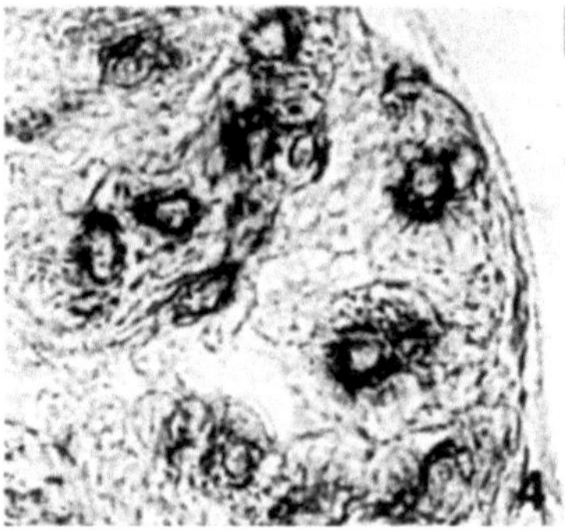

Fig. 4. αANP immunoreactivity in atrial cardiocytes of *N. neglecta*. Immunoreactivity is concentrated in the perinuclear region. Peroxidase – antiperoxidase (PAP). x 400

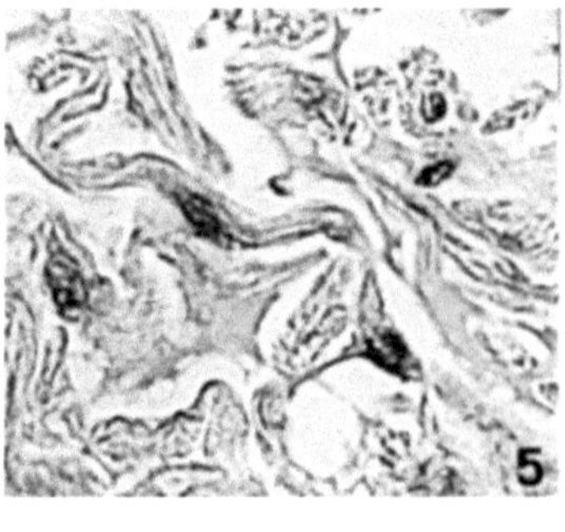

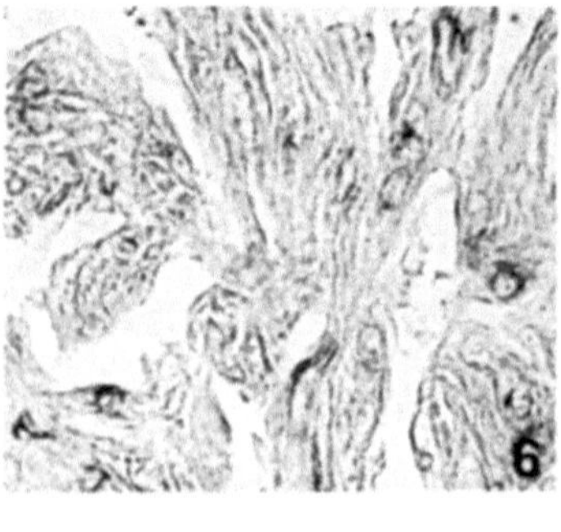

Figs. 5,6. α-rANP immunoreactivity in ventricular myocytes of *C. gunnari* (**Fig. 5**) and *N. neglecta* (**Fig. 6**). PAP. x 350

The low or absent ANP-like immunostaining in the ventricles of Antarctic teleosts resembles the observations made in mammals, in which the concentration of ANP in the ventricles is very low compared with that in the atria [25]. It may be that ANP, synthesized in the ventricle, is secreted without being concentrated in granules, while atrial cardiocytes store most of the ANP that is synthesized. However, ultrastructural studies showed in the ventricle of the Antarctic icefish *Chaenocephalus aceratus* numerous granulated cells [26]. Two molecular forms of natriuretic peptides have been identified in the eel ventricles and named ventricular natriuretic peptides (VNP 1-36 and VNP 1-25); the two forms slightly differ from the atrial natriuretic peptide [27]. The low immunoreactivity observed in the ventricular chamber of the teleosts investigated may reflect the presence of a VNP with a lower crossreactivity to the atrial natriuretic peptide antibody.

All the blood from the heart courses forward into a ventral aorta through the conus arteriosus that possesses a myocardial musculature. The bulbus arteriosus follows the conus and develops as a proximal part of the ventral aorta. In Antarctic teleosts the conus is much reduced. In the Antarctic fishes examined, the myocytes of the conus showed a very clear ANP immunoreactivity, particularly intense in the perinuclear zone. There was less intense staining in the bulbus arteriosus [24].

Ir-Like ANP in Kidney, Gills, Urinary Bladder and Skin

In the kidney of *N. neglecta* and *C. antarcticus*, the basal regions of the proximal tubule cells gave positive immunoreactions for α-rANP antibodies (Fig. 7). Positively stained cells were also found in the epithelium of the filaments in the gills. A weaker or completely absent immunoreactivity was observed in *C. hamatus*.

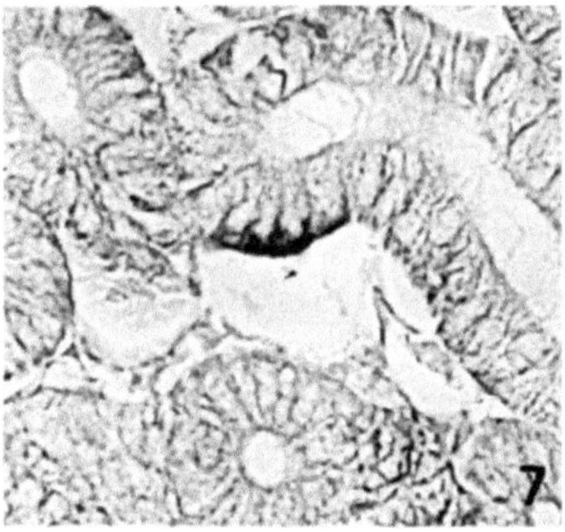

Fig. 7. α-rANP immunoreactivity in cells of the mesonephros in *N. neglecta*. PAP. x 300

Distribution of ANP Specific Binding Sites

α-rANP binding sites were examined by in vitro autoradiography of Bouin-fixed sections using 79 pM ^{125}I ANP (specific activity 1397 Ci/mmol, Peninsula) LM-1 emulsion (Amersham, Buckinghamshire, UK) [22].

ANP binding sites were localized in the atrium, ventricle, conus and bulbus arteriosus in the heart of *C. hamatus* and *T. bernacchii*. Labeling was stronger in the subepicardial region of all the cardiac chambers, but was also present in the atrial and ventricular trabeculae (Fig. 8).

In vitro quantitative autoradiography [18,28], using the same iodinated ANP, revealed in the heart of *Trematomus* one class of receptor in the atrium, ventricular endocardium and outer bulbar layer, and two classes of receptor in the inner layer of the bulbus arteriosus; the difference may underline the different embryological origin and functional activity of the bulbus arteriosus. ANP binding sites were demonstrated in the aglomerular kidney of both *C. hamatus* and *T. bernacchii*. Labeling was detected in the apical brush border of the cells in the mesonephric proximal segments.

ANP labeling was also found in the endothelial layer and in the smooth muscle of renal vessels. The kidney is therefore a potential target organ for ANP in Antarctic teleosts. Binding also occurred in the filaments of the gills, presumably in close contact with chloride cells, both in *C. hamatus* and *T. bernacchii* [22]. ANP receptors have been described in isolated branchial cells of both freshwater and seawater eels and it was postulated that the binding was in close contact with the sodium-chloride transporting elements of the gills [29].

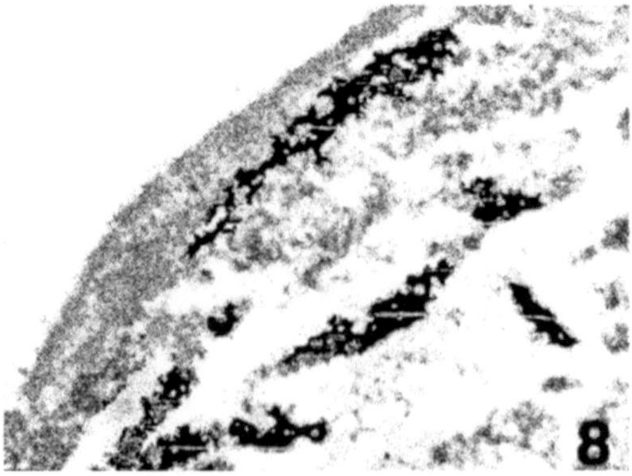

Fig. 8. Autoradiographical localization of ANP binding sites in the atrium of *C. hamatus*. x 250

Ir-Like ANP Immunoreactivity in Antarctic Teleost Larvae

The presence of ANP was studied in larvae of *Pleuragramma antarcticum* (Nototheniidae) about 2 cm long. Immunoreactivity for α-rANP was present in the thin wall of the atrium and in the trabeculae of the ventricle (Fig. 9), in the conus, and in the bulbus arteriosus, where the immuno-staining was particularly concentrated in the inner layer (Fig. 10). Immuno-reactivity was also present in the walls of the ventral aorta and was localized in cells of the branchial epithelium (Fig. 11) and of the archinephric duct (Fig. 12). ANP is present during fetal life in mammals and the presence of immunoreactive ANP granules was found in both atrial and ventricular cardiocytes, contrasting with its restriction to the atrium of the adult heart [30]. In the frog *Rana dalmatina*, ANP-like immuno-reactivity was observed in both the ventricle and the two atria from larval stages 22/23. At stages 25-27 ANP was confined to the atria while the ventricle was completely devoid of immunoreactivity [31]. The α-rANP cardiac localization seems to change during development in mammals and frogs, immunoreactivity concentrates in the atria as maturation progresses and is absent from ventricular cardiocytes. The presence of α-rANP in the ventricles of adult Antarctic teleosts is controversial; there is ventricular localization in *N. neglecta* [32] and *Champsocephalus* [24] but not in *Chionodraco* and *Trematomus* [22]. It seems that during ontogeny storage and release of ANP changes in some teleosts as in frogs and mammals.

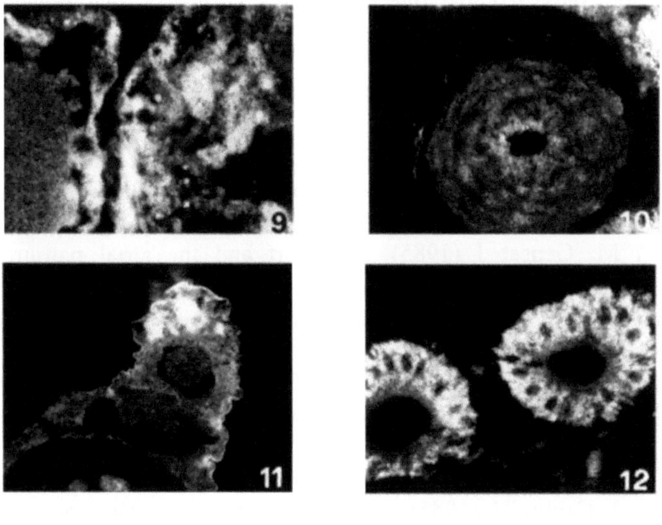

Figs. 9-12 Larvae of *Pleuragramma antarcticum*. Immunofluorescence for α-rANP in atrium and ventricle (**Fig. 9**), in bulbus arteriosus (**Fig. 10**), in the gill lamellae (**Fig. 11**) and in archinephric tubules (**Fig. 12**). x 400

Concluding Remarks

Atrial natriuretic peptides are generally considered to be regulators of body fluid volume and electrolyte balance, and to control hemodynamics in vertebrates. Studies on natriuretic peptides from the eel indicate that they play a role in the adaptations to different enviromental salinities and cardiovascular activity. Antarctic teleosts are very specialized fishes and atrial natriuretic peptides may be essential for their survival in their extreme environment. Further work is necessary to characterize the molecular sequence of the peptides, and the affinity and structure of the receptor subtypes. The administration of homologous ANPs may clarify the physiological role played by this family of peptides in the different environmental and hemodynamic conditions encountered by the Antarctic teleosts.

Acknowledgments

This research has been supported by the Italian Programme for Antarctic Research (PNRA).

References

1. Takei Y (1992) Vasoactive hormones and body fluid homeostasis: a phylogenetic consideration. In: Kuwasawa K, Hill RB, McMahon BR, T Kuramoto (eds) Comparative physiology, vol 11. Karger Flynn, Basel, pp 239-249
2. DeBold ML, DeBold AJ (1983) The amino acid sequence of an atrial peptide with potent diuretic and natriuretic properties. Biochem Biophys Res Commun 117:859-865
3. Currie MG, Geller DM, Cole BR, Siegel NR, Fok KS, Adams SP, Eubanks SR, Galluppi GR, Needleman P (1984) Purification and sequence analysis of bioactive atrial peptides (atriopeptins). Science 223: 67-69
4. Cantin M, Genest J (1985) The heart and the atrial natriuretic factor. Endocrinol Rev 6: 107-127
5. Goetz KL (1988) Physiology and pathophysiology of atrial peptides. Am. J Physiol 254:E1-E15
6. Gray DA, Schatz H, Gerstberger R (1991) Interaction of Atrial Natriuretic Factor and osmoregulatory hormones in the Peking duck. Gen Comp Endocrinol 81, 246-255
7. Chapeau C, Gutkowska J, Sciller PW, Milne RW, Thibault G, Garcia R, Genest J, Cantin M (1985) Localization of immunoreactve synthetic atrial natriuretic factor (ANF) in the heart of various animal species. J Histoch Cytochem 33:541-550

8. Kim SH, Cho KW, Koh GY, Seul KH, So JN, Ryu H (1989) Phylogenetic study on the immunoreactive atrial natriuretic peptide in the heart. Gen Comp Endocrinol 74:127-135
9. Netchitailo P, Feuilloley M, Pellettier G, De Lean A, Ong H, Cantin M, Gutkowska J, Leboulenger F, Vaudry H (1988) Localization and identification of immunoreactive atrial natriuretic factor (ANF) in the frog ventricle. Peptides 9:1-6
10. Ryu H, Cho KW, Kim SH , Kim SZ, Oh SH, Hwang YH, Lee GY (1992) Frog lymph heart synthesizes and stores immunoreactive atrial natriuretic peptide. Gen Comp Endocrinol 87:171-177
11. Kim SZ, Cho KW, Kim SH, Ryu H, Hwang YH, Lee GY, (1991) Identification of immunoreactive atrial natriuretic peptide in bulbus arteriosus of freshwater teleostean fish. Comp Biochem Physiol 11B:575-578
12. Masini MA, Sturla M, Napoli L, Uva BM (1996) Immunoreactive localization of vasoactive hormones (atrial natriuretic peptide and endothelin) in the heart of *Protopterus annectens*, an African lungfish. Cell Tissue Res 284:501-507
13. Reinecke M, Betzler D, and Forsmann WG (1987) Immunocytochemistry of cardiac polypeptide hormones (cardiodilatin/atrial natriuretic polypeptide) in brain and hearts of *Myxine glutinosa* (cyclostomata). Histochem 86:233-239
14. Duff DW, Olson KR (1986) Trout vascular and renal responses TO atrial natriuretic factor and heart extracts. Am J Physiol 251:R639-R642
15. Uemura H, Naruse M, Takei Y, Nakamura S, Hirohama T, Ando K, Aoto T (1991) Immunoreactive and bioactive atrial natriuretic peptide in the carp heart. Zool Science 8:885-891.
16. Sheide JI, Zadunaiski JA (1988) Effect of atriopeptin II on isolated opercular epithelium of *Fundulus heteroclitus*. Am J Physiol 254:R27-R32
17. Feller G, Bassleer R, Goessens G, Hamoir G (1983) Relative size and myocardial structure of the heart of an Antarctic fish devoid of haemoglobin and myoglobin, *Channichthys rhinoceratus*. J Zool London 199:51-57
18. Tota B, Agnisola C, Schioppa M, Acierno R, Harrison P, Zummo G (1991) Structural and mechanical characteristics of the heart of the icefish *Chionodraco hamatus* (Lönnberg). In: di Prisco G, Maresca B, Tota B (eds) Biology of Antarctic Fishes. Springer Verlag, Berlin, pp 204-219
19. Hemmingsen EA, Douglas EL (1972) Respiratory and circulatory responses in a hemoglobin-free fish *Chaenocephalus aceratus*, to changes in temperature and oxygen tension. Comp Biochem Physiol 43A:1031-1043
20. Hemmingsen EA, Douglas EL (1970) Respiratory characteristics of the hemoglobin-free fish *Chaenocephalus aceratus*. Comp Biochem Physiol 33:733-744
21. Takei Y, Takahashi A, Watanabe TX, Nakajima K, Sakakibara S (1989) Amino acid sequence and relative biological activity of eel atrial natriuretic peptide. Biochem Biophys Res Commun 164:537-543
22. Uva BM, Masini MA, Napoli L, Devecchi M (1993) Immunoreactive atrial natriuretic-like peptide in Antarctic teleosts. Comp Biochem Physiol 104A:291-297

23. Hemmingsen EA (1991) Respiratory and cardiovascular adaptations in hemoglobin-free fishes: resolved and unresolved problems. In: di Prisco G, Maresca B, Tota B (eds) Biology of Antarctic fishes. Springer Verlag, Berlin, pp 191-203
24. Masini MA, Sturla M, Uva B (1997) Vasoactive peptides in the heart of *Chamsocephalus gunnari*. Comp Biochem Physiol 118A:1083-1086
25. Arai H, Nakao K, Saito Y, Morii N, Sugawara A, Yamada T, Itoh H, Shiono S, Mukoyama M, Ohkubo H, Nakanishi S, Imura H (1987) Simultaneous measurement of natriuretic polypeptide (ANP) messenger mRNA and ANP in rat heart - evidence for preferentially increased synthesis and secretion of ANP in left atrium of spontaneously hypertensive rats (SHR). Biochem Biophys Res Commun 148:239-245
26. Tota B, Farina F, Zummo G (1988) Ultrastructural aspects of functional interest in the ventricular myocardial wall of the Antarctic icefish *Chaenocephalus aceratus*. Comp Biochem Physiol 90B:561-566
27. Takei Y, Takahashi A, Watanabe TX, Nakajima K, Sakakibara S (1991) A novel natriuretic peptide isolated from eel cardiac ventricles. FEBS Letters 282:317-320
28. Cerra MC, Acierno R, Tota B (1996) Cardiac natriuretic peptide receptors in teleosts: comparison between eurytherm and stenotherm Antarctic species. Society for Experimental Biology Meeting, Canterbury, UK
29. Broadhead CL, O'Sullivan UT, Deacon CF, Henderson IW (1992) Atrial natriuretic peptide in the eel, *Anguilla anguilla*. Its cardiac distribution, receptors and actions on isolated branchial cells. J Mol Endocrinol 9:103-114
30. Smith FG, Sato T, Varille VA, Robillard JE (1989) Atrial natriuretic factor during fetal and postnatal life: a review. J Develop Physiol 12:55-62
31. Ghiani P, Uva B, Mandich A, Masini MA (1991) Atrial natriuretic hormone and its receptors during development of the frog *Rana dalmatina*: preliminary observations. Acta Embryol Morphol 12:33-34
32. Sturla M, Masini MA, Uva B Distribution of vasoactive peptides in the Antarctic teleost *Notothenia neglecta*. Annales d'Endocrinologie, 57:P36

Exercise in the Cold: High Energy Turnover in Antarctic Fish

Peter L.M. van Dijk, Iris Hardewig, and Hans Otto Pörtner

Alfred Wegener Institute for Polar and Marine Research, Columbusstrasse, 27568 Bremerhaven, Germany

Introduction

Antarctic fishes inhabit one of the world's coldest marine habitats. At high latitudes the temperatures in the Southern Ocean are close to the freezing point of seawater at -1.86 °C and display little seasonal variation. The thermal conditions in the Antarctic have been relatively constant for several million years. The inhabiting fish fauna has therefore become highly specialized under the permanent cold conditions of this habitat which is reflected in a generally low upper lethal temperature of Antarctic fish at 5-6 °C [1]. In contrast to cold stenothermal polar fish, eurythermal boreal species must be able to survive a broader temperature range. While these species may be exposed to similarly cold water temperatures around 0 °C during winter, thermal conditions during summer require an elevated upper lethal temperature limit. On the other hand, eurythermal species are able to confine high cost metabolic activities like growth and reproduction to more favorable warmer seasons. Hence, acclimation to seasonal cold in eurythermal temperate species as opposed to adaptation to cold stenothermal conditions at high latitudes is associated with different requirements for the organism and may therefore result in distinct physiological features.

Recent histological and biochemical investigations suggest that, despite relatively low resting rates of metabolism, the capacity for aerobic energy production is enhanced in Antarctic species. Therefore, it may be more meaningful to investigate situations of high energy flux to determine whether metabolic cold adaptation does occur in Antarctic fish. The highest energy flux in fish is observed during burst swimming activity. Maximum speed is an important factor determining the success of prey capture and predator avoidance. Therefore, it is crucial for the survival of individuals and presumably subject to high selective pressures. In the following review we will summarize the available data on the ability of cold adapted fish to perform strenuous exercise and to recover from this

G. di Prisco, E. Pisano, A. Clarke (Eds)
Fishes of Antarctica. A biological overview
© Springer-Verlag Italia 1998

situation of high energy turnover. Special attention will be directed towards potential differences between seasonal cold *acclimation* of eurythermal temperate species as opposed to *adaptation* to the cold environment on the evolutionary time scale in Antarctic fish.

Effect of Temperature on the Ability to Perform Exercise

The actual activity pattern and the mode of locomotion observed in fishes depend upon the species' lifestyle. Two main types of swimming can be distinguished. The marathon-type of performance is powered mainly by red musculature and can be sustained for hours or even days. The maximal *sustainable* speed, which is the maximum speed that can be maintained for more than 200 min [2], is largely determined by the aerobic capacity of the red musculature of the fish. Maximal *swimming* speed is reached only for a short period of time during burst swimming when white musculature becomes active. The white muscle relies mostly on glycolytic energy production and therefore this fiber type is rapidly fatigued.

Acute cold exposure reduces the ability to exercise, but the negative effect of low temperature on aerobic and anaerobic swimming performance is compensated to various degrees during cold acclimation. While in goldfish acclimation to 5 °C leads to almost complete compensation of the maximal (anaerobic) swimming velocity compared to animals acclimated to 35 °C, killifish show only moderate compensation and no acclimatory effects were observed in trout [3,4].

Compensatory effects on the aerobic swimming capacity (maximal sustainable speed) have been observed in several species such as goldfish, carp and striped bass [5-7]. However, cold acclimation does not lead to complete compensation since warm acclimated fish are able to maintain higher swimming speeds than cold acclimated animals at their respective acclimation temperature (Table 1). Evolutionary adaptation to the polar environment, however, may have led to a higher degree of compensation as evidenced by a comparison of the maximal sustainable speeds from Antarctic and temperate fish. Maximum sustainable speeds of active pelagic *Pagothenia borchgrevinki* range between 1.8 and 2.15 bl sec^{-1} [8,9], while benthic *Notothenia coriiceps* can only maintain 0.8-1.0 bl sec^{-1} [10,11]. Table 1 shows that these values are well in the range of the maximum sustainable speeds of temperate fish species.

Burst swimming performance appears to be more constrained in polar fish, especially in larvae and juvenile animals as opposed to adult stages. No thermal compensation is observed in the startle response of cold adapted Antarctic fish larvae [12].

Table 1. Maximum sustainable swimming speeds in Antarctic and temperate fish. Fish swam at their respective acclimation temperature given in the Table

Species	V_{max} (bl sec^{-1})	Acclimation Temperature (°C)	Lifestyle	Source
Pagothenia borchgrevinki	1.8- 2.15	0	moderately active, pelagic	8,9
Notothenia coriiceps	0.8- 1.0	2	benthic	10,11
Gadus morhua	2.1 2.6	5 8	moderately active, pelagic	14
Morone saxatilis	1.9 2.8	9 25	migrator, pelagic	7
Oncorhynchus nerka	2.8 3.1	10 18	migrator, pelagic	15
Coregonus clupeaformis	1.4 1.7	5 12	moderately active, pelagic	16

Their maximum speed at 0 °C is 2 - 3 times lower than that of temperate and tropical larvae at their habitat temperatures as would be predicted from a Q_{10} of 2. Juvenile *N. coriiceps* show maximum swimming speeds lower than those of juveniles from a variety of temperate fish at higher temperatures [11]. Maximal velocities of adult *N. coriiceps* and *P. borchgrevinki*, however, are in the lower range of values for temperate fish [11,13]. It should be noted that maximum swimming speeds are influenced by several factors such as body shape, swimming mode and life style of a particular fish species. Therefore, comparisons between Antarctic and temperate fish species may be biased.

From the limited amount of data, one may conclude that cold compensation does occur for aerobic and to a lesser degree for anaerobic swimming performance in Antarctic fish. High speeds, however, are obviously not achievable at subzero temperatures, reflected in the generally slow movements and only moderately active life styles of Antarctic fish.

Energy Provision During Exercise

Aerobic Capacity

Sustained swimming is powered by slow aerobic muscle fibers that rely almost exclusively on aerobic energy provision. An increase of the maximum sustainable speed during cold adapation should therefore go along with an increase of the aerobic capacity. How do fish enhance their aerobic capacity in the cold? Several studies on temperate fish have shown a significant increase of the mitochondrial volume density in muscle cells during cold acclimation [17,18]. A similar trend is observed in Antarctic fish. Mitochondrial densities in slow fibers of several notothenioids are higher than those in temperate species at warm acclimation temperatures [19,20]. Besides a compensatory increase in oxidative capacity a second functional implication has been ascribed to changes in mitochondrial density. Increased number of mitochondria and concomitant reduction of the diffusion pathlength between capillaries or cytosolic compartments and these organelles may compensate for the reduced diffusion coefficients of metabolites and oxygen at low temperatures [17,21]. A recent study on diffusion limitations in fish muscle has shown, however, that diffusional fluxes of neither ATP nor phosphocreatine limit muscular function at low temperatures even in the absence of increased mitochondrial densities [22].

Only few studies have investigated the impact of cold acclimation or adaptation on the functional properties of mitochondria. There is some evidence that both cold acclimation and adaptation lead to an increase in mitochondrial aerobic capacity [23]. In the short-horned sculpin *Myxocephalus scorpius* a decrease of the acclimation temperature from 15 °C to 5 °C doubled maximal rates of pyruvate oxidation per milligram of mitochondrial protein [24]. A comparison of oxidative capacities of Antarctic, temperate and tropical fish showed only moderate cold compensation [25]. While respiration rates of mitochondria from tropical tilapia extrapolate to zero at around 2 °C, mitochondria of the Antarctic *N. coriiceps* still oxidize substrates at considerable rates at -1.5 °C. However, the aerobic capacity of tropical and temperate fish was considerably higher than that of Antarctic fish when compared at their habitat temperatures.

The mechanisms that underlie the increase of oxidative capacity of mitochondria are still the subject of investigation. Cristae density does not change during thermal acclimation of fish [17,21] and is probably not enhanced in Antarctic fish species [26].

A compensatory increase of mitochondrial enzyme activities after cold acclimation has been reported for many temperate fish like goldfish, carp and green sunfish [27-29] but may be absent in salmonid species like trout and lake whitefish [30,31]. Higher activities of oxidative enzymes per gram of tissue may be caused by an increased number of mitochondria, a higher amount of enzyme per mitochondrion or modulation of the specific activity of the individual enzyme molecule. Several studies of Wodtke [28,32] provide evidence that the activity of membrane bound mitochondrial enzymes - like cytochrome c oxidase (COX) and succinate dehydrogenase (SDH) - is enhanced by changes of the fluidity and lipid composition of the membranes during cold acclimation. While the increases of COX activity in seasonally cold acclimated fish range between 30 and 100% [27-29], *Gobionothoten gibberifrons* displays 5 times higher COX activities in the white muscle than a temperate zone fish with similar lifestyle [33]. Oxidative enzymes, located in the mitochondrial matrix as opposed to membrane bound enzymes, generally show only a moderate increase in cold acclimated or adapted animals. While COX and SDH are significantly enhanced in cold acclimated green sunfish, malate dehydrogenase remains unchanged [29]. Crockett and Sidell [33] found only 1.4 - 2.8 times increased citrate synthase activities in *Trematomus newnesi* and *G. gibberifrons* when compared to temperate species. Thus, a rise in mitochondrial density during cold acclimation or adaptation leads to increased activities of oxidative enzymes. The activity of membrane bound enzymes, such as COX and SDH, may be further increased by an activating effect of homeoviscous response of mitochondrial membranes.

Shifts in the saturation and fluidity of mitochondrial membranes during thermal acclimation have been documented for several species, such as goldfish and carp [32,34]. No data are available, however, on the membrane composition of mitochondria from Antarctic fish. Cytoplasmic membranes of Antarctic fish contain high percentages of polyunsaturated fatty acids and show up to 100% homeoviscous response [35]. Cold adapted organisms commonly show higher efficiencies of homeoviscous adaptation than eurythermal animals after cold acclimation [35]. Therefore, it can be assumed that mitochondrial membranes of Antarctic fish show at least the same degree of homeoviscous response as mitochondria from cold acclimated animals. At a given temperature a higher fraction of unsaturated fatty acids increases the permeability of membranes to ions especially to protons [36]. High rates of proton leakage through the inner mitochondrial membrane increase the oxygen consumption of mitochondria without a concomitant rise in ATP synthesis. Therefore, lower ADP/O values may be expected in cold adapted animals when acutely exposed to warmer temperatures. We suggest that an increase of aerobic capacity in Antarctic fish tissues brought about by an increase of the number of mitochondria in combination with homeoviscous adaptation of mitochondrial membranes may set the upper temperature limit at which oxidative phosphorylation can work efficiently [37]. The low tolerance of Antarctic animals towards temperature elevation may be due to increased proton leakiness of the inner mitochondrial membranes at higher temperatures.

Anaerobic Capacity

During burst swimming total ATP turnover may increase over 100-fold compared to resting rates [12]. Oxidative metabolism alone is not able to meet these high rates of energy utilization. In fast twitch white muscles anaerobic glycolysis is the major source of energy during burst swimming in fish. High glycolytic rates lead to an accumulation of lactate and protons in the muscle tissue.

Rainbow trout acclimated to 5 °C or 18 °C accumulated the same amounts of lactate in the white muscle during exhaustive swimming [38]. These results are in contrast to data on roach [39,40] in which a change of acclimation from 20 °C to 4 °C leads to 50% reduction of lactate formation during exhaustive exercise. Data on changes of glycolytic enzyme activities are also contradictory. While trout and lake whitefish (*Coregonus clupeaformis*) show a compensatory increase of hexokinase and phosphofructokinase activities during cold acclimation [30,31], these enzymes are even less active in cold acclimated goldfish and striped bass [27,41]. Guderley [23] suggested that glycolytic enzymes may increase in species which do not change mitochondrial enzyme activity during cold acclimation (as in salmonids), while they may tend to decrease in species which exhibit a compensatory rise in oxidative enzyme activity.

A number of studies on capture and exercise stress in Antarctic notothenioids suggest that the anaerobic capacity of these animals is greatly reduced [42]. Exhaustive exercise induced only a minor increase of lactate in the white muscle of about 1-3 μmol g^{-1} in *N. coriiceps* and *P. borchgrevinki* [9,43]. Dunn and Johnston [43] examined maximal activities of glycolytic enzymes in *N. coriiceps* and concluded that the absence of lactate production is caused by a reduced glycolytic capacity in these fish. Enzyme profiles for several notothenioids show the same trend: generally, oxidative enzymes and creatine kinase show high degrees of cold compensation, while glycolytic enzymes do not [19,33]. The reduced glycolytic capacity has been interpreted as a special adaptation to cold environment [43]. However, since absence of lactate formation has only been shown in notothenioids the question arises if indeed the low glycolytic capacity has an adaptional value or if it is rather a special phylogenetic trait of this fish family [9,44]. To shed light on this question we investigated the effect of exhaustive exercise on an Antarctic nonnotothenioid fish, the zoarcid *Pachycara brachycephalum* [45]. In this benthic sluggish fish we found relatively high amounts of lactate (11.5±0.7 μmol g^{-1} muscle tissue), which is in the range found in flounder acclimated to 11 °C [46]. Based on these data, it may be concluded that a low glycolytic capacity is not a general phenomenon in Antarctic fish.

Interestingly, we did not find significant amounts of lactate in the plasma of *P. brachycephalum* after exhaustive exercise. In contrast to mammals, the bulk of lactate is retained in the white musculature of fish for glyconeogenesis in situ [47]. Nevertheless, significant increases in lactate concentrations are found in the plasma after severe exercise in temperate species. However, the muscle-to-blood gradient of lactate is dependent on the acclimation temperature [38]. Cold acclimated trout had lower blood lactate levels although intracellular concentrations were the same in both groups. The authors explain this phenomenon by a reduction of lactate diffusion from the intracellular space at low temperatures. It should be noted, therefore, that in cold adapted fish the onset of anaerobic metabolism must be established by the analysis of lactate levels in the muscle tissue rather than in the plasma, as is frequently done in temperate species.

Recovery from Exhaustive Exercise

During recovery from exhaustive exercise the restoration of pre-exercise conditions must be achieved within reasonable time to enable successive bouts of activity which may be crucial during predator-prey interactions. Despite only partial cold compensation of locomotory performance in fish, recovery processes seem to be largely independent of temperature [38,48]. Metabolic recovery from escape swimming occurred at identical rates in herring larvae reared at 5 and 12 °C [48]. In trout rephosphorylation of phosphocreatine and lactate clearance from blood and muscle occurred at the same rates in fish acclimated to 5 and 18 °C [38]. This shows that although resting levels of oxygen consumption are reduced in cold acclimated trout [49] the capacities for high power output during exercise and for postexercise recovery metabolism are largely compensated in cold acclimated animals. Again, the roach *Rutilus rutilus* shows a different response to cold acclimation. While most other eurythermic species show perfect compensation of recovery processes, roach not only produce less lactate during burst activity at 4 °C, but also need about 4 times longer to regain pre-exercise lactate levels than their counterparts acclimated to 20 °C. Obviously, no compensatory mechanisms are functioning in this species despite an enhanced aerobic capacity [40].

In Antarctic fish the time required for recovery of blood composition and rates of oxygen consumption after exercise and capture stress is similar to temperate species. For most parameters this is essentially achieved within 8-24 h [8,9,42].

The rate of lactate clearance [45] in the eelpout *P. brachycephalum* (3.3 μmol g^{-1} h^{-1}) was even faster than that observed in flounder acclimated to 11 °C (0.62 μmol g^{-1} h^{-1}) and comparable to that observed in juvenile trout at 15 °C [46,50]. Also phosphocreatine is rephosphorylated within 3 hours [45], which is in the range of recovery times observed in trout [47].

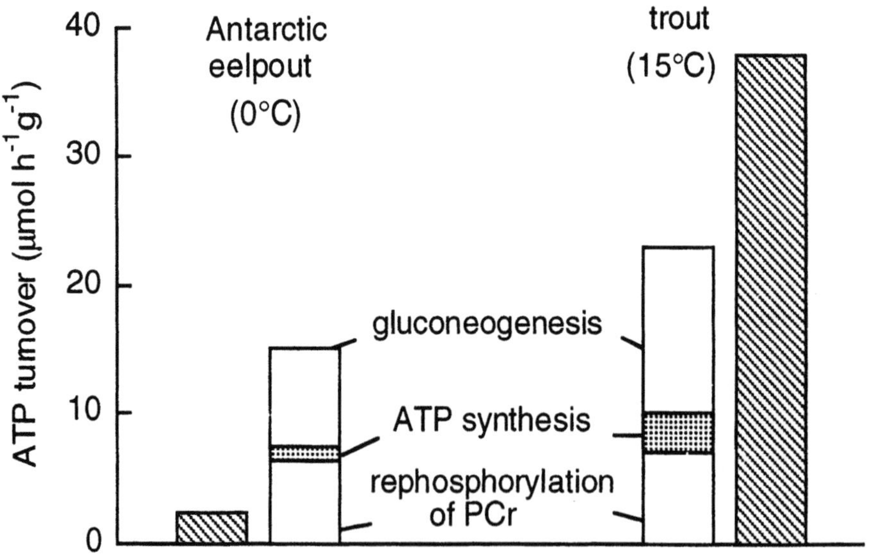

Fig. 1. The excess postexercise ATP production (on top of resting rates) during the first hour of recovery [45] from exhaustive [50] swimming in Antarctic eelpout *P. brachycephalum* at 0 °C and trout at 15 °C based on metabolite data (stacked bars) compared to resting levels of ATP turnover calculated from oxygen consumption rates (hatched bars)

A fast recovery from exhaustive swimming requires largely increased postexercise metabolic rates to pay off the oxygen debt. Based on metabolite data we calculated the ATP turnover necessary for gluconeogenesis, rephosphorylation of PCr and ATP synthesis during the first hour of recovery and compared the results with data on trout (Fig. 1). This comparison reveals that the ATP requirement during the first hour of recovery is about the same in trout at 15 °C and Antarctic eelpout at 0 °C! On the basis of the low resting metabolic rate in the latter (0.38 µmol O_2 g^{-1} h^{-1} compared to 6.3 µmol O_2 g^{-1} h^{-1} in trout; I. Hardewig, P. van Dijk and H.O. Pörtner, unpublished results; [50]) the factorial increase of ATP production required for recovery is much higher in eelpout. These data strongly suggest that Antarctic fish show metabolic cold adaptation. This adaption is not reflected in elevated resting metabolic rates, but becomes evident during situations of high energy turnover.

Conclusions

Antarctic fish have obviously developed mechanisms to compensate for the adverse effects of low temperatures on locomotory performance. Aerobic metabolic capacity and, accordingly, aerobic swimming velocities exhibit clear cold adaptation in Antarctic species. However, the increase of aerobic capacity may cause a high sensitivity towards temperature increases. This may be tolerated in cold stenothermal Antarctic species that do not experience large temperature variations in their habitat. Eurythermal organisms, however, must be able to maintain vital functions over a broad temperature range. A trade off between an increase in metabolic capacity and tolerance towards shifts in habitat temperature may have led to different degrees of cold compensation in stenothermal versus eurythermal species: While both groups generally appear to employ the same qualitative mechanisms to survive in the cold, the degree of cold compensation with respect to maximum rates of energy turnover during exercise and recovery seems to be more pronounced in stenothermal Antarctic fishes.

References

1. Somero GN, DeVries AL (1967) Temperature tolerance in some Antarctic fishes. Science 156:257-258
2. Beamish FWH (1978) Swimming capacity. In: Hoar WS, Randall DJ (eds) Fish physiology. Academic Press, New York, pp 101-187

3. Johnson TP, Bennet AF (1995) The thermal acclimation of burst escape performance in fish: an integrated study of molecular and cellular physiology and organismal performance. J Exp Biol 198:2165-2175
4. Johnson TP, Bennet AF, McLister JD (1996) Thermal dependence and acclimation of fast start locomotion and its physiological basis in rainbow trout (*Oncorhynchus mykiss*). Physiol Zool 69(2):276-292
5. Fry FEJ, Hart JS (1948) Cruising speed of goldfish in relation to water temperature. J Fish Res Bd Can 7:175-199
6. Rome LC, Loughna PW, Goldspink G (1984) Muscle fiber activity in carp as a function of swim speed and muscle temperature. Am J Physiol 247:R272-R279
7. Sisson JE, Sidell B (1987) Effect of thermal acclimation on muscle fiber recruitment of swimming striped bass (*Morone saxatilis*). Physiol Zool 60:310-320
8. Forster ME, Franklin CE, Taylor HH, Davison W (1987) The aerobic scope of an Antarctic fish *Pagothenia borchgrevinki* and its significance for metabolic cold adaptation. Polar Biol 8:155-159
9. Davison W, Forster ME, Franklin CE, Taylor HH (1988) Recovery from exhausting exercise in an Antarctic fish, *Pagothenia borchgrevinki*. Polar Biol 8:167-171
10. Johnston IA, Clarke A, Ward P (1991) Temperature and metabolic rate in sedentary fish from the Antarctic, North Sea and Indo-West Pacific Ocean. Mar Biol 109:191-195
11. Archer SD, Johnston IA (1989) Kinematics of labriform and subcrangiform swimming in the Antarctic fish *Notothenia neglecta*. J Exp Biol 143:195-210
12. Johnston IA, Johnson TP, Battram JC (1991) Low temperature limits burst swimming performance in Antarctic fish. In: di Prisco G, Maresca B, Tota B (eds) Biology of Antarctic fish. Springer Verlag, Berlin, pp 179-190
13. Montgomery JC, MacDonald JA (1984) Performance of motor systems in Antarctic fishes. J Comp Physiol 135:157-164
14. Beamish FWH (1966) Swimming endurance of some Northwest Atlantic fishes. J Fish Res Bd Can 23:341-347
15. Brett JR (1964) The respiratory metabolism and swimming performance of young sockeye salmon. J Fish Res Bd Canada 21:1183-1226
16. Bernatchez L, Dodson JJ (1985) Influence of temperature and current speed on the swimming capacity of lake whitefish (*Coregonus clupeaformis*) and cisco (*C. artedii*). Can J Fish Aquat Sci 42:1522-1529
17. Egginton S, Sidell BD (1989) Thermal acclimation of subcellular structures of fish muscle. Am J Physiol 256:R1-R10
18. Johnston IA (1993) Phenotypic plasticity of fish muscle to temperature change. In: Rankin JC, Jensen FB (eds) Fish ecophysiology. Chapman Hall, London, pp 322-340
19. Johnston IA (1987) Respiratory characteristics of muscle fibres in a fish (*Chaenocephalus aceratus*) that lacks haem pigment. J Exp Biol 133:415-428

20. Dunn JF, Archer SD, Johnston IA (1989) Muscle fiber types and metabolism in post-larval and adult stages of notothenioid fish. Polar Biol 9:213-223
21. Tyler S, Sidell BD (1984) Changes in mitochondrial distribution and diffusion distances in muscle of goldfish upon acclimation to warm and cold temperatures. J Exp Zool 232:1-9
22. Hubley MJ, Locke BR, Moerland TS (1997) Reaction-diffusion analysis of the effects of temperature on high-energy phosphate dynamics in goldfish skeletal muscle. J Exp Biol 100:975-988
23. Guderley H (1997) Temperature and growth rates as modulators of the metabolic capacities of fish muscle. In: Pörtner HO, Playle R (eds) SEB Series 66, Cold ocean physiology. Cambridge University Press
24. Guderley H, Johnston IA (1996) Plasticity of fish muscle mitochondria with thermal acclimation. J Exp Biol 199:1311-1317
25. Johnston I, Guderley H, Franklin CE, Crockford T, Kamunde C (1994) Are mitochondria subject to evolutionary temperature adaptation? J Exp Biol 195:293-306
26. Archer SD, Johnston IA (1991) Density of cristae and distribution of mitochondria in the slow muscle fibers of Antarctic fish. Physiol Zool 64:242-258
27. Sidell BD (1980) Responses of goldfish (*Carassius auratus*, L.) muscle to acclimation temperature: alterations in biochemistry and proportions of different fiber types. Physiol Zool 53:98-107
28. Wodtke E (1981) Compensation of the molar activity of cytochrome c oxidase in the mitochondrial energy-transducing membrane during thermal acclimation of the carp (*Cyprinus carpio* L.). Biochim Biophys Acta 640:710-720
29. Shaklee JB, Christiansen JA, Sidell BD, Prosser CL, Whitt GS (1977) Molecular aspects of temperature acclimation in fish: contributions of changes in enzyme activities and isozyme patterns to metabolic reorganization in green sunfish. J Exp Zool 201:1-20
30. Blier P, Guderley H (1988) Metabolic responses to cold acclimation in the swimming musculature of lake whitefish, *Coregonus clupeaformis*. J Exp Zool 246:244-252
31. Guderley H, Gawlicka A (1992) Qualitative modification of muscle metabolic organization with thermal acclimation of rainbow trout, *Onchorynchus mykiss*. Fish Physiol Biochem 10:123-132
32. Wodtke E (1981) The effects of acclimation temperature on the unsaturation of the main neutral and charged phospholipids in mitochondrial membranes of the carp (*Cyprinus carpio* L.). Biochim Biophys Acta 640:698-709
33. Crockett EL, Sidell BD (1990) Some pathways of energy metabolism are cold adapted in Antarctic fishes. Physiol Zool 63:472-488
34. Van den Thillart G, de Bruin G (1981) Influence of environmental temperature on mitochondrial membranes. Biochim Biophys Acta 640:439
35. Cossins AR (1994) Homeoviscous adaptation of biological membranes and its functional significance. In: Cossins AR (ed) Temperature adaptation of biological membranes. Portland Press, London, pp 63-76

36. Brand MD, Steverding D, Kadenbach B, Stevenson PM, Hafner RP (1992) The mechanism of the increase in mitochondrial proton permeability induced by thyroid hormones. J Biochem 206:775-781
37. Pörtner HO, Hardewig I, Sartoris FJ, van Dijk PLM (1997) Energetic aspects of cold adaptation: critical temperatures in metabolic, ionic and acid-base regulation. In: Pörtner HO, Playle R (eds) SEB Series 66, Cold ocean physiology. Cambridge University Press
38. Kieffer JD, Currie S, Tufts BL (1994) Effects of environmental temperature on the metabolic and acid-base responses of rainbow trout to exhaustive exercise. J Exp Biol 194:299-317
39. Wieser W, Koch F, Drexel E, Platzer U (1986) "Stress" reactions in teleosts: effects of temperature and activity on anaerobic energy production in roach (*Rutilus rutilus* L.). Comp Biochem Physiol 83A:41-45
40. Dalla Via J, Huber M, Wieser W, Lackner R (1989) Temperature related responses of intermediary metabolism to forced exercise and recovery in juvenile *Rutilus rutilus* (L.) (Cyprinidae: Teleostei). Physiol Zool 62:964-976
41. Sidell BD, Moerland TS (1989) Effects of temperature on muscular function and locomotory function in teleost fish. Adv Comp Env Physiol 5:115-156
42. Egginton S, Davison W (1997) Effects of environmental and experimental stress on Antarctic fishes. In: Pörtner HO, Playle R (eds) Cold ocean physiology. Cambridge University Press (in press)
43. Dunn JF, Johnston IA (1986) Metabolic constraints on burst-swimming in the Antarctic teleost *Notothenia neglecta*. Mar Biol 91:433-440
44. Egginton S (1997) A comparison of the response to induced exercise in red- and white-blooded Antarctic fishes. J Comp Physiol 167:129-134
45. Hardewig I, van Dijk PLM, Pörtner HO (1997) High energy turnover at low temperatures: recovery from exhaustive exercise in Antarctic and common eelpout (Zoarcidae). Am J Physiol, submitted
46. Milligan CL, Wood CM (1987) Muscle and liver intracellular acid-base and metabolite status after strenuous activity in the inactive, benthic starry flounder (*Platichthys staelatus*). Physiol Zool 60:54-68
47. Milligan CL (1996) Metabolic recovery from exhaustive exercise in rainbow trout. Comp Biochem Physiol 113A:51-60
48. Franklin CE, Johnston IA, Batty RS, Yin MC (1996) Metabolic recovery in herring larvae following strenuous activity. J Fish Biol 48:207-216
49. Beamish FWH (1964) Respiration of fishes with special emphasis on standard oxygen consumption. II. Influence of weight and temperature on respiration of several species. Can J Zool 42:177-188
50. Scarabello M, Heigenhauser GJF, Wood CM (1992) Gas-exchange, metabolite status and excess post-exercise oxygen consumption after repetitive bouts of exhaustive exercise in juvenile rainbow trout. J Exp Biol 167:155-169

Ion and Metabolite Transport Through the Intestinal Luminal Membranes of the Antarctic Fish *Trematomus bernacchii*

Michele Maffia, Raffaele Acierno, Mariella Rollo, and Carlo Storelli

Laboratory of General Physiology, Department of Biology, University of Lecce, Via Provinciale Lecce-Monteroni, 73100 Lecce, Italy

Introduction

Because they live in an aquatic environment, fish are continuously challenged with preserving the internal fluid osmolarity from an increase (sea water fish) or a lowering (fresh water fish) and with maintaining an invariant ion composition despite great differences from the external milieu. The intestinal, renal and gill epithelia are the main physiological structures involved in these homeostatic processes since with their passive and active transport mechanisms they maintain ion concentrations, organic metabolites and H_2O in the physiological range. In particular the intestinal epithelium is a selective barrier that controls ion and water absorption from sea water introduced into the gut and accomplishes the uptake of nutrients and metabolites needed for life and growth. Transport phenomena, either active or passive, through morphologically and functionally polarized cells of this epithelium are regulated by the components of their apical and basolateral plasma membranes such as lipids (phospholipids and cholesterol) and protein (enzymatic and transport proteins). Because fishes are essentially ectothermic animals, all these components are affected by the environmental temperature, the main effects being on membrane fluidity, permeability, and protein activity.

Between the late Oligocene and early Miocene (about 25 million years ago) the complete separation of Antarctica from other continents took place [1]. After this separation, the climatic history of Antarctica shows a constant trend of temperature decrease that has brought sea water temperature from ~5 to ~-2 °C (present day). In this condition, all ectothermic organisms that continued to live in Antarctic waters dealt with the thermodynamically adverse effect of low temperature on biochemical constraints and were able to survive in this cold environment.

Our group has been involved, for several years, in studying the adaptation of epithelial absorbing or secreting membranes to cold [2-4].

Data presented here represent some of the results that we have collected, using the intestine of the Antarctic teleost, *Trematomus bernacchii*, in the

G. di Prisco, E. Pisano, A. Clarke (Eds)
Fishes of Antarctica. A biological overview
© Springer-Verlag Italia 1998

study of ion and metabolite transport through the enterocyte cell membrane. These data, which are part of a larger scientific project related to the study of biochemical and physiological adaptation to low Antarctic temperature, deal with the investigation of molecular adaptation mechanisms which allow these teleosts to keep the activities of their carrier proteins compatible with maintenance of basic homeostatic processes at below zero temperatures.

Material and Methods

Specimens of *Trematomus bernacchii* were collected in the Ross Sea (Antarctica) and kept unfed in open circulating seawater aquaria of the Italian Station in Terra Nova Bay at 0 °C for at least 2 weeks before sacrifice.

Fish were killed with a sharp blow on the head and intestines were rapidly used for brush border membrane vesicles (BBMV) prepared by selective Mg^{2+}-precipitation [3]. These vesicles were employed for indirect transport measurements through plasma membranes with fluorescent dyes and for enzymatic determinations [3]. Some intestines were immediately frozen in liquid nitrogen, stored at -80 °C, and used for BBMV preparations that were subsequently employed in uptake experiments with radioactive D[^{14}C]-glucose in the Laboratory of General Physiology (Lecce, Italy).

European eels (*Anguilla anguilla*) at the yellow stage were purchased from the commercial pond Ittica Ugento (Lecce, Italy) and kept without feeding in recirculating seawater aquaria at 18 °C for at least 2 weeks All reagents were purchased from Sigma Chemical Co (St. Louis, MO). The Na^+-coupled cotransport activities of isolated BBMV were measured at -1 °C in Terra Nova Bay by using a fluorescent dye sensitive to electrical membrane potential changes, DiS-C_2(5), and in the Laboratory of General Physiology (Lecce) by using radioactive D-glucose, according techniques described in detail elsewhere [3]. The enzymatic activity of Na^+/K^+-ATPase of entire mucosal homogenate was measured by following the appearance of inorganic phosphate [3]. Statistical analysis of data was performed by ANOVA and Student's t-test (★=p<0.05, ★★=p<0.01).

Results

Na^+-Dependent Transport Systems of *T. bernacchii* Intestinal BBMV

Figure 1 shows a representative experiment performed at Terra Nova Bay using the electrical membrane potential sensitive fluorescent dye, DiS-C_2(5), as previously described [3]. The upper trace represents nonspecific fluorescent quenching of the dye induced by injection of BBMV in an extravesicular medium containing the same concentration of KCl (100 mM) as of the intravesicular medium and valinomycin (0.005 mM). Under these experimental conditions the symmetrical distribution of KCl in the intra- and extravesicular media and the presence of K-specific ionophore valinomycin, eliminated a membrane potential (short-circuit conditions) and the fluorescence quenching of the dye was the result of its redistribution inside and outside the membrane vesicles [5]. When in the extravesicular medium KCl was replaced by NaCl (100 mM), a much higher initial fluorescence quenching of the dye was observed as a first effect (lowest trace: sharp decrease).

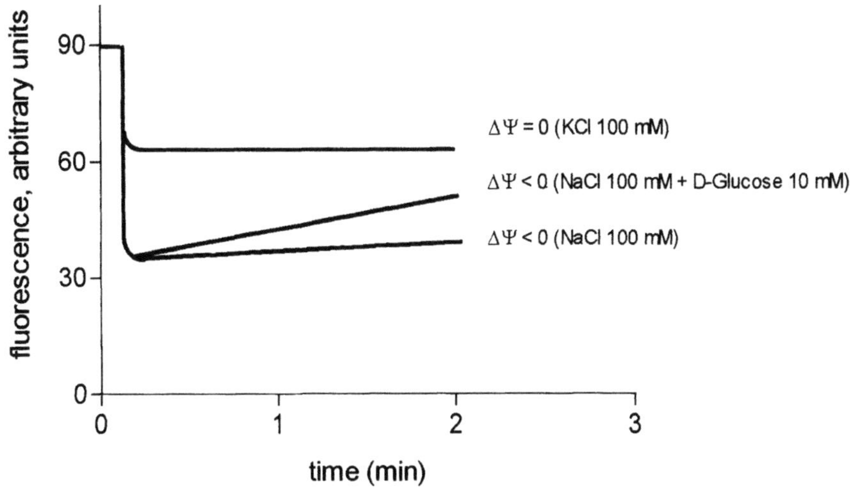

Fig. 1. Fluorescent traces of a typical experiment with the electrical potential sensitive dye, DiS-C_2(5), on BBMV of *T. bernacchii.* Vesicles were resuspended in 100 mM mannitol + 100 mM KCl + 20 mM hepes adjusted with tris at pH 7.8. The external media were those reported in figure + valinomycin 0.005 mM + mannitol 100 mM or 90 mM when D-glucose 10 mM was present

This phenomenon is explained by the higher distribution of the positively charged fluorescent dye as a dimeric nonfluorescent form at the internal side of the vesicular membrane, correlated to the formation of an inside negative electrical membrane potential due to K^+ diffusion through the plasma membranes.

In the experimental conditions K^+ permeability was artificially increased by valinomycin producing a much higher response with respect to that of other permeant ionic species present in the incubation media (Na^+, Cl^-). Starting from this point, the inside-negative electrical potential began to dissipate mainly by Na^+ entry, that followed both its favorable concentration and electrical potential gradients (lowest trace: slow recovery). Under these experimental conditions a substrate such as D-glucose significantly increased the dissipation of fluorescence quenching (intermediate trace). This phenomenon, explained by a more rapid depolarization of the initial electrical membrane potential, should be due to a higher entry rate of Na ions presumably mediated by Na^+-D-glucose cotransport activity. The activity of this putative Na^+-D-glucose cotransport, expressed as the decay of fluorescence quenching within 15 sec after the lowest fluorescence value recorded, is reported in Fig. 2, where are also reported experimental values obtained by applying the same experimental procedure to substrates as aminoacids. All substrates analyzed induced a fluorescence quenching dissipation significantly higher than that recorded in presence of mannitol (used as a control value), thus suggesting that Na^+-coupled transporters located on the apical membranes of *T. bernacchii* enterocyte work actively at -1 °C.

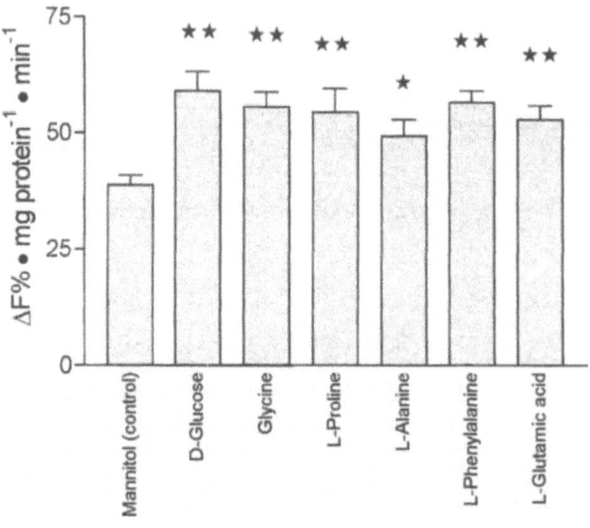

Fig. 2. Na^+-coupled transport of the intestinal brush border membrane in the Antarctic teleost *T. bernacchii*. ΔF% corresponds to the changes in fluorescence expressed in arbitrary units. ★= $p<0.05$, ★★= $p<0.01$

Na^+-D-Glucose Cotransport of *T. bernacchii* Intestinal BBMV

Na^+-D-glucose cotransport characteristics were further investigated by directly evaluating the uptake of radioactive D-glucose. Figure 3 illustrates a typical time-course of D-[^{14}C]-glucose uptake in *T. bernacchii* intestinal BBMV under different ionic gradient conditions. An inwardly directed Na^+-gradient induced D-glucose uptake after 2 min that was significantly higher than the equilibrium value reached after 90 min. This transient overshoot phenomenon, always correlated to the activity of a Na^+-coupled-D-glucose transport mechanism, was completely inhibited by phloridzin, a Na^+-D-glucose cotransport inhibitor, or by substitution of NaCl with choline Cl. This result confirms the presence of a Na^+-D-glucose cotransport mechanism on the apical membranes of the Antarctic fish enterocyte.

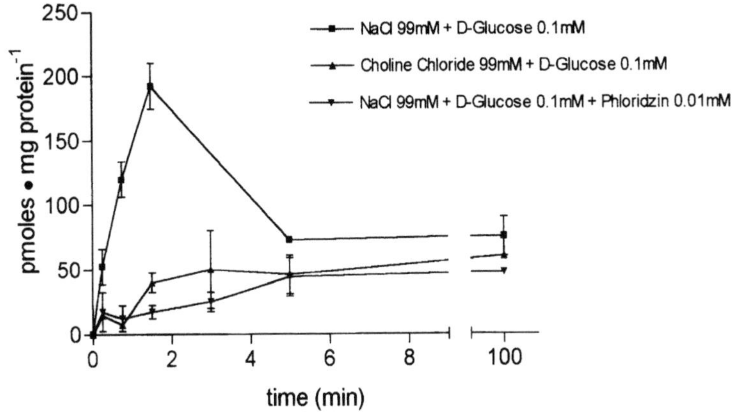

Fig. 3. Time-course of D-[14C]-glucose uptake in *T. bernacchii* intestinal BBMV. Vesicles were resuspended in the same solution as Fig. 1. A valinomycin concentration equal to 0.005 mM was present in all the external media

Table 1. Main characteristics of Na^+-D-glucose cotransport in fish and other vertebrates

	T. bernacchii	*A. anguilla*	Other vertebrates
Na-dependence	+	+	+
Electrogenicity	+	+	+
Substrate specificity	β-Me-glucose ≥ α-Me-glucose ≥ D-glucose ≥ D-galactose >3-O-Me-glucose ≥ D-fructose > 2-deoxy-D-glucose	same	
K_m (mM)	0.16 ± 0.03	0.10 ± 0.02	0.03 - 4.0
J_{max} (nmol/mg protein/min)	0.23 ± 0.009	6.87 ± 0.89	0.65 - 11.0
Na^+-D-glucose stoichiometric ratio	?	2:1	2:1; 1:1
Temperature optimum	-2 / 0 °C	> 12 °C	

The main characteristics of this transport system, obtained by using the above described radioactive technique in kinetic and specificity experiments reported elsewhere [3], are summarized in Table 1.

Temperature Effect on Transport Systems

An investigation of the temperature effects on plasma membrane transport systems was performed by analyzing at different temperatures, both in *T. bernacchii* and *A. anguilla* intestine, activity of the luminal Na^+-dependent

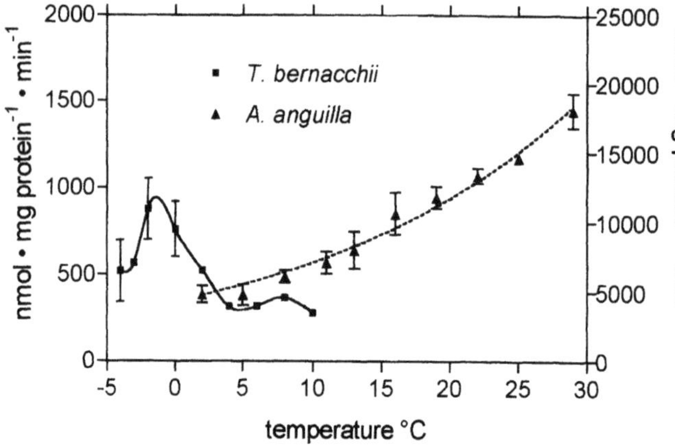

Fig. 4. Temperature dependence of Na^+- D[^{14}C]-glucose uptake (expressed as nmoles of D-glucose x mg proteins^{-1} x min^{-1}) in *T. bernacchii* and *A. anguilla* brush border membrane vesicles. Left and right ordinate pertain to the Antarctic and the temperate species respectively.

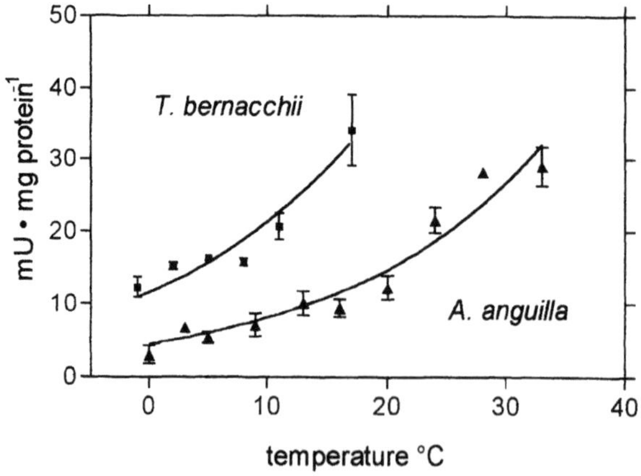

Fig. 5. Effect of temperature on Na^+/K^+-ATPase activity in *T. bernacchii* and *A. anguilla.* mU correspond to nmoles of inorganic phosphate for minute deriving from the Na^+/K^+-ATPase catalysed reaction.

sugar transport and of the ouabain-sensitive basolaterally located Na^+/K^+-ATPase. Figure 4 shows that while Na^+-dependent D[^{14}C]-glucose uptake of eel intestine increases exponentially with temperature in the full range tested (3 - 28 °C), the sugar transport activity of the Antarctic fish reached its maximum at -2/0 °C and slowly decreased at higher temperatures. On the other hand, Fig. 5 shows that the activity of Na^+/K^+-ATPase of both species' enterocytes exponentially increased with temperature over the range of experimental temperature used.

Discussion and Conclusions

The success of notothenioids in surviving in the Antarctic Sea seems to involve, among other things, adaptive modifications of intrinsic proteins involved with the transport of water and electrolytes through the plasma membranes. Although many researchers have concentrated their efforts on the analysis of various enzymes involved in different cellular functions, reaching many interesting conclusions regarding the molecular adaptive features of these enzymes at low temperatures, very few data exist on membrane transport proteins, which have the peculiarity of needing to function in a lipid microenvironment with a specific physical state of fluidity. Since 1983 our group has studied the characteristics of various transport systems in different epithelia of teleosts. One of the first epithelia used for these studies was the intestine, because a variety of different transport systems for ions, nutrients and other substrates are localized in this tissue. By using brush border and basolateral membranes isolated as vesicles, in association with radioactive and fluorescent tracers, we have demonstrated in the intestine of carnivorous and herbivorous teleost species the existence of transport systems for ions and metabolites with characteristics similar to those previously described in higher vertebrates [6-10]. Recently we have applied similar experimental procedures in the study of intestinal transport systems of *Trematomus bernacchii.* The main conclusions deriving from these studies are summarized in Fig. 6. This Figure suggests that on the apical membranes of *T. bernacchii* enterocytes are present: 1) the Na^+-dependent, electrogenic transport mechanisms for sugars and amino acids [3]; 2) a H^+/dipeptide cotransport (Rollo et al. unpublished data) with characteristics similar to those previously reported in *A. anguilla* [10] and 3) an Na^+/H^+ exchanger (Acierno et al. unpublished data) that does not exist on eel intestinal BBMV [11]. This exchanger, however, has been found in other fish species, confirming that an exchanger commonly present on apical membranes of mammalian intestine [12] is also expressed in teleosts. The Na^+/K^+-ATPase enzymatic activity (mU/mg protein) of the Antarctic fish intestine is comparable, at -1 °C, to that measured in eel, at 18 °C, suggesting that one of the main homeostatic

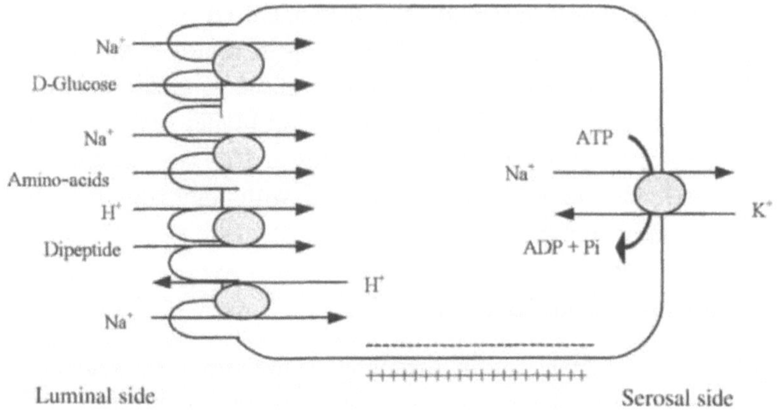

Fig. 6. Transport systems of *T. bernacchii* enterocyte

mechanisms for the regulation of intracellular K^+ and Na^+ concentrations is functionally active at low temperatures in maintaining transmembrane ionic gradients and electrical potential. These so-called driving forces can energize Na^+-dependent, electrogenic transporters such as Na^+-D-glucose cotransport of Antarctic fish intestine whose main characteristics are reported in Table 1, in comparison with data previously obtained on eel [13] and higher vertebrates [14]. Carrier protein binding properties of this transporter family seem to be strongly conserved since substrate specificity and affinity (apparent K_m) are similar in all species investigated. As a consequence, Na^+-D-glucose cotransport of the Antarctic fish could be structurally homologous to the mammalian small intestine SGLT-1 protein, whose mRNA seems to be highly conserved through evolution [15]. However a modification in a few amino acids of the carrier, essential in regulating its transport activity, could be hypothesized on the basis of the low J_{max} value, a kinetic parameter dependent on carrier density and catalytic rate (k_{cat}), and of the temperature optimum (-2/0 °C) reported for Na^+-D-glucose transport of *T. bernacchii*. As suggested by the large J_{max} range reported for other vertebrates, this parameter is dependent on diet habit, size of animals, seasonality, food supply, etc. and for these reasons it is not easily comparable among vertebrates. However a good sugar absorptive capacity could be maintained by a longer food residence in the intestinal tract of *T. bernacchii* or by other unknown events regarding sugar basolateral transport or the action of metabolic enzymes, etc., within the intact cell. It cannot be excluded that this carnivorous teleost could have a low sugar need, as also suggested by recent studies of Sidell et al. [16] showing that for oxidative metabolism of cold water species, monounsaturated fatty acids (MUFA) represent a preferred substrate compared to glucose. As confirmation of this observation, the intestinal mucosa of the Antarctic fish possess a much higher amount of MUFA with respect to that of the European eel [4]. Temperature dependence (Fig. 4) of Na^+-D-glucose cotransporter of *T. bernacchii* intestine is not only completely different from that of eel intestinal sugar transporter but also different from those previously reported for other functionally extrinsic enzymatic proteins such as alkaline phosphatase and leucine aminopeptidase of the same Antarctic species [2]. An exponential increase of activity with temperature has also been found for the intestinal Na^+/K^+-ATPase (Fig. 5) similar to that previously reported in kidney of *T. bernacchii* [17]. However both these experimental observations concern only the effects of temperatures on ATP-hydrolysis catalysed by Na^+/K^+-ATPase while no information is at the moment available on the ion transport capacity mediated by this active pump at different temperatures. In light of these considerations, the most important characteristics of the Na^+-D-glucose transport activity of *T. bernacchii* intestine are the quantification of the true substrate transport and the inactivation temperature limit (that intriguingly corresponds to that lethal for these animals). A protein unfolding, but more likely a modification of lipid-protein interaction forces, could be possible explanations of this temperature effect. In this regard *T. bernacchii* intestinal brush border membranes have an amount of poly- and monounsaturated fatty acids, significantly higher than that reported for the European eel (Acierno et al. unpublished data), in accordance with the phenomenon of homeoviscus adaptation, characteristic of cold-acclimated poikilotherms [18]. In principle this phenomenon, significantly influencing the physical state of the membrane by increasing its fluidity at low temperatures, could induce, at temperatures higher than those of *T. bernacchii* environment, a lipid-protein disassembling leading to a negative effect on functionality of intrinsic proteins such as Na^+-D-glucose cotransporter. However, also if influences of lipid modification on transport protein activity were previously observed [19], other studies are needed to better clarify this aspect.

References

1. Eastman JT (1993) Antarctic fish biology. Academic Press, San Diego
2. Maffia M, Acierno R, De Ceglie G, Vilella S, Storelli C (1993) Adaptation of intestinal cell membrane enzymes to low temperatures in the Antarctic teleost *Pagothenia bernacchii.* J Comp Physiol B163:265-270
3. Maffia M, Acierno R, Cillo E, Storelli C (1996) Na^+-D-glucose cotransport by intestinal BBMVs of the Antarctic fish *Trematomus bernacchii.* Am J Physiol 271:R1576-R1583
4. Acierno R, Maffia M, Sicuro P, Fiammata L, Rollo M, Ronzini L, Storelli C (1996) Lipid and fatty acid composition of intestinal mucosa of two Antarctic

5. Cassano G, Maffia M, Vilella S, Storelli C (1988) Effects of membrane potential on Na^+ cotransporters in eel intestinal brush border membrane vesicles: studies with a fluorescent dye. J Membr Biol 101:225-236
6. Ahearn GA, Storelli C (1994) Use of membrane vesicle techniques to characterize nutrient transport processes of the teleost gastrointestinal tract. In: Hochachka P, Mommsen T (eds) Biochemistry and molecular biology of fishes. Elsevier, Amsterdam, pp 513-524
7. Storelli C, Vilella S, Romano MP, Maffia M, Cassano G (1989) Brush border amino acid transport mechanisms in carnivorous eel intestine. Am J Physiol 257:R506-R510
8. Vilella S, Reshkin SJ, Storelli C, Ahearn GA (1989) Brush border inositol transport by intestines of carnivorous and herbivorous teleosts. Am J Physiol 256:G501-G508
9. Maffia M, Ahearn GA, Vilella S, Zonno V, Storelli C (1993) Ascorbic acid transport by intestinal brush-border membrane vesicles of the teleost *Anguilla anguilla*. Am J Physiol 264:R1248-R1253
10. Maffia M, Verri T, Danieli A, Thamotharan M, Pastore M, Ahearn GA, Storelli C (1997) H^+-glycyl-L-proline cotransport in brush-border membrane vesicles of eel *(Anguilla anguilla)* intestine. Am J Physiol 272:R217-R225
11. Cassano G, Vilellà S, Zonno V, Storelli C (1985) Na^+/H^+ exchange activity is not present on eel intestinal brush border membrane vesicles. In: Alvarado F, Van Os CH (eds) Ion-gradient coupled transport. Elsevier, Amsterdam, pp 282-286
12. Wakabayashi S, Munekazu S, Poussegur J (1997) Molecular physiology of vertebrate Na^+/H^+ exchangers. Physiol Reviews 77:51-74
13. Storelli C, Vilella S, Cassano G (1986) Na^+-dependent D-glucose and L-alanine transport in eel intestinal brush border membrane vesicles. Am J Physiol 251:R463-R469
14. Wright EM (1993) The intestinal Na^+-glucose cotransporter. Ann Rev Physiol 55:575-589
15. Hediger MA, Rhoads DB (1994) Molecular physiology of sodium-glucose cotransporters. Physiol Rev 74:993-1026
16. Sidell BD, Crockett EL, Driedzic WR (1995) Antarctic fish tissues preferentially catabolize monoenoic fatty acids. J Exp Zool 271:73-81
17. Gonzalez-Cabrera PJ, Dowd F, Pedibhotla VK, Rosario R, Stanley-Samuelson D, Petzel D (1995) Enhanced hypo-osmoregulation induced by warm-acclimation in Antarctic fish is mediated by increased gill and kidney Na^+/K^+-ATPase activities. J. Exp. Biol. 198:2279-2291
18. Sinensky M (1974) Homeoviscous adaptation - a homeostatic process that regulates the viscosity of membrane lipids in *Escherichia coli.* Proc. Natl. Acad Sci USA 71:522-525
19. Hesket KR, Smith GA, Houslay MD, Mc Gill KA, Birdsall NJM, Metcalfe JC, Warren GB (1976). Annular lipids determine the ATPase activity of a calcium transport protein complexed with dipalmitoyllecithin. Biochemistry 15:4145-4151.

The Neuroendocrine System in the Intestinal Tract and Pancreas of Antarctic Fish

Grazia Tagliafierro, Marina Carlini, Gabriella Faraldi, and Lorenzo Gallus

Istituto di Anatomia Comparata, Università di Genova, 16132 Genova, Italy

Introduction

The neuroendocrine system (NES) is composed of endocrine cells and nerve elements containing a large variety of low molecular weight peptides; they are generally called neuropeptides or regulatory peptides and are contained in secretory granules. The endocrine cells can be grouped to form endocrine glands such as the pancreas, or they can be found, as single elements, within the epithelial component of different organs, such as the gut and the respiratory tract. The epithelial endocrine cells lie on the basement membrane and frequently have a cytoplasmic process extending to the lumen; this basal-granulated cell is called the open type endocrine cell. Other basal-granulated cells do not reach the lumen, and are called the closed type endocrine cell; the latter cell type lies flat on the basement membrane [1]. In both cases the release of the granule content occurs at the base of the cell near a blood vessel [2].

There is also the possibility that a small portion of the product of secretion is released into the gut lumen. The apical release of regulatory peptides can be compared to the dendritic release of messengers in bipolar neurons [3]. The peptide-secreting nerve elements, both neuron somata and nerve fibers, are distributed throughout the central and the peripheral nervous system. No anatomical boundaries are present between the endocrine and the nerve compartment, often sharing identical, or nearly identical, bioactive substances, viewed as the gene products of neuro-endocrine cells. The bioactive substances released by the nervous and the endocrine compartments may function as neurotransmitters and/or neuromodulators, or they may diffuse to the blood and act as hormones, or they can also diffuse locally, acting as paracrine messengers on target neighboring cells.

The Enteropancreatic Neuroendocrine System in Teleosts

The NES of the enteropancreatic tract is relatively well known, being the object of research for about 30 years. Most results have been gained on

G. di Prisco, E. Pisano, A. Clarke (Eds)
Fishes of Antarctica. A biological overview
© Springer-Verlag Italia 1998

mammals; however in recent years, increasing data are also available for fish, particularly teleosts whose NES appears no simpler than that of mammals.

Teleosts are characterized by great variability in the number, distribution and localization of neuroendocrine elements. This variability could be related to that of some morphological gut characters such as presence or absence of the stomach, the number of pyloric ceca, or the kind of nutrients, and environmental conditions [4-6].

Immunohistochemistry has shown the presence of a number of peptides and of serotonin in nerves and endocrine cells of the fish gut and pancreas. Most of the peptides have been grouped into a few families on the basis of their similarity in amino acid sequence, shown by biochemical methods [7-10]; for example, vasoactive intestinal peptide (VIP), peptide histidine isoleucine (PHI), pituitary adenylate cyclase activating peptide (PACAP), glucagon and secretin are grouped into the glucagon/VIP family. The distribution and localization of these bioactive substances in the gastro-enteropancreatic (GEP) tract of fish are currently detected by means of immunohistochemical methods although recently more resolution has been gained through the in situ hybridisation technique [11]. In both cases a major problem is that of specificity, because the first method utilizes antibodies raised against mammalian peptides, and the second one utilizes mammalian cDNA probes to detect mRNAs encoding a particular peptide or its precursor. For this reason the immunodetected substances in fishes can be only partly similar to those used to produce the antisera or the cDNA probes. The utilization of a panel of different specificity controls can overcome this problem [12].

The physiological interpretation of these substances in fish is in progress, and numerous experiments are in favor of their role in the control of gastrointestinal absorption, secretion, motility, blood flow, growth and differentiation, and furthermore in the regulation of the mucosal immune and inflammatory response [13-17]. Some of these functions are greatly influenced by temperature, thus the enteropancreatic neuroendocrine system of Antarctic fish could show significant deviations from that of other teleosts [18,19].

The Enteropancreatic Tract of Antarctic Notothenioid Teleosts

There are only few papers concerning the morphology and histology of the digestive system in Antarctic teleosts. In these fish there is a simplification of intestinal loops and a reduction of the number of pyloric ceca. The intestinal wall is generally similar to that of other teleosts [20-22] but the enteric submucous plexus seems to be reduced and is lacking neuron cell bodies [23]. The pancreas is diffuse, as in most actinopterygians, and the endocrine portion forms a large aggregation of islet cells, the Brockmann body, with some smaller accessory islets [22]. Only one Brockmann body was observed in *Cryodraco antarcticus*, *Pagetopsis macropterus* and *Pleuragramma antarcticum* but in *Notothenia coriiceps* three islets of 1 cm diameter were found [24].

The enteropancreatic NES has been studied in three channichthyid species: *Chionodraco hamatus*, *Cryodraco antarcticus*, *Pagetopsis macropterus* and in four nototheniids: *Notothenia coriiceps*, *Pleuragramma antarcticum*, *Trematomus bernacchii*, *Trematomus newnesi* [23-26]. The data concerning the enteropancreatic NES of non-Antarctic teleosts are also reported for comparison.

Intestinal Endocrine Cells

Numerous immunoreactive (IR) endocrine cells were seen in the intestinal epithelium of *C. hamatus*, *C. antarcticus*, *P. macropterus, N. coriiceps*, *P. antarcticum*, *T. bernacchii* and *T. newnesi*; their immunopositivity to the different antisera slightly varied according to the species [23-26].

Gut endocrine cells are always of the open type (Fig. 1). Generally at the base of the cell there is a more intense immunostaining, which corresponds to a greater concentration of the immunodetected substance.

Most showed some basal cytoplasmic extensions that make contact with neighboring cells (Fig. 2). Endocrine cells positive to serotonin, substance P (sub P), cholecystokinin octapeptide (CCK 8), cholecystokinin decapeptide (CCK 10), glucagon, gastrin, pancreatic polypeptide (PP), VIP and PHI antisera were observed in all species. Sub P-, CCK 8- and serotonin-IR cells were equally frequent along the gut, while gastrin-, glucagon-, VIP-, PACAP- and PHI-IR cells seemed to be more numerous, sometimes exclusive, in the proximal intestinal portion.

In the intestinal epithelium of *T. bernacchi* and *T. newnesi* we also found big-gastrin- and bombesin-IR cells. Fewer somatostatin-, serotonin-, gastrin/CCK-IR cells were seen, and, on the contrary, more numerous VIP- and PP-IR cells were present.

Somatostatin-IR cells could be found in both channichthyid species, but were particularly abundant in *C. antarcticus*; on the contrary, serotonin-IR cells appeared more frequent in *C. hamatus*. In *C. antarcticus* numerous insulin immunoreactive endocrine cells were found along the entire gut (Fig. 3). This endocrine cell type was instead rarely, if ever, seen in the nonAntarctic perciformes *Serranus cabrilla* (Serranidae) and *Diplodus annularis* (Sparidae) [25,26]. The pattern of distribution of peptides in these temperate water teleosts is more similar to that of nototheniids than that of channichthyids.

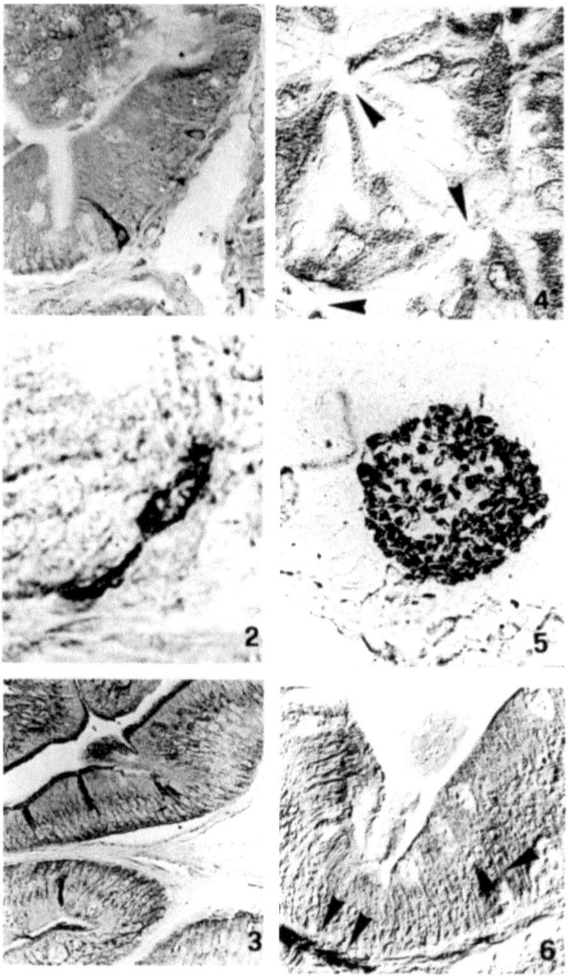

Fig. 1. *C. antarcticus*. Open type CCK 10-IR cell in the intestinal epithelium. Immunoperoxidase x 250. **Fig. 2.** *T. newnesi*. IR cells showing a basal cytoplasmatic process. Immunoperoxidase x 700. **Fig. 3.** *C. antarcticus*. Insulin-IR cells in the intestinal epithelium. Immunoperoxidase x 250. **Fig. 4.** *P. macropterus*. Brockmann body. Somatostatin-IR cells are polarized towards blood capillaries (arrow). Immunoperoxidase x 500. **Fig. 5.** *P. antarcticum*. In the 2 cm long larvae insulin-IR cells are already detected in the central region of the Brockmann body. Immunoperoxidase x 200. **Fig. 6.** *P. macropterus*. VIP immunoreactivity is localized in both endocrine cells (arrow) and nerve elements (double arrow). Immunoperoxidase. Nomarski optic x 300

Pancreatic Endocrine Cells

The pancreatic endocrine cells are always of the closed type, and often polarized towards blood vessels (Fig. 4). The topographical distribution of insulin-, glucagon-, PP- and somatostatin-IR cells in the Brockmann body in *C. antarcticus*, *P. macropterus*, *N. coriiceps* and *P. antarcticum* was similar to that found in most other teleosts [6,24,27]. Glucagon-IR cells are located at the periphery while somatostatin 14-IR and insulin-IR are situated in the central portion. Different endocrine cell associations can be seen in smaller islets: insulin-IR cells can be found at the periphery, while somatostatin- and PP-IR cells can be detected throughout the islet. In the large islet of *C. antarcticus*, PP-IR cells are fewer than that observed in nothotheniids and in *Serranus cabrilla* [24]. The pancreatic endocrine cells appear very early during the embryonic development and a large islet with insulin, glucagon and somatostatin-IR cells can be observed in 20 mm long *P. antarcticum* larvae (Fig. 5; Tagliafierro et al., unpublished).

Intestinal and Pancreatic Nerves

In the gut of *C. hamatus*, *C. antarcticus*, *P. macropterus*, *N. coriiceps*, *P. antarcticum*, *T. bernacchii* and *T. newnesi* and in the pancreas of *C. antarcticus*, *P. macropterus*, *N. coriiceps* and *P. antarcticum*, many peptidergic nerve elements were seen [22,23,25,26]. Their presence, distribution and frequency are only partially similar to those found in other teleosts [27].

VIP, PHI and PACAP immunoreactivity was present both in the endocrine cells and nerve elements (Fig. 6); nevertheless, VIP- PHI- and PACAP-IR nerve elements are less common than in the gut and pancreas of some other teleosts such as *S. cabrilla* and *D. annularis* (Figs. 7,8) [5,26]. In the blood vessel wall of the Antarctic species no VIP immunoreactivity was found, while numerous CCK- and serotonin-IR nerves were instead seen. It has been demonstrated that gastrin/CCK and serotonin stimulate smooth muscle motility and exert a constrictor role on blood vessels influencing the blood flow [28]. Sub P-IR fibers were particularly numerous in the myenteric plexus (Fig. 9) often surrounding nonIR neuron cell bodies; serotonin-IR fibers were present in the blood vessel wall and in the lamina propria beneath the intestinal epithelium. Most of these IR nerve fibers were of the extrinsic type, at least in channichthyids where few IR neuron somata were seen. In *C. antarcticus* insulin-IR nerve terminals and neuron cell bodies were also detected (Fig. 10). Insulin immunoreactivity in *C. antarcticus* has been confirmed by the in situ hybridization method (Tagliafierro, unpublished). In the Brockmann body of the notothenioids, we detected only few VIP-, PACAP- and PHI-IR fibers, while we saw many in *S. cabrilla* and in *D. annularis*, as was the case in other teleosts and in mammals [29,30]. VIP, PACAP and PHI belong to the same family and are candidate messengers

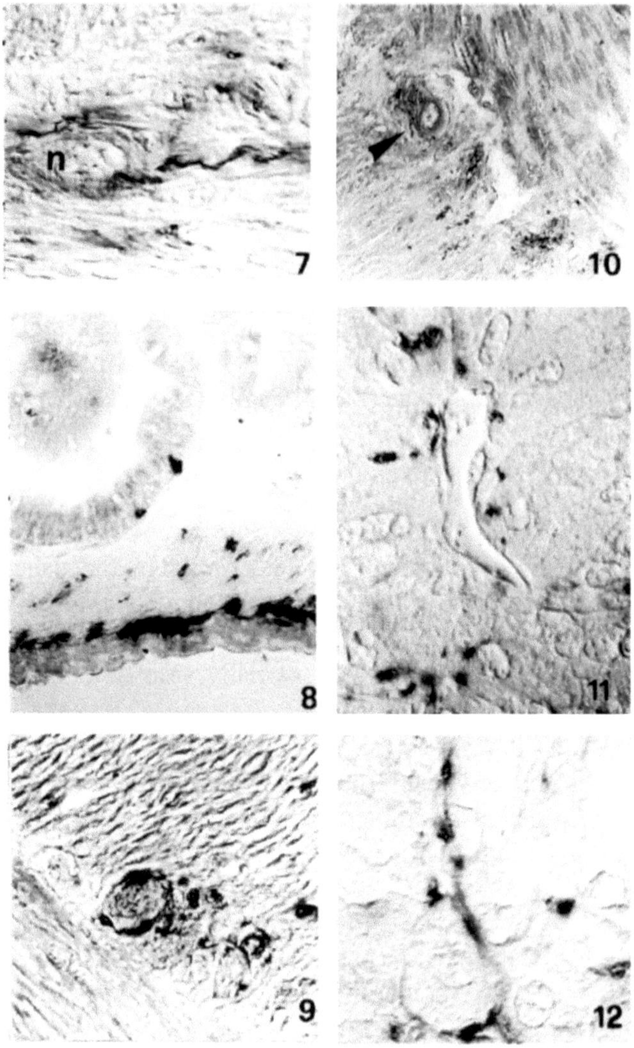

Fig. 7. *C. antarcticus.* VIP-IR nerve fibers surrounding a nonIR neuron cell body (n). Nomarski optic. Immunoperoxidase x 700. **Fig. 8.** *S. cabrilla.* A strong VIP immunoreactivity is present in epithelial endocrine cells and in numerous nerve elements. Immunoperoxidase x 100. **Fig. 9.** *P. macropterus.* Sub P-IR nerve fibers surrounding nonIR neuron cell bodies in the myenteric plexus. Immunoperoxidase. Nomarski optic x 600. **Fig. 10.** *C. antarcticus.* Insulin-IR nerve fibers in the muscular layer. Some of them surround a faintly IR neuron cell body (arrow). Immunoperoxidase x 200. **Fig. 11.** *P. macropterus.* Brockmann body. Sub P-IR fibers near a blood vessel in the central region. Immunoperoxidase. Nomarski optic x 700. **Fig. 12.** *P. macropterus.* Brockmann body. Sub P-IR fibers surrounding a positive glucagon cell in the peripheral region. Immunoperoxidase. Nomarski optic x 700

in nonadrenergic, noncholinergic nervous control of pancreatic and intestinal secretion, motility and blood flow; they appear to stimulate the release of islet hormones and the flow of pancreatic juice [31].

Surprisingly, in our Antarctic specimens, many sub P-IR fibers were seen, especially in the outer portion of the large islet, near glucagon- and PP-IR cells (Figs. 11,12). No sub P-IR fibers were until now seen in the pancreatic endocrine portion of *S. cabrilla, D. annularis* and of other teleosts [26,29], and only few sub P-IR fibers seem to be present in the endocrine pancreas of mammals [32].

Sub P enhances the blood flow but does not seem to have any influence on the release of mammalian pancreatic hormones. No data are available about the role of this peptide on fish pancreatic function. Nevertheless the finding of sub P-IR nerve fibers surrounding the endocrine cells of the Antarctic fish pancreatic islet suggests that this peptide might play some role in the release of pancreatic hormones.

Conclusions

A general conclusion can be drawn from the data reviewed here: a great variability is present in the NES of teleosts, and, furthermore, a peculiar pattern of peptide immunoreactivity is present in the digestive system of Antarctic notothenioid teleosts. This probably means that peptides different from those usually found in temperate teleosts are involved in the regulation of intestinal motility, blood flow and pancreatic hormone release.

The channichthyids, which are the most phyletically derived among notothenioids, are those presenting the great differences concerning in particular the innervation of the blood vessel wall, the muscular layers, and the endocrine islets.

The enteric nervous system seems to be less developed than that of nonAntarctic teleosts. VIP- and PACAP-IR nerve fibers were less frequently seen and they are reduced in the Brockmann body where numerous sub P-IR nerve fibers were instead found. On the contrary a greater frequency of CCK/gastrin-, serotonin- and somatostatin-IR cells and nerve elements was found throughout the intestinal wall. The finding of insulin immunoreactivity in the gut endocrine cells and nerve elements of *C. antarcticus* deserves further investigation.

At present it is difficult to explain the meaning of these results; nevertheless, we should remember that the NES may play an important role in the control of digestive functions, and the NES itself is under the control of many factors. One of these could be the environmental conditions, such as the low sea water temperature. This can affect the blood flow and the digestive functions, and these problems can be overcome utilizing a different pattern of regulatory peptides for the control of intestinal wall motility or the calibration of blood vessels. Another contributing factor could be the numerous parasites always present in the gut of Antarctic fishes [33].

Parasites enhance the immune response and produce a chronic inflammatory status. The NES and the immune system can cross talk using different kinds of messengers, such as citokines, hormones, and neuropeptides. The neuropeptides more involved in the gut immune regulation, sub P, VIP, CCK and somatostatin [15], are among those which present a different distribution. At present no data about NES modifications by intestinal parasites are available in nonAntarctic teleosts to support this hypothesis.

Acknowledgments

We acknowledge the financial support from the Italian National Programme for Antarctic Research (PNRA).

References

1. Fujita T, Kobayashi S (1977) Structure and function of gut endocrine cells. Int Rev Cytol Suppl 6:187-233
2. Vigna SR (1986) Gastrointestinal tract. In: Pang PKT, Schreibman MP (eds) Vertebrate endocrinology: fundamentals and biomedical implications. Academic Press, Orlando, pp 261-278
3. Fujita T, Kanno T, Kobayashi S (1988) The paraneuron. Springer-Verlag, Tokyo
4. Langer M, Van Noorden S, Polak JM, Pearse AGE (1979) Peptide hormone-like immunoreactivity in the gastrointestinal tract and endocrine pancreas of eleven teleost species. Cell Tissue Res 199:493-508
5. Bjenning C, Holmgren S (1988) Neuropeptides in the fish gut. An immunohistochemical study of evolutionary patterns. Histochemistry 88:155-163
6. Groff KE, Youson JH (1997) An immunohistochemical study of the endocrine cells within the pancreas, intestine, and stomach of the gar (*Lepisosteus osseus* L.). Gen Comp Endocrinol 106:1-16
7. Plisetskaya EM (1989) Pancreatic peptides. In: Holmgren S (ed) The comparative physiology of regulatory peptides. Chapman and Hall, London, New York, pp 174-202
8. Conlon JM (1989) Biosynthesis of regulatory peptides: evolutionary aspects. In: Holmgren S (ed) The comparative physiology of regulatory peptides. Chapman and Hall, London, New York, pp 344-369
9. Uesaka T, Yano K, Yamasaki M, Ando M (1995) Somatostatin-vasoactive intestinal peptide- and granulin-like peptides isolated from intestinal extracts of goldfish, *Carassius auratus*. Gen Comp Endocrinol 99:298-306

10. Wang Y, Conlon JM (1995) Purification and structural characterization of vasoactive intestinal polypeptide from the trout and bowfin. Gen Comp Endocrinol 98:94-101
11. Larsson LI, Hougaard DM (1993) Sensitive detection of rat gastrin mRNA by in situ hybridization with chemically biotinylated oligodeoxynucleotides: validation, quantitation, and double-staining studies. J Histochem Cytochem 41:157-163
12. Watson SJ, Sherman TG, Kelsey JE, Burke S, Akil H (1987) Anatomical localization of mRNA: in situ hybridization of neuropeptide systems. In: Valentino KL, Eberwine JH, Barchas JD (eds) In situ hybridization. Application to neurobiology. Oxford University Press, New York, pp 126-145
13. Holmgren S (1989) Gut motility. In: Holmgren S (ed) The comparative physiology of regulatory peptides. Chapman and Hall, London, New York, pp 231-255
14. Jonsson AC, Holmgren S (1989) Gut secretion. In: Holmgren S (ed) The comparative physiology of regulatory peptides, Chapman and Hall, New York, pp 256-2781
15. Shanahan F, Anton PA (1994) Role of Peptides in the regulation of the mucosal immune and inflammatory response. In: Walsh JH, Dockray GJ (eds) Gut peptides: biochemistry and physiology. Raven Press, New York, pp 851-867
16. Hill AM, McCune SK, Alvero RJ, Glazner GW, Brenneman DE (1996) VIP regulation of embryonic growth. In: Arimura A, Said SI (eds) VIP, PACAP and related peptides. Ann NY Acad Sci 805, NY Acad Sci New York, pp 259-268
17. Tang H, Sun L, Xin Z, Ganea D, (1996) Down regulation of cytokine expression in murine lymphocytes by PACAP and VIP. In: Arimura A, Said SI (eds.) VIP, PACAP and related peptides. Ann NY Acad Sci 805, NY Acad Sci New York, pp 768-778
18. Macdonald JA, Montgomery JC, Wells RMG (1988) The physiology of Mcmurdo Sound fishes: current New Zealand research. Comp Biochem Physiol 90 B:567-578
19. Eastman JT (1993) Antarctic fish biology. Evolution in a unique environment. Academic Press, San Diego
20. Korovina VM, Prirodina VP (1986) Anatomical-histological features of the intestine of the sculpin spinecheek *Cottoperca gobio* (Bovichthyidae, Notothenioidei). J Ichthyol 26:130-136
21. Korovina VM, Neyelov AV, Bondarenko YP, (1991) Intestinal anatomy and histology of the marbled notothenia *Notothenia rossi marmorata.* J Ichthyol 3:79-90
22. Eastman JT, DeVries AL (1997) Morphology of the digestive system of Antartic nototheniid fishes. Polar Biol 17:1-13
23. Tagliafierro G, Faraldi G, Delù M, Morescalchi MA (1995) Gut regulatory peptides in some Antarctic notothenioids. Polar Biol 15:429-435

24. Tagliafierro G, Carlini M, Delù M, Faraldi G. (1996) Organizzazione del pancreas endocrino in alcuni pesci antartici. Abstracts 57° Congresso Unione Zoologica Italiana, S. Benedetto del Tronto, September 1996
25. Tagliafierro G, Faraldi G, Morescalchi A (1996) GEP neuroendocrine system in fishes: distribution and ontogeny. Regul Peptides 64:186
26. Tagliafierro G, Carlini M, Faraldi G, Gallus L (1997) The neuroendocrine system in the gut and pancreas of Antarctic fish. Abstracts, Final Conference Network "Fishes of the Antarctic Ocean" Pontignano, Siena, Italy
27. Nilsson S, Holmgren S (1993) Autonomic nerve functions. In: Evans DH (ed) The physiology of fishes. CRC Press Inc, Boca Raton, London, pp 279-313
28. Jonsson AC, Holmgren S, Holstein B (1987) Gastrin/CCK-like immunoreactivity in endocrine cells and nerves in the gastrointestinal tract of the cod, *Gadus morhua*, and the effect of peptides of the gastrin/CCK family on cod gastrointestinal smooth muscle. Gen Comp Endocrinol 66:190-202
29. Jonsson AC (1991) Regulatory peptides in the pancreas of two species of elasmobranchs and in the Brockmann bodies of four teleost species. Cell Tissue Res 266:163-172
30. Katsoulis S, Schmidt WE (1996) Role of PACAP in the regulation of gastrointestinal motility In: Arimura A, Said SI (eds) VIP, PACAP and related peptides. Ann NY Acad Sci 805, NY Acad Sci New York, pp 364-378
31. Dockray GJ (1994) Vasoactive intestinal polypeptide and related peptides. In: Walsh JH, Dockray GJ (eds) Gut peptides. Raven Press, New York, pp 447-472
32. Larsson LI (1979) Innervation of the pancreas by substance P, enkephalin, vasoactive intestinal polypeptide and gastrin/CCK immunoreactive nerves. J Histochem Cytochem 27: 1283-1284
33. Zdzitowiecki K (1997) Diversity of Digenea, parasites of fishes in various areas of the Antarctic. Abstracts, Final Conference Network "Fishes of the Antarctic Ocean" Pontignano, Siena, Italy

Part 3

Evolution and Phylogeny

Four Years in Notothenioid Systematics: A Molecular Perspective

Luca Bargelloni[1] and Guillaume Lecointre[2]

[1]Dipartimento di Biologia, Università di Padova, via U. Bassi, 58/B 35131 Padova, Italy
[2]Service de systématique moleculaire et Laboratoire d'Ichtyologie, Muséum National d'Histoire Naturelle, 43 rue Cuvier, 75231 Paris Cedex 05, France

Why Notothenioids?

The perciform suborder Notothenioidei is a highly diversified group of fish inhabiting the waters around the Antarctic continent. This is a unique environment, characterized, especially in the high Antarctic Zone, by temperatures as low as -2 °C, presence of sea ice, and large seasonal fluctuations of the primary production. At the Oligocene-Miocene boundary, about 25-22 million years ago, with the opening of the Drake Passage and the formation of a circumpolar hydrographic barrier, the Polar Front Zone, the Antarctic marine environment became effectively isolated.

This isolation greatly limited migration, therefore forcing the shallow-water fish fauna either to adapt to the harsh environmental conditions or to become extinct. Notothenioids were able to survive and successfully occupied a large variety of ecological niches. In terms of numbers of species and biomass, they represent the dominant element of the Antarctic fish fauna [1]. Notothenioids also show a remarkable degree of endemism (97% of the species), and are distributed nearly exclusively in the Antarctic marine waters. Compared to all the other groups of fishes in Antarctica, notothenioids evolved several, remarkable adaptations to the Antarctic conditions. Among the physiological and molecular adaptations displayed by notothenioid fish, the most striking one is certainly the presence of the antifreeze glycopeptides (AFGPs) which are responsible for the depression of the freezing point [2].

Also some relevant physiological features, such as blood viscosity, oxygen transport and cardiovascular system, significantly differ in these fishes compared with temperate species [3]. In some Antarctic species hemoglobins have reduced affinity for oxygen; the extreme case is represented by the icefish (family Channichthyidae) that lacks

G. di Prisco, E. Pisano, A. Clarke (Eds)
Fishes of Antarctica. A biological overview
© Springer-Verlag Italia 1998

hemoglobins [4]. Among morphological modifications, certainly remarkable is the evolution of neutral buoyancy, in the absence of a swim bladder, as shown by some notothenioid species, which can be likely considered an adaptation to the pelagic niche [5].

All these peculiar features as well as their ecological success make the study of notothenioid evolution extremely interesting.

In recent years, the necessity of using phylogenetic trees in comparing physiological, behavioral and ecological traits among different taxa has become evident [6]. Hence, a solid knowledge of notothenioid systematics is important not only for taxonomic purposes but also as an essential tool for any evolutionary study on this group of organisms.

Until recently, reconstruction of notothenioid relationships has been based on morphological characters, with "conservative" results. Iwami in 1985 [7] and Hastings in 1993 [8] produced notothenioid cladograms where each notothenioid family appeared as monophyletic (except bathydraconids which were paraphyletic in Hasting's cladogram).

Since 1993 our two groups have undertaken a long-term project aimed at reconstructing the phylogenetic relationships among and within notothenioid families based on DNA sequence data. These studies have been focused on two mitochondrial genes, respectively encoding for the large (16S) and the small (12S) subunit of ribosomal RNA [9], as well as on the nuclear gene for the 28S subunit of ribosomal RNA [10]. Sequence information from these ribosomal genes, in fact, has proved to be very useful in the reconstruction of the phylogenetic relationships in various taxonomic groups.

In the present paper we summarize previous data on notothenioid systematics, in the light of new results stemming from the analysis of a larger number of taxa.

Phylogenetic Methods

Using a methodological approach already described [9], 33 notothenioid species were examined as listed in Table 1. The zoarcids *Lycodichthys dearborni* and *Pachycara brachycephalum* were included in the study as outgroup. Phylogenetic analyses were performed by maximum parsimony (MP), using the PAUP 3.1.1 computer program [11].

The reliability of the heuristic searches was improved by using the option "random addition of taxa" with 100 replications in PAUP. Several different character weighting schemes were used [9].

Statistical confidence of MP trees was assessed using bootstrapping [12] with 200 replications each. The obtained topology was also compared to previously proposed phylogenetic hypotheses using the Kishino and Hasegawa test and Wilcoxon signed-rank test.

Table 1. List of the species examined[a]

Nototheniidae	Harpagiferidae
Nototheniinae	*Harpagifer antarcticus*
Notothenia coriiceps	
Notothenia rossii	Artedidraconidae
Indonotothenia cyanobrancha	*Histiodraco velifer*
Gobionotothen gibberifrons	*Pogonophryne scotti*
Gobionotothen acuta	
Lepidonotothen squamifrons	Bathydraconidae
Patagonotothen tessellata	*Parachaenichthys charcoti*
Lepidonotothen nudifrons	*Cygnodraco mawsoni*
	Gymnodraco acuticeps
Trematominae	
Trematomus newnesi	Channichthyidae
Trematomus pennellii	*Chionodraco hamatus*
Trematomus scotti	*Cryodraco antarcticus*
Trematomus bernacchii	*Chaenocephalus aceratus*
Trematomus nicolai	*Pagetopsis macropterus*
Trematomus eulepidotus	*Champsocephalus gunnari*
Trematomus hansoni	
Pagothenia borchgrevinki	Bovichtidae
	Bovichtus variegatus
Pleuragramminae	
Aethotaxis mitopterix	
Pleuragramma antarcticum	
Eleginopinae	
Eleginops maclovinus	
Dissostichus mawsoni	
Dissostichus eleginoides	

[a]Species names are as in [1] pp 61-62, with a single modification (*Indonotothenia cyanobrancha* for *Notothenia cyanobrancha*)

The homogeneity of the rate was estimated using the relative rate test [13]. The number of transversional substitutions per site was estimated using Kimura's formulas [14]. Variances and covariances were determined according to Wu and Li [15]. Using the bovichtid *Bovichtus variegatus* as reference taxon, the homogeneity of the rate was statistically tested for all possible pairs of remaining notothenioids; levels of significance were calculated using a standardized normal test (t-test), with infinite degrees of freedom. Divergence times between taxa were calculated as the ratio of genetic distance (or mean genetic distance, when comparing clades with more than one taxon) and divergence rate. A more detailed description of methods will be presented elsewere (Bargelloni et al., unpublished).

Phylogenetic Evidence from Molecular Data: Monophyly, Paraphyly and Artefacts

MP phylogenetic analyses of the mitochondrial gene sequences yielded a topology (Fig. 1) in overall agreement with the phylogenetic scheme derived from cladistic analysis of morphological characters [1].

B. variegatus appears to be the sister group to the other families of notothenioids being highly divergent from them. This species, however,

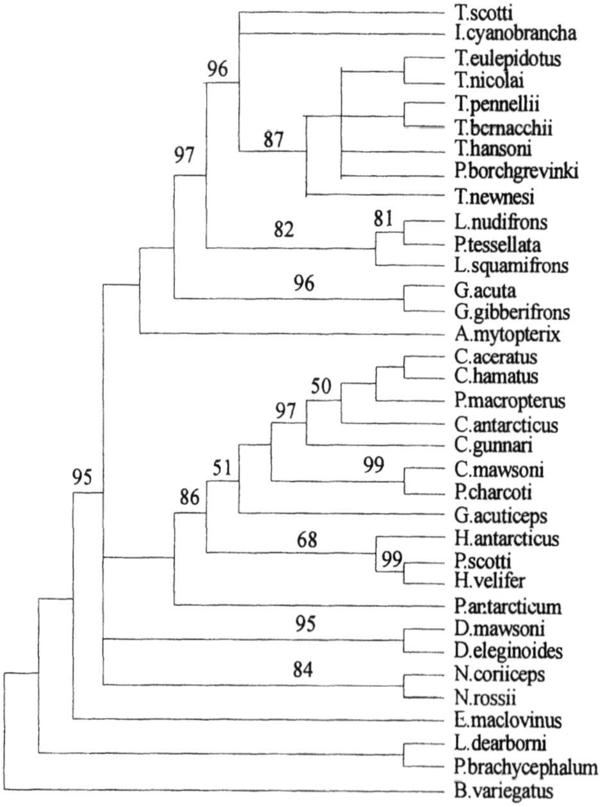

Fig. 1. Strict consensus tree of 16 most parsimonious trees found with maximum parsimony (transversions:transitions weight 2:1). Numbers above branches refer to bootstrap values after 200 replicates. Only values greater than 50% are shown

cannot be considered to be representive of the family Bovichtidae. We know, in fact, from both nuclear [10] and mitochondrial [16] ribosomal DNA data that bovichtids are not a natural group. In both cases it has been shown with good statistical support that bovichtids are paraphyletic, with *Pseudaphritis* closer to the rest of the notothenioids than are *Cottoperca* and *Bovichtus*. The two latter genera are so close to perciform outgroups that it is difficult to recover monophyletic notothenioids (Fig. 2).

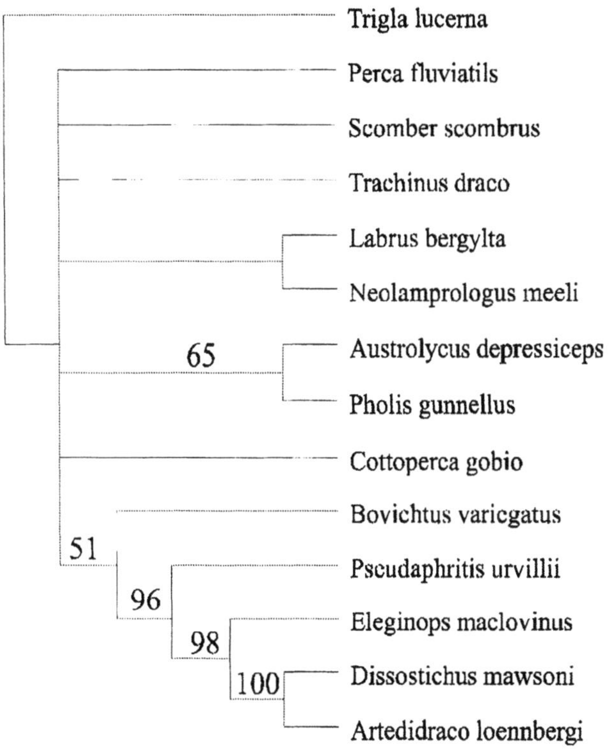

Fig. 2. Phylogenetic relationships of the three bovichtid genera as resulted from the analysis of nuclear ribosomal DNA (28S) dataset [10]. This tree is a strict consensus tree of six equiparsimonious trees. Numbers above branches are bootstrap proportions obtained after 1000 replicates analyzed by the parsimony method as implemented in PAUP [11]

With regard to the remaining notothenioid families, all our analyses suggest that *Eleginops maclovinus* should not be included in the family Nototheniidae (Fig. 1), since it branches out before the divergence of all the other species (excluding bovichtids). It has been recently proposed [17] based on morphological data that this species should form a monospecific family (Eleginopsidae), and should be considered the sister group to all the other families (again excluding bovichtids). In our opinion, the concordance of molecular and morphological analyses on the new phylogenetic position of *E. maclovinus* is extremely interesting for the study of notothenioid evolution. In fact, this species might be regarded as a much less distant outgroup (compared to bovichtids) and could be utilized as such in future evolutionary studies.

The evidence for a revised systematic position of *E. maclovinus* brings up the problem of the taxonomic status of the family Nototheniidae. This is the most speciose notothenioid family, with more than 49 species. So far, nototheniids have been assumed to be monophyletic and divided into four subfamilies (see Table 1). In order to test the monophyly of the entire family as well as the intrafamily relationships, we conducted a comparison between the most parsimonious phylogenetic tree (Fig. 1) and the tree(s) which can be obtained forcing the topology to follow the phylogenetic relationships within nototheniids which have been previously proposed. As shown in Table 2, neither the family Nototheniidae nor two nototheniid subfamilies (Nototheniinae and Pleuragramminae) can be considered monophyletic groups.

Table 2. Statistical comparisons of different tree topologies

Topology	Tree length	Kishino-Hasegawa test[a]	Wilcoxon test[a]
Most parsimonious tree (Fig. 1)	828	best	best
Nototheniidae monophyletic	843	$p<0.005$	$p<0.005$
Pleuragramminae monophyletic	838	$p<0.05$	$p<0.05$
Nototheniinae monophyletic	849	$P<0.005$	$p<0.005$
Nototheniidae cladogram (Fig. 3)	808	$p<0.0001$	$p<0.0001$

[a]*p* values refer to the approximate probability of getting a more extreme test statistic under the null hypothesis of no difference between the two trees (two-tailed test)

The tree with all nototheniids in a single clade is significantly less supported ($p<0.01$) than the MP tree; a similar result is observed when Nototheniinae or Pleuragramminae are constrained to monophyly. An even less robust tree (Table 2) is obtained when the tree topology is forced to reflect the "unsupported" cladogram (Fig. 3) describing the relationships among subfamilies [1].

Taken together, the above analyses suggest that the previous hypotheses on the phylogenetic relationships within the Nototheniidae family as well as the monophyly of the family itself are questionable, and several taxonomic groups are in need of revision. This is not surprising since nototheniid systematic relationships based on morphology were considered unsatisfactory [1] due to the lack of clearly defined synapomorphies.

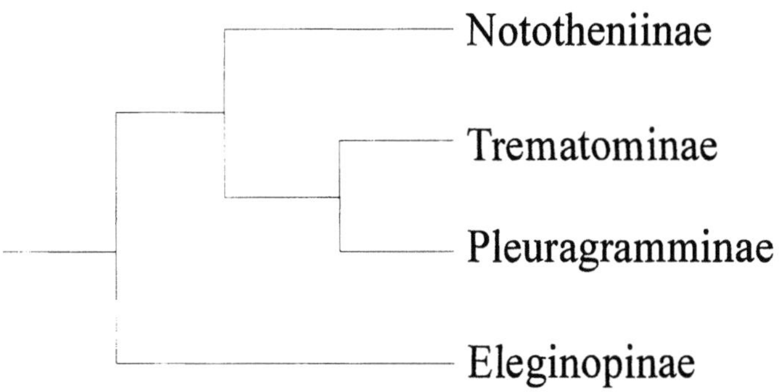

Fig. 3. Cladogram showing the phylogenetic relationships among nototheniid subfamilies as traditionally proposed. Note that based on molecular evidence at least three of these subfamilies (Eleginopinae, Nototheniinae, and Pleuragramminae) are not natural groups

As for the other notothenioid families (Artedidraconidae, Harpagiferidae, Bathydraconidae, Channichthyidae), the species we analyzed are all clustered in a single clade, with good bootstrap support (Fig. 1). Within this clade, we observe a strong indication of monophyly for Artedidraconidae (represented by *Histiodraco velifer* and *Pogonophryne scotti*) and Channichthyidae (with five species). Within the latter family, *Champsocephalus gunnari* is the sister taxon to the other species as previously hypothesized [7] and found in the complete phylogeny of this family based on molecular data (Chen et al., this Vol.). Other results, though not in contrast with morphological evidence, receive less strong support.

Harpagiferidae, whose position was unresolved based on morphology [1], appear to be the sister taxon to Artedidraconidae. The position of these two families as sister group to Bathydraconidae and Channichthyidae was expected, but the bootstrap analysis does not rule out other possibilities.

Also unresolved is the taxonomical status of the family Bathydraconidae. Molecular analyses seem to indicate paraphyly as already suggested by Hastings [8], but when this hypothesis is tested against the traditional one (bathydraconids monophyletic), the latter is not significantly less supported.

The paraphylies detected by molecular studies are to be evaluated in the light of the widely spread long-branch attraction artefact [18,19]. The notothenioid tree has a general shape that can be suspected to be due to such an artefact: an asymmetric part (bovichtids and nototheniids) and a symmetric part, with crown groups (artedidraconids, channichthyids). In molecular systematics, outgroups often have longer branches than members of the ingroup. When one or several member(s) of the ingroup has a higher rate of molecular evolution, the lineages implied are successively and artefactually branched at the base of the tree, producing a "Hennig's comb shape" of the branching pattern, obscuring possible monophylies for some groups. Longer branches are not directly detectable from the first glance at the tree because the taxa implied are basal. This is the reason why some authors [20,21] consider the asymetric part of a molecular phylogenetic tree as not reliable, even when the basal position of a lineage is supported with high bootstrap proportions. Indeed, long branch attraction artefact is a pitfall that bootstrapping cannot reveal [22,23]. In the case of the notothenioids, the question is far from being meaningless: the two most basal families, i.e. the bovichtids and the nototheniids, all appear as paraphyletic. Are these paraphylies artefactual? There are two means to refute the possible nonreliability of basal paraphylies. First, the use of several genes to confirm the same basal paraphylies is the most powerful tool to detect such an artefact, because two genes different in nature (for instance one mitochondrial and one nuclear) have poor chance to concomitantly accelerate their mutational rate in the same lineages. The second mean is the intrinsic quality of each data set. In absence of mutational saturation, in absence of detectable acceleration as shown with a relative rate test, and in absence of excessive amounts of homoplasy [high retention index (RI)], there is little reason to suspect such a long-branch attraction artefact.

The long-branch attraction artefact as initially described by Felsenstein [18] needs some random noise in the data to occur. For a few taxa (for instance in four or five-species trees), some homoplasy will very easily provoke the artefact, i.e. grouping the species with the closest mutational rates together with high bootstrap proportions [22,23]. The artefact will be less likely to happen when the number of taxa increases, because in principle increasing the number of taxa breaks long branches. In other words, when the homoplasy content of a data set is low and when the number of species is high (usually more than 15), the artefact will be unlikely.

These elements led us to conclude that the paraphylies of most of the notothenioid families revealed by the recent molecular studies are not artefactual. First of all, both the paraphyly of the Bovichtidae (with *Pseudaphritis* as the sister group of the rest of the nonbovichtid notothenioids) and the paraphyly of the Nototheniidae are supported both by the nuclear 28S rDNA [10] and the mitochondrial 12S-16S rDNA

[9,16] with high bootstrap proportions. In each of these data sets, no mutational saturation was detectable among members of the ingroup, and no significant deviation of base composition shared by some taxa and the outgroup could be suspected to provoke a false grouping of these taxa closer to the outgroups. Last but not least, retention indices were rather good (0.73 in [16]; 0.704 within the ingroup for Lecointre et al. [10]), showing that homoplasy was not predominant in these diverse data.

The mitochondrial and nuclear sequence data for notothenioids therefore reliably demonstrate the paraphyly of bovichtids and the paraphyly of nototheniids. These basal paraphylies are not artefactual and one can notice that, in the mtDNA tree, a symmetric topology appears relatively early in the topology because of the numerous members of the genus *Trematomus* clustered with *Pagothenia*. Once these paraphylies are technically demonstrated, the epistemological causes can be raised. Most authors will readily admit that bovichtids and nototheniids are paraphyletic: these families were morphologically very poorly defined or were defined on the basis of symplesiomorphies (i.e. shared primitive characters [1,8,10]). This feature is very common in the history of systematics. Whatever the characters phylogenies are based on, a cladistic procedure will more often demonstrate the paraphyly of a group previously considered as monophyletic (in the sense of Mayr [24]) or merely justified as useful (prokaryotes, bryophytes, pterydophytes, invertebrates, agnathans, fishes, reptiles, prosimians, etc.) rather than establishing the monophyly of previously unclassified or dispersed lineages (Otocephala, Archonta). This is only due to the fact that, between Darwin and Hennig, we have spent a century to define groups on diagnoses made of a mixture of shared primitive characters (characters too general for the scope of the study, i.e. already present in outgroups) and shared derived characters or, more trivially, diagnoses based on global similarity from which finalism and anthropocentrism were not excluded [25]. Defining a group on the basis of shared primitive characters is surely making a grade, i.e. a paraphyletic unit [26]. A technical consequence of this is that, since a method, cladistics, proposes that the only way to create strictly monophyletic groups (holophyletic groups of Mayr) is to define them exclusively on the basis of common derived characters (i.e. synapomorphies), the destiny of a huge number of taxa is to explode into a Hennig's comb. It was the case, for instance in actinopterygian phylogeny, for paleoniscides, holosteans, protacanthopterygians, malacopterygians, etc. This is also true at lower taxonomic levels and also true when revealed by molecular studies: only two out of six notothenioid families remained monophyletic; and only one out of seven to eight cyprinid subfamilies remained monophyletic [27].

In summary, molecular evidence from mitochondrial and nuclear ribosomal genes gives indications that roughly parallel the phylogenetic picture based on the analysis of morphological characters, if we take into account those morphologists that have recognized the paraphyly of some families (i.e. Balushkin [17] for *Eleginops*). Nevertheless, one exception exists if we add to this data set the three bovichtid genera. The highly supported relationships of *Pseudaphritis* found from two different genes [10,16] strongly contradict both the poorly argumented monophyly of bovichtids [8] and the paraphyly of this family as presented by Balushkin [17].

Phylogenies, Speciation Patterns and Molecular Clocks

As already mentioned, the importance of phylogenies in comparative and evolutionary studies cannot be overlooked. For instance, the distribution of present-day species can be superimposed on well-established phylogenetic trees to infer patterns of speciation [28]. Since Darwin, the formation of species has represented one of the most important, yet also one of the most elusive subjects in evolutionary biology, and a large number of studies have been aimed at understanding speciation. As a result of these efforts, a basic model of speciation arose, which is now named "allopatric speciation." The "allopatric" scenario is rather simple: a large, continuous population is divided in smaller subunits by the presence of extrinsic barriers, which reduce the genetic exchange among separated subpopulations. This favors the creation of genetic differences, ultimately leading to reproductive isolation. In marine systems, however, barriers to gene flow are far less common if compared to terrestrial or freshwater environments. In addition, marine organisms often show a high dispersal potential, further reducing the opportunity for genetic divergence among populations. It is not surprising, therefore, that the Polar Front Zone (PFZ) has attracted the attention of several authors, as a unique hydrographic barrier which isolates the Southern Ocean, playing a pivotal role in the evolution of the Antarctic marine fauna. Vicariance (i.e. "passive" allopatric speciation), due to the formation of the PFZ, has in fact been invoked as the most important mode of species formation in several Antarctic taxa. Evidence from molecular data casts some doubt on this scenario. The vicariant hypothesis stems from the observation that the area cladogram of Nototheniidae shows that the more phyletically derived taxa (Trematominae and Pleuragramminae) have a High Antarctic distribution, while less derived groups (Eleginopinae and Nototheniinae) present a relevant number of subAntarctic or lesser-Antarctic species. However, the cladogram itself seems to be seriously flawed, if molecular results are taken into account. As shown before, neither the placement of Trematominae and Pleuragramminae together nor the existence of the clade Pleuragramminae are supported. Similarly the taxonomic status of the subfamily Nototheniinae is uncertain. At best, the vicariant scenario should be reevaluated in light of the new phylogenetic relationships among nototheniids.

Phylogenetic trees based on sequence data can also give informations on the tempo and mode of evolution of the species examined. For instance, the low level of resolution at the interfamilial level seems not to be due to the use of inappropriate molecular tools, since a reliable phylogenetic signal is recovered for deep branching events (divergence of *B. variegatus* and *E. maclovinus*) as well as for more recent ones. This might be, instead, the phylogenetic mark of a radiation that occurred early on in the evolution of notothenioids.

Moreover, genetic distances between species can be used to measure the time since their divergence; this is possible when the rate of genetic divergence is linear with time and homogeneous between taxa (molecular clock hypothesis). If the clock is valid, molecular data provide a valuable tool in timing evolutionary events. However, even accepting the clock hypothesis, a problem still remains: calibrating the clock with absolute timing. This calibration can be accomplished when geological records for cladogenetic events are available; dates are usually provided by good fossil records of the taxa under investigation and/or by specific, well-established geological events associated with separation of taxa [29].

With regard to our data, evolutionary distances were calculated on all possible changes (since no mutational saturation is present either for transitions or for transversions [16]). Divergence estimates were also based on transversions (TV) only, since transversions seem to evolve linearly with time. A few differences in the rate were detected by the relative rate test. However, rates were considered sufficiently homogeneous to be used in estimating time of divergence. Two different estimates of divergence rates were applied to our data in order to calculate time of divergence between notothenioid species. A rate of 0.3-0.7% substitutions per million years (Myr) has been proposed by Rand [30] for mtDNA of ectotherms, while a rate of 0.14% TV/Myr for ribosomal mitochondrial genes (12S and 16S) was calculated in mountain newts [29] and bovids [31]. When the separation of *E. maclovinus* and the rest of nonbovichtid notothenioids is considered, our estimate of the divergence time ranges between 23 and 22 Myr, which is roughly the time of the formation of the Antarctic Convergence, suggesting that the Polar Front possibly played a role as a vicariant factor, separating the lineage that led to *E. maclovinus* from the ancestor to the rest of the notothenioid taxa.

From this common ancestor, according to our estimate, the major radiation of notothenioids probably occurred 16-10 Ma, when a consistent sea ice formation, drop in water temperatures and expansion of ice sheet played an important role in modifying the Antarctic climate [1].

These major changes in the Antarctic environment might have created empty niches to which very few nonnotothenioid fish were able to adapt. If antifreeze glycopeptides (AFGPs) evolved only once in notothenioid evolution, after the separation of *B. variegatus* and *E. maclovinus* and before the diversification of all the other notothenioids, as proposed by Bargelloni et al. [9], AFGPs might have played a key role, among other features, in the adaptive radiation of notothenioid fishes. Recently, this hypothesis has found direct confirmation, with the evidence of a recent, common evolution of the ancestral AFGP gene from a trypsinogen-encoding gene [32].

Conclusive Remarks

Molecular studies seem to corroborate the hypothesis that notothenioids underwent an adaptive radiation, possibly quite recently (after the formation of the PFZ). The ecological plasticity and the ability to survive the freezing conditions through AFGPs might well have been the key factors which allowed the remarkable diversification of the notothenioid fish.

Regarding the notothenioids from the taxonomic point of view, it should be noticed that molecular phylogenies helped to define new morphological synapomorphies supporting some nodes of the notothenioid tree. According to Patterson et al. [33], such interactions are not so common. Most of molecular phylogenies limit their conclusions in a statement of disagreement, but will rarely reconsider morphological characters. The clade rendering bovichtids paraphyletic (*Pseudaphritis* plus the rest of nonbovichtid notothenioids) is highly supported by two molecular studies based on different genes extracted from different structures of the cell. Such strong evidence eliminates hesitations in reconsidering morphological characters. Lecointre et al. [10] proposed that this node could be supported by six synapomorphies: (1) the second (the uppermost free) pectoral radial in connection both with the scapula and the coracoid (the first being fused with the coracoid), (2) last second dorsal fin ray divided to the base, (3) the ventral process of the hyomandibula shortened, (4) the loss of the connection between the ectopterygoid and the metapterygoid, (5) appearance in the opercular bones of connective tissues, (6) major Hb1 component in the hemoglobin, loss of all hamoglobins being a more derived state. In the future, if *Bovichtus* and

Cottoperca can be shown to be the sister group of one of the perciform outgroups with a significant support, then these apomorphies could provide a new definition of the Notothenioidei. For the moment, although the monophyly of the notothenioids (i.e. including *Cottoperca* and *Bovichtus*) is difficult to recover from molecular studies, the definition of the Notothenioidei in its classical sense seems to lie on a single synapomorphy, the presence of three platelike pectoral radials [the first (upper) one being fused with the coracoid], admitting a convergence in the Callionymidae.

Future Studies

In spite of important recent progress, the work on notothenioid systematics is not finished. Further phylogenetic research is needed to get a more precise picture of the phylogeny within some families. Although channichthyids show very little genetic divergence, a new molecular phylogeny of channichthyid genera would be extremely useful (Chen et al., this Vol.). Interrelationships of harpagiferids, artedidraconids and the paraphyletic bathydraconids are not yet based on robust nodes. Much work remains to be done also within nototheniids. It would also be very interesting to test in a cladistic framework Voskoboynikova's [34] hypothesis on the interrelationships of the nototheniid genera based on the structure of the visceral skeleton. This hypothesis assumed the monophyly of the nototheniids; it would be fruitful to reconsider Voskoboynikova's topology in the light of the paraphyly of the family. Finally, another topic for further research is the identification of the sister group of the notothenioids, a very difficult task because of the absence of definition for the perciforms.

Acknowledgments

We would like to thank all the members of the Coordination Committee of the "Fishes of the Antarctic Ocean" Network and the European Science Foundation itself for providing such an invaluable opportunity for meeting and discussing about biology and evolution of notothenioid fish. L.B. has received support from the Italian PNRA (Programma Nazionale di Ricerche in Antartide).

References

1. Eastman JT (1993) Antarctic fish biology. Academic Press, San Diego
2. Cheng CC, DeVries AL (1991) The role of antifreeze glycopeptides and peptides in the freezing avoidance of cold water fish. In: di Prisco G (ed) Life under extreme conditions: biochemical adaptation. Springer-Verlag, Berlin, pp 1-14

3. Wells RMG (1987) Respiration of Antarctic fish from McMurdo Sound. Comp Biochem Physiol 88A: 417-424
4. Ruud JT (1954) Vertebrates without erythrocytes and blood pigment. Nature 173: 848-850
5. Eastman JT (1985) The evolution of neutrally buoyant notothenioid fishes: their specialization and potential interactions in the Antarctic marine food web. In: Siegfried WR, Condy PR, Laws RM (eds) Antarctic nutrient cycles and food webs. Springer-Verlag, Berlin Heidelberg, pp 430-436
6. Harvey PH, Pagel MD (1991) The comparative method in evolutionary biology. Oxford University Press, Oxford
7. Iwami T (1985) Osteology and relationships of the family Channichthyidae. Memoirs of National Institute of Polar Research Ser E: 1-69
8. Hastings PA (1993) Relationships of the fishes of the perciform suborder Notothenioidei. In: RG Miller (ed) A history and atlas of the fishes of the Antarctic Ocean. Foresta Institute for Ocean and Mountain Studies, Carson City, pp 99-107
9. Bargelloni L, Ritchie PA, Patarnello T, Battaglia B, Lambert DM, Meyer A (1994) Molecular evolution at subzero temperature: mitochondrial and nuclear phylogenies of fishes from Antarctica (suborder Notothenioidei), and the evolution of antifreeze glycopeptides. Mol Biol Evol 11: 854-863
10. Lecointre G, Bonillo C, Ozouf-Costaz C, Hureau JC (1997) Molecular phylogeny of the Antarctic fishes: paraphyly of the Bovichtidae and no indication for the monophyly of the Notothenioidei (Teleostei). Polar Biology 18: 193-208
11. Swofford DL (1993) Phylogenetic analysis using parsimony (PAUP). Illinois Natural History Survey, Champaign
12. Felsenstein J (1985) Confidence limits on phylogenies: an approach using the bootstrap. Evolution 39: 783-791
13. Wilson AC, Carlson SS, White TJ (1977) Biochemical evolution. Ann Rev Biochem 46: 573-639
14. Kimura M (1980) A simple method for estimating evolutionary rate of base substitutions through comparative studies of nucleotide sequences. J Mol Evol 16: 111-120
15. Wu CI, Li WH (1985) Evidence for higher rates of nucleotide substitution in rodents than in men. Proc Natl Acad Sci USA 82: 1741-1745
16. Ritchie PA, Lavoue S, Lecointre G (1997) Molecular phylogenetics and the evolution of Antarctic notothenioid fishes. Comp Biochem Physiol 118A:1009-1026
17. Balushkin AV (1992) Classification, phylogenetic, and origins of the families of the suborder Notothenioidei (Perciformes). J Ichthyol 32: 90-110
18. Felsenstein J (1978) Cases in which parsimony or compatibility methods will be positively misleading. Syst Zool 27: 401-410
19. Kim J (1996) General inconsistency conditions for maximum parsimony: effects of branch lengths and increasing numbers of taxa. Syst Biol 45:363-374

20. Philippe H (1997) Communication at the international symposium “Molecules and morphology in systematics”, Paris, March 1997
21. Philippe H, Adoutte A (1997) The molecular phylogeny of Protozoa: solid facts and uncertainties. In: Coombs GH, Vickerman K, Sleigh MA, Warren A (eds) Evolutionary relationships among Protozoa Systematics Association
22. Lecointre G, Philippe H, Lê HLV, Le Guyader H (1993) Species sampling has a major impact on phylogenetic inference. Mol Phyl Evol 2: 205-224
23. Philippe H, Douzery E (1994) The pitfalls of molecular phylogeny based on four species, as illustrated by the Cetacea/Artiodactyla relationships. J Mam Evol 2: 133-152
24. Mayr E (1982) The growth of biological thought. Diversity, evolution and inheritance. The Bellknap Press of Harvard University Press, Harvard
25. Romer AS (1967) Major steps in vertebrate evolution. Science 158: 1629-1637
26. Dupuis C (1978) Permanence et actualité de la systématique: la “systématique phylogénétique” de W Hennig (historique, discussion, choix de références). Cahiers des Naturalistes Bull N P n s 34: 1-69
27. Gilles A, Lecointre G, Faure E, Chappaz R, Brun G (in press) Mitochondrial phylogeny of the European cyprinids: implications for their systemics, reticulate evolution and colonisation time. Mol Phyl Evol
28. Brooks DR, McLennan DA (1991) Phylogeny, ecology and behaviour: a research program in comparative biology. Chicago University Press, Chicago
29. Caccone A, Milinkovitch MC, Sbordoni V, Powell JR (1994) Molecular biogeography: using the Corsica-Sardinia microplate disjunction to calibrate mitochondrial rDNA evolutionary rates in mountain newts (*Euproctus*). J Evol Biol 7: 227-245
30. Rand DM (1994) Thermal habit, metabolic rate and the evolution of mitochondrial DNA. Trends Ecol Evol 9: 125-131
31. Allard MW, Miyamoto MM, Jarecki L, Kraus F, Tennant MR (1992) DNA systematics and evolution of the arctiodactyl family Bovidae. Proc Natl Acad Sci USA 89: 3972-3976
32. Chen L, DeVries AL, Cheng CC (1997) Evolution of antifreeze glycoprotein gene from a trypsinogen gene in Antarctic notothenioid fish. Proc Natl Acad Sci USA 94: 3811-3816
33. Patterson C, Williams DM, Humphries CJ (1993) Congruence between molecular and morphological phylogenies. Annu Rev Ecol Syst 24: 153-188
34. Voskoboynikova OS (1993) Evolution of the visceral skeleton and phylogeny of Nototheniidae. J Ichthyol 33: 23-47

Chromosome Diversification in Antarctic Fish (Notothenioidei)

Eva Pisano[1], Catherine Ozouf-Costaz[2], and Valentina Prirodina[3]

[1]Institute of Comparative Anatomy, University of Genova, Italy
[2]Museum of Natural History, Paris, France
[3]Zoological Institute, Russian Academy of Sciences, St. Petersburg, Russia

Introduction

During recent years cytogenetic studies on Antarctic fish have received increasing attention [1-4]. Representatives from all the families of the Notothenioidei suborder have been studied in High Antarctic, SubAntarctic and cold-PeriAntarctic regions (Fig. 1).

The diploid number and chromosomal morphology are known for 63 species of Notothenioidei. These data represent a relatively high percentage of the cytogenetic information available for perciform fish [5].

Patterns of chromosomal differentiation have been described and interpretations of the karyological data have been attempted. However, it is clear that for phylogenetic analyses classical data are limited [4,6,7]. More recently, the development of banding studies and the introduction of molecular cytogenetic methods promise to add a new dimension to analyses of a wide range of evolutionary problems [7,8].

In this report we summarize the broad patterns of chromosomal diversification of Notothenioidei. We focus on cytogenetic evidence in "bovichtids," on intraspecific polymorphism, and on phylogenetic inference of chromosome character analysis.

The monophyly of the Notothenioidei is still an unproven hypothesis that merits further investigations (Bargelloni and Lecointre, this Vol.). Therefore, the chromosomal data are discussed in relation to the classification of the suborder in the eight families of Bovichtidae, Pseudaphritidae, Eleginopsidae, Nototheniidae, Harpagiferidae, Artedidraconidae, Bathydraconidae, Channichthyidae [9].

Karyotypic Patterns

Among the suborder Notothenioidei the most frequent diploid number is 48. This diploid number is also found in 211 of 660 species of Perciformes so far karyotyped [5].

G. di Prisco, E. Pisano, A. Clarke (Eds)
Fishes of Antarctica. A biological overview
© Springer-Verlag Italia 1998

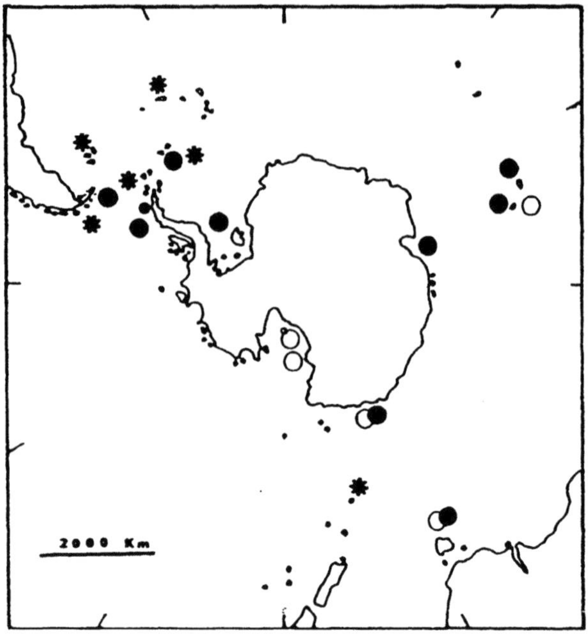

Fig. 1. Sampling areas for cytogenetic research by Russia (asterisks), France (black circles) and Italy (white circles). The small black circle refers to a single study by Brazil

A reasonable hypothesis is that the common ancestor of Notothenioidei also had a karyotype made of 48 all acrocentric chromosomes [1,2]. Then, various pathways of chromosomal differentiation were followed in the eight families (Fig. 2).

The types of chromosomal modifications, leading to the now existing karyotypes, are the same as those seen in the cytogenetic differentiation of other vertebrates, including fish. Most rearrangments are probably Robertsonian (centric fission or fusion) or pericentric inversions. Additions and amplifications of heterochromatin have also been hypothesized [1,6,10,11].

A set of chromosomes apparently identical to that of the presumed ancestor (Fig. 2,1) is found in the bovichtids *Cottoperca gobio* [3] and *Bovichtus angustifrons* (Ozouf-Costaz et al., unpublished).

Pericentric inversions in the ancestral karyotype could have given rise to the two small metacentric pairs found in the karyotype of the family *Pseudaphritidae* (Fig. 2,2) and in the karyologically very similar *Eleginops maclovinus* (Ozouf-Costaz et al., unpublished), representing the monotypic family *Eleginopsidae* (Fig. 2,3). Pericentric inversions could have been an initial step in karyological diversification among Notothenioidei, because two pairs of small meta-submetacentric chromosomes are present in most karyotypic lines.

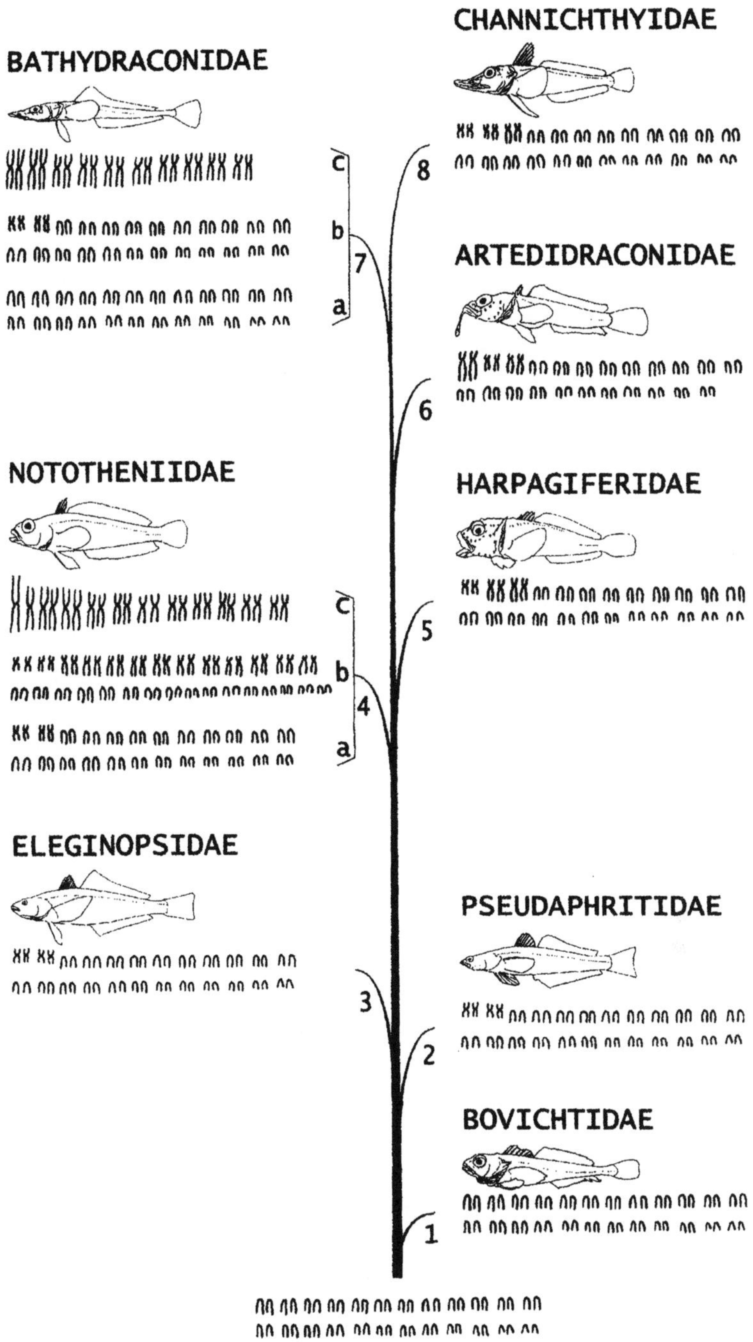

Fig. 2. Major patterns of karyotypic diversification in the notothenioid families from a presumed ancestral chromosome set

An alternative, less parsimonious, hypothesis is that pericentric inversions are a frequent mutation which became independently fixed in various phylogenetic lines. The fact that two apparently similar small meta-submetacentric elements are present in other Antarctic perciforms [4] gives some support to the second hypothesis. However, until chromosomal homologies can be established by molecular methods neither of these two hypotheses can be tested.

Nototheniidae and Bathydraconidae appear to be the most heterogeneous families (Fig. 2, Table 1). Three main pathways in chromosome differentiation were found in Nototheniidae. In one, the species have not diverged very much from the ancestral karyotype and has maintained the same diploid number (Fig. 2,4a). With the exception of the more derived *Pleuragramminae,* at least one species in each nototheniid subfamily conserves this pattern, including species of the genera *Dissostichus, Trematomus, Pagothenia, Lepidonotothen, Patagonotothen.* In the second lineage (Fig. 2,4b), additional rearrangements apparently gave rise to a higher chromosomal number, as in *Trematomus nicolai* [12]. The third group of species has highly rearranged karyotypes with a reduced number of large metacentric elements (Fig. 2,4c), as it is found in some *Trematomus* and typically in *Notothenia* and *Paranotothenia* species [2].

The karyotypic diversity found in Nototheniidae is not surprising considering that they are the most diverse family in the suborder in terms of species number, geographic distribution and in ecological characteristics [13,14].

The seven bathydraconid species studied so far have notably different karyotypes [2,10]. Here the main three karyotypic lines are illustrated.

One is represented only by the species *Psilodraco breviceps* and apparently has retained a primitive condition of 48 acrocentric chromosomes (Fig. 2,7a). However, further study and banding is needed to confirm if its karyotype is really plesiomorphic.

A slightly divergent set of chromosomes characterizes a second karyotypic line in dragon fish (Fig. 2,7b). The more derived line leads to a symmetrical karyotype made of all metacentric elements (Fig. 2,7c), as found in *Prionodraco evansii*.

The presence of intraspecific sex-linked heteromorphism is an additional feature of high karyotypic variability in both Nototheniidae and Bathydraconidae (Table 1).

The high karyotypic diversity found in both Nototheniidae and Bathydraconidae is consistent with various molecular and morphological results suggesting the paraphyletic nature of these two notothenioid families (Bargelloni and Lecointre, this Vol.).

In contrast, the family Channichthyidae has very homogeneous karyotypes. All the 13 studied species have 47-48 chromosomes which have apparently diverged little from the supposed ancestral condition (Fig.

2,8). The karyotypic stability on chromosome number of the icefish has already been discussed as well as the limited number of derived karyological features which contrasts with their great phenotypic differentiation and with their derived phyletic position among Notothenioidei [6,10].

The chromosomal characteristics of Channichthyidae could indicate that their morphophysiologically derived features are not correlated with and evolved at a much higher rate than the karyotype. However, this conclusion does not mean that these fish are not undergoing genomic changes. A minority of channichthyids has rearranged karyotypes with a higher number of submetacentric chromosomes. Both intraspecific variations in genome size and banding studies indicate minor differences concerning the amount and distribution of heterochromatin and the location of ribosomal genes [6,8]. Moreover, chromosomal rearrangements producing heteromorphic sex-linked chromosomes are operative in most of these species and they can also be considered a source of genomic reorganization.

The karyotypes of the eight artedidraconid fish species studied are also homogeneous and had not diverged very much from the primitive condition, while very little cytogenetic information is available on the Harpagiferidae [2,10] (Fig. 2,5,6).

Table 1. Karyotypic characters of the notothenioid families

	No. of species	2n	NF	XY	kd
Bovichtidae	2	48	48	-	0
Pseudaphritidae	1	48	52	-	+
Eleginopsidae	1	48	52	-	+
Nototheniidae	30	22-58	44-88	+	+++
Harpagiferidae	2	48	50-54	-	+
Artedidraconidae	8	46	50-54	-	+
Bathydraconidae	7	20-48	40-58	+	+++
Channichthyidae	13	47-48	52-62	+	+

2n: diploid number, NF: fundamental number; XY: sex-linked chromosomes; kd: karyotypic diversity estimate

The Position of "Bovichtids"

Most authors have included the taxa *Cottoperca*, *Bovichtus*, and *Pseudaphritis* in the family Bovichtidae, the temperate thornfish [13-16]. This family has been widely regarded as the most primitive among Notothenioidei and is the basal clade in phylogenetic reconstructions [14,17,18]. On the basis of osteological characters, Balushkin [9] separated the new family Pseudaphritidae (comprising only the monotypic

Pseudaphritis) leaving together in the family Bovichtidae the genera *Cottoperca* and *Bovichtus*. Phylogenetic analysis of various data sets has shown that the Bovichtidae are not monophyletic. The analysis of hemoglobin sequence places *Pseudaphritis* closer to non-bovichtid notothenioids [19] and analyses based on 28S rDNA and mitochondrial DNA sequencing clearly separate *Cottoperca-Bovichtus* from the more derived *Pseudaphritis* [20,21]. Cytogenetic evidence indicates different pathways of karyological differentiation in *Cottoperca gobio* and *Pseudaphritis urvillii*, a karyologically derived taxon [3]. Moreover, the two pairs of small metacentric elements in the karyotype of *Pseudaphritis* seem to indicate a common karyotypic pathway in this species and other notothenioid phyletic lineages (Fig. 2). Karyotypic features of *B. angustifrons* mark the taxonomic distance between *Pseudaphritis* and the taxa *Cottoperca* and *Bovichtus*. Although cytogenetic data cannot help to estimate divergence times, they strongly support the separation of the taxon *Pseudaphritidae*. Because *Cottoperca* and *Bovichtus* both retain a primitive karyological condition in chromosome number, morphology and banding patterns, no reliable cytotaxonomic hypotheses can be developed for these two taxa.

Intraspecific Polymorphism

As the amount of cytogenetic research on Antarctic fish increased and new methods have been applied, more intraspecific polymorphisms of both autosomes and sex-linked chromosomes have been found.

One of the most common source of intraspecific polymorphism seems to be linked to the amount of ribosomal genes per haploid set, giving rise to two homologous autosomes bearing different sized nucleolar organizing regions (NORs) [8]. In this respect, the Antarctic species do not differ very much from temperate and tropical fishes where intra- and interindividual variation of these DNA regions has been widely reported [22].

Several species of channichthyids and nototheniids and one bathydraconid species possess heteromorphic sex-linked chromosomes with male heterogamety (Table 1). In most cases a multiple Y and an odd (2n-1) number of chromosomes is present in the males. This feature seems to have been both independently and recently acquired by notothenioid taxa and it has been considered a convergent apomorphic character in karyological differentiation of Notothenioidei [1,6,12].

Although sex chromosomes are present in some non-Antarctic perciform species [5,23], the frequency of sex-linked heterochromosomes seems higher among Antarctic fish. For this reason it has sometimes been considered as an adaptation of sex-determining mechanisms to the permanently cold Antarctic waters, where thermal invariability prevents a temperature-dependent sex determination [24].

Intraspecific Polymorphism in *Trematomus hansoni*

The karyotype of the circumAntarctic nototheniid *Trematomus hansoni* was studied from the Antarctic Peninsula [25], the Weddell Sea [1], the Ross Sea [12] and Adélie Land [7]. Specimens from the Atlantic Sector have 48 chromosomes. The population of the Ross Sea has 45/46 chromosomes and heterogonosomes while the population from Adélie Land has 46 chromosomes, but no heterogonosomes. Moreover, the chromosomes of the Adélie Land sample have a very unusual pattern of Ag-NOR banding (Table 2).

The intraspecific variability at various levels of chromosomal organization in *T. hansoni* makes it possible to expect similar diversification in other species, especially those with a wide geographic distribution and could indicate an active process of genomic restructuring. Moreover, if we simply look at the chromosomes as an expression of cellular morphology, the observed chromosomal polymorphism could even be viewed as an aspect of the phenotypic plasticity of this species.

Table 2. Chromosomal characters in *Trematomus hansoni*

Sampling area	N	sex	2n	XY	NOR
Antactic Peninsula	6	?	48	?	?
Weddell Sea	1	F	48	?	?
Ross Sea	13	M/F	45/46	YES	2
Adélie Land	10	M/F	46	NO	42

N: number of specimens; F/M: female/male; 2n: diploid number; XY: sex-linked chromosomes; NOR: number of Ag-NOR (nucleolar organizing regions) positive chromosomes

Phylogenetic Analysis of Chromosomal Characters

Although traditional karyology has made considerable contributions to our present cytogenetic knowledge, it has limited capacity to identify chromosome homologies. Moreover, apparently similar rearrangments seem to have occurred in the various notothenioid taxa. This situation makes classical karyotypic characters a poor data base for comparative analysis. Even the presumed ancestral notothenioid chromosome set, critical for phylogenetic reconstruction, and any other hypothesis, will remain speculative if not confirmed by a detailed resolution understanding of the features of chromosomes and their unambiguous homologies.

Recently, the pattern of the NOR banding was used as new chromosomal character to clarify the phylogenetic relationships between *Notothenia* and *Paranotothenia* species, with *C. gobio* and *B. angustifrons* as outgroups.

The analysis produced a phylogenetic tree where the taxa *coriiceps*, *rossii, magellanica* and *microlepidota* are clustered together (and could be subgenera of the genus *Notothenia*) while *cyanobrancha* is clustered with other Nototheniidae [7] (Fig. 3).

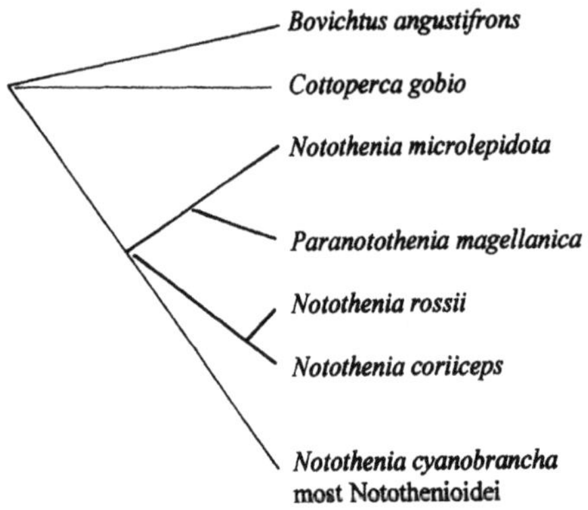

Fig. 3. Phylogenetic relationships among *Notothenia/Paranotothenia* species, based on Ag-NOR chromosome banding (from [7])

The topology of this tree makes it possible to form hypotheses concerning chromosome rearrangements leading to changes in NOR-band position, in Nototheniodei. The interstitial location of this band typically found in *Notothenia* and *Paranotothenia* species (Fig. 4c) is more derived and should have arisen from the translocation of a previously submetacentric NOR bearing chromosome on an acrocentric pair, and the NOR located on a submetacentric chromosome in many species (Fig. 4b) could have originated from a pericentric inversion of an ancestral acrocentric chromosome (Fig. 4a).

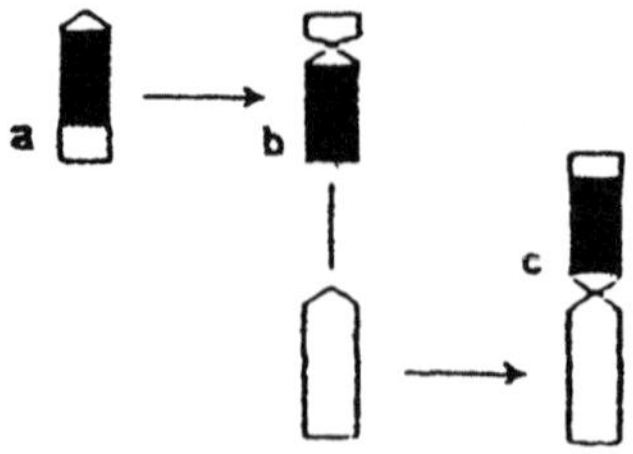

Fig. 4. Supposed chromosomal rearrangements leading to changes of NOR bands in Antarctic fish, from the ancestral (a) to the more derived (b) and (c) position (from [7])

Molecular Cytogenetics and Future Studies

The technology now exists which will permit major advances in genetics and cytogenetics of fishes (and Antarctic fishes) at the molecular level. One promising molecular cytogenetic method with broad application is fluorescence in situ hybridization (FISH) [26]. This technique has been widely used in human genetics to identify chromosomes and to localize genes on the chromosome arms [27]. Recent results using FISH have resolved the problem of species chromosome homology between many mammals and has defined the chromosomal syntenies and major translocations involved in the genome evolution of higher vertebrates, especially primates [28]. Genomic specific sequences corresponding to highly repetitive ribosomal genes (rDNAs) were obtained from Antarctic fish and then used as a probe to map the chromosome location of the 28S rDNA in *Champsocephalus gunnari* using FISH. The structural information derived has implication for general knowledge of the genome in teleosts [8]. Using similar methods a satellite DNA was isolated from *Chionodraco hamatus* and mapped to that species' chromosomes. The interstitial location of this satellite in the Y-chromosome supports the hypothesis about the origin of the sex-linked heteromorphic chromosomes in this species [29]. Specific centromeric and telomeric probes could be used to examine intraspecific chromosome rearrangements and painting with chromosome probes could also be used to assess intraspecific homologies and to provide unambiguous mapping information. Such advances would overcome many of the limits of conventional karyotyping for phylogenetic analysis. Moreover information on chromosome changes in various species would help to test more general assumptions regarding the role of chromosomes in speciation and evolution [11,30].

Finally, various molecular cytogenetic techniques are now available which permit the visualization of smaller DNA sequences and even single genes directly on chromosomes, in interphase nuclei or on stretched DNA filaments. These methods would allow the mapping of genes important in the diversification and adaptation of the Antarctic species such as globin or antifreeze protein genes. Such information would provide new perspective on understanding how changes in genome organization and function are correlated with critical events in the evolution of Antarctic fish.

Acknowledgments

The Italian National Programme for Antarctic Research, the Institut Français pour la Recherche et la Technologie Polaire, the Australian Antarctic Division, the CNRS GDR 1005 and the Muséum National d'Histoire Naturelle of Paris provided financial support. The European Science Foundation Network "Fishes of the Antarctic Ocean" also supported both collaborative exchanges and field work. R.

Stanyon kindly provided helpful comments on the manuscript. Artwork in Fig. 2 was produced by F. Iecle.

References

1. Ozouf-Costaz C, Hureau J-C, Beaunier M (1991) Chromosome studies on fish of the suborder Notothenioidei collected in the Weddell Sea during EPOS 3 cruise. Cybium 15(4):271-289
2. Prirodina VP (1994) Review of karyotypic and taxonomic diversity in the suborder Notothenioidei (Perciformes). J Ichthyol 34(6):1-13
3. Pisano E, Ozouf-Costaz C, Hureau J, Williams R (1995) Chromosome differentiation in the subAntarctic Bovichtidae species *Cottoperca gobio* (Günther, 1861) and *Pseudaphritis urvillii* (Valencienne, 1832) (Pisces, Perciformes). Antarct Sci 7(3):381-386
4. Morescalchi A, Morescalchi MA, Odierna G, Stingo V, Capriglione T (1996) Karyotype and genome size of zoarcids and notothenioids (Teleostei, Perciformes) from the Ross Sea: cytotaxonomic implications. Polar Biol 16:559-564
5. Klinkhardt M, Tesche M, Greven H (1995) Database of fish chromosomes. Westarp Wissenschaften, Magdeburg
6. Pisano E, Ozouf-Costaz C, Bonillo C, Caimo A, Rossetti S, Williams R (1997) Cytogenetics of the Antarctic icefish *Champsocephalus gunnari* Lönnberg, 1905 (Channichthyidae, Notothenioidei). Comp Biochem Physiol 118A:1087-1094
7. Ozouf-Costaz C, Pisano E, Thaeron C, Hureau J-C (1997) Antarctic fish chromosome banding: significance for evolutionary studies. Cybium 21(4):399-410
8. Ozouf-Costaz C, Pisano E, Bonillo C, Williams R (1996) Ribosomal DNA location in the Antarctic fish *Champsocephalus gunnari* (Notothenioidei, Channichthyidae) by banding and fluorescence in situ hybridization. Chrom Res 4:557-561
9. Balushkin AV (1992) Classification, phylogenetic relationships, and origins of the families of the suborder Notothenioidei (Perciformes). J Ichthyol 32(7):90-110
10. Prirodina VP (1997) The direction of the karyotype specialization in the suborder Notothenioidei (Pisces, Perciformes). Cybium 21(4):393-398
11. White MJD (1973) Animal cytology and evolution. Cambridge University Press, Cambridge
12. Morescalchi A, Pisano E, Stanyon R, Morescalchi MA (1992) Cytotaxonomy of Antarctic teleosts of the *Pagothenia/Trematomus* complex (Nototheniidae, Perciformes). Polar Biol 12:553-558
13. Gon O, Heemstra PC (eds) (1990). Fishes of the Southern Ocean. JLB Smith Institute of Ichthyology, Grahamstown, South Africa
14. Eastman JT (1993) Antarctic fish biology: evolution in a unique environment. Academic Press, San Diego
15. Hureau J-C (1986) Relations phylogénétiques au sein des Notothenioidei. Océanis 12(5):367-376

16. Andriashev AP (1987) A general review of the Antarctic bottom fish fauna. In: Kullander SO, Fernholm, B (eds) Proc 5th Congr Eur Ichthyol, Stockholm, Swedish Museum of Natural History, Stockholm, pp 357-372
17. Iwami T (1985) Osteology and relationships of the family Channichthyidae. Mem Natl Inst Polar Res, Tokyo Ser E 36:1-69
18. Bargelloni L, Ritchie PA, Patarnello T, Battaglia B, Lambert DM, Meyer A (1994) Molecular evolution at subzero temperatures: mitochondrial and nuclear phylogenies of fishes from Antarctica (suborder Notothenioidei), and the evolution of antifreeze glycopeptides. Mol Biol Evol 11:854-863
19. Stam WT, Beintema J, D'Avino R, Tamburrini M, di Prisco G (1997) Molecular evolution of hemoglobins on Antarctic fishes (Notothenioidei). J Mol Evol 45:437-445
20. Lecointre G, Bonillo C, Ozouf-Costaz C, Hureau J-C (1997) Molecular phylogenetics and the origins of Antarctic fishes: paraphyly of the Bovichtidae and no indication for the monophyly of the Notothenioidei (Teleostei). Polar Biol 18:193-208
21. Ritchie PA, Lavoué S, Lecointre G (1997) Molecular phylogenetics and evolution of Antarctic notothenioid fishes. Comp Biochem Physiol 118A:1009-1026
22. Foresti F, Almeida Toledo LF, Toledo SA (1981) Polymorphic nature of nucleolus organizer regions in fishes. Cytogenet Cell Genet 31:137-144
23. Koehler MR, Neuhaus D, Engel W, Schartl M, Schmid M (1995) Evidence for an unusual ZW/ZW'/ZZ sex-chromosomes in *Scardinius erythrophthalmus* (Pisces, Cyprinidae), as detected by cytogenetic and H-Y antigen analyses. Cytogenet Cell Genet 71:356-362
24. Morescalchi A (1992) Chromosomes, sex determination and environment in teleosteans. In: Dallai R (ed) Sex origin and evolution. Selected Symposia and Monographs UZI, 6, Mucchi, Modena, pp 137-149
25. Phan VN, Gomes V, Suzuki H (1986) Estudios Citogenéticos de Peixes Antarticos. I. Cariotipos de *Notothenia gibberifrons* (Lönnberg, 1905), *Trematomus bernacchii* (Boulenger, 1902) e *T. hansoni* (Boulenger, 1902) Perciformes, Nototheniidae. Jpn J Ichthyol 33(4):384-387
26. Phillips RB, Reed KM (1996) Application of fluorescence in situ hybridization (FISH) techniques to fish genetics: a review. Aquaculture 140:197-216
27. Trask BJ (1991) Fluorescence in situ hybridization: applications in cytogenetics and gene mapping. Trends Genet 7:149-154
28. Wienberg J, Stanyon R (1995) Chromosome painting in mammals as an approach to comparative genomics. Current Opinion in Genetics and Development 5:792-797
29. Capriglione T, Morescalchi A, Olmo E, Rocco L, Stingo V, Manzo S (1994) Satellite DNAs, heterochromatin and sex chromosomes in *Chionodraco hamatus* (Channichthyidae, Perciformes). Polar Biol 14:285-290
30. King M (1993) Species evolution: the role of chromosome change. Cambridge University Press, Cambridge

Phylogeny of the Channichthyidae (Notothenioidei, Teleostei) Based on Two Mitochondrial Genes

Wei-Jen Chen, Céline Bonillo, and Guillaume Lecointre

Laboratoire d'Ichtyologie générale et appliquée et Service de Systématique moléculaire, Muséum National d'Histoire Naturelle, 43 rue Cuvier, 75231 Paris Cedex 05, France

Introduction

The family Channichthyidae contains 11 genera and 15 species [1] and is considered monophyletic [2,3-5]. Interrelationships of channichthyids were first investigated by Iwami [2] based on morphological characters.

Some parts of Iwami's tree remained obscure, for instance the enigmatic position of *Dacodraco*. The other study based on the osteological development [6] failed to recover the same tree as in Iwami's. Therefore, another marker which is independent from morphological characters is necessary for this phylogenetic study. We explored the variability of four different genes for several channichthyid species in order to investigate the complete phylogenetic picture of the family. Among the genes studied, three were mitochondrial: 16S mtDNA, cytochrome b, D-loop and one was nuclear: 28S ribosomal DNA in its most variable domains D2 and D8.

To obtain the phylogeny of the family, we performed a total evidence approach involving 1135 nucleotides from the the two informative genes cytochrome b and the D-loop for 14 channichthyid species and one outgroup.

Material and Methods

DNA sequences were obtained from the 11 channichthyid genera currently recognized and 14 out of the 15 species of the family, plus one Bathydraconidae as an outgroup. Details about their sampling are available on request (Chen [7]). In order to examine intraspecific variation in the most variable gene, the D-loop, three D-loop sequences from individuals of *Chionodraco hamatus* from the population of Ross Sea, taken from Bargelloni's unpublished data were used. To carry out the calibration of molecular rate of the cytochrome b, available sequence of *Trachurus trachurus* (Perciformes) was taken from GenBank under the accession number of X81568.

G. di Prisco, E. Pisano, A. Clarke (Eds)
Fishes of Antarctica. A biological overview
© Springer-Verlag Italia 1998

DNA extraction was performed following Lecointre et al. [8]. DNA amplification was performed by polymerase chain reaction (PCR) [9,10]. Double-stranded PCR products were sequenced in two ways: direct sequencing or sequenced after a cloning step. Primer sequences, PCR protocols and sequencing protocols are available on request (Chen [7]).

Obtained sequences were read and entered twice using the computer package MUST [11], and aligned using the facilities of ED, within MUST. MUST includes the neighbor-joining (NJ) method [12] and allows very fast bootstrap analyses with this tree-construction method using the NJBOOT program. The main phylogenetic analyses were performed using PAUP 3.1.1. [13]. Branch and Bound searches were performed with the whole species set (15 species). Relative and intrinsic saturation were evaluated for transitions and transversions separately using the methods described in Lecointre et al. [8]. No weighting schemes were used because of absence of transitional saturation. Bootstrap analyses [14] were performed with PAUP, using Branch and Bound searches and 500 iterations, and with NJ using 1000 iterations. Morphological data were evaluated using PAUP and MacClade [15].

Results

Sequence Analysis

Sequences are available in GenBank under accession numbers from AF037109 to AF037138.

Both the alignment of 16S mtDNA sequences (500 bp) and D2 (611 bp) and D8 (443 bp) of the 28S rDNA showed that these genes were not variable enough to allow the inference of channichthyid phylogeny. For the cytochrome b data, DNA sequences from 15 species (including one out-group taxon) were aligned according to comparison with *Trachurus trachurus* (Teleostei, Perciformes). Of the total 752 sites, 161 (21%) are variable within the ingroup and 89 (12%) are informative for parsimony (at least two types of nucleotides each represented at least twice). When codon positions are considered, the third codon position provides 88% of the variability. A test of absolute saturation of transitional substitution in the third codon position showed no saturation (not shown, the regression coefficient of the plot is 0.91). For the D-loop data, the size of the D-loop sequences are not the same through taxa because of numbers of deletion/insertion events. After alignment, 383 positions were obtained for all species studied including 187 (49%) variable sites in which 99 (26%) sites are informative for parsimony. *Chionodraco rastrospinosus* has the longest sequence owing to a significant insertion for 12 cytosines in the 5’ end region. Noncorrected pairwise distances between the 14 ingroup species ranged from 4.07% to 19.97% (not shown). Hence, D-loop is potentially more variable than cytochrome b gene for which pairwise differences ranged from 1.86% to 9.31%. Intraspecific distances were evaluated by comparing our individual of *Chionodraco hamatus* from Terre Adélie with three individuals of *C. hamatus* from the Ross Sea (sequences taken from L. Bargelloni's unpublished data). The pairwise differences ranged from 0.82% to 1.73% within the populations from the Ross Sea and from 1.36% to 1.64% between the populations from two different localities. These values of intraspecific diversity are smaller than those of interspecific ones. Anyway, the topologies obtained from these two different mitochondrial genes showed very few differences, suggesting that possible bias from intraspecific variability should be negligible here and permit us to reconstruct a robust phylogeny for channichthyids at the species level. Moreover, no saturation was detected when the pairwise observed transitional differences were plotted against the pairwise inferred transitional substitutions (the regression coefficient is 0.93).

Phylogenetic Analyses

Branch and Bound searches were performed on the 15 species for the two mitochondrial genes (cytochrome b and D-loop respectively) separately, each yielding only one most-parsimonious tree [tree length (l) = 299, retention index (RI) = 0.546, consistency index (CI) = 0.639 for the cytochrome b; l = 354, R.I. = 0.612, C.I. = 0.698 for the D-loop]. These two trees shared the same topology excepting the positions of three species: *Chaenodraco wilsoni, Chaenocephalus aceratus* and *Channichthys rhinoceratus*, but all having their branching point in the same area. The NJ method yielded the same topologies as above. It must be noted that differences in topology only affected nodes with short internal branches and low bootstrap proportions. Anyway, we consider that, when differences in topology occur, the best way to evaluate the relative strength of the two phylogenetic signals is to perform a total evidence approach, because the congruence of characters is more important than the congruence of trees [16]. Thus, we combined the two data sets in a single one. An interesting feature of this study is that the phylogenetic analysis based on the 1135 sites from the two mitochondrial genes resulted in a better resolution of the channichthyid relationships in terms of bootstrap proportions. A Branch and Bound search yielded two equiparsimonious trees (659 steps, CI = 0.665, RI = 0.570, Fig. 1). The NJ method yielded one of these two (Fig. 1, left). These trees were very similar to the ones derived from the two genes separately, but bootstrapping showed higher bootstrap proportions: one node over 13 showed a bootstrap proportion below 50% and 9 nodes above 70%.

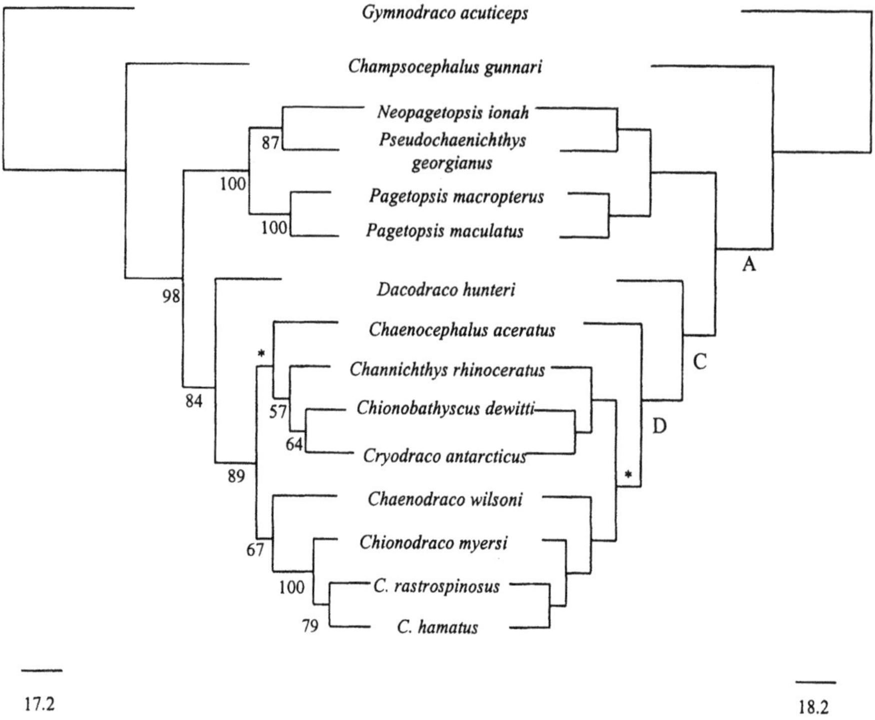

Fig. 1. Two equiparsimonious trees obtained from a Branch and Bound search without weighting schemes (PAUP 3.1.1) on 1135 sites of two mitochondrial genes (cytochrome b and D-loop). Each tree has a length of 659 steps, CI of 0.665 and RI of 0.570. Branch lengths are proportional to the number of steps affected to the branch (ACCTRAN). The numbers below branches refer to bootstrap proportions (500 iterations). Asterisks refer to bootstrap proportions inferior to 50%. The difference between these two trees is the relative position of *Chaenocephalus aceratus*

Considering the three species that were previously instable, we now confirm the position of *Chaenodraco wilsoni* as the sister-group of *Chionodraco* species (bootstrap proportion of 67%) and *Channichthys* as the sister-group of the clade *Chionobathyscus* + *Cryodraco*. The total evidence approach fails to clearly identify the relationships of *Chaenocephalus*. This tree from the total evidence (Fig. 1) is strikingly similar to the one proposed by Iwami [2], excepting the relationships of *Channichthys*.

Discussion

Morphological Data Sets

Phylogenetic analyses based on morphological characters in notothenioid fishes often expressed the intrinsic difficulties in assessing character polarization because of the pedomorphic nature of most of them. Pedomorphosis plays an important role in the speciation of this group of fishes, particularly in the Channichthyidae [6,17]. Moreover, channichthyid adaptation for diverse niches to the extreme cold Antarctic environment provoked phenotypic alterations corresponding to ecomorphological forms, but has entailed little alteration of the basic perciform body. According to Eastman [1], this mode of diversification does not favor the emergence of morphological novelties that could constitute detectable synapomorphies. For this reason, morphological studies attempting to reconstruct the channichthyid phylogeny were difficult to carry out and left some unresolved nodes. The two available studies [2,6] failed to recover the same tree but this difference can easily be explained by their respective methodologies.

Iwami [2] used 17 morphological characters studied in a "cladistic" framework. However, the principle of parsimony was not followed. We entered Iwami's data matrix in PAUP 3.1.1 and have obtained four equiparsimonious trees (l = 19, CI = 0.895, RI = 0.889) while Iwami presented only one (Fig. 2, left). He therefore just mapped his 17 characters

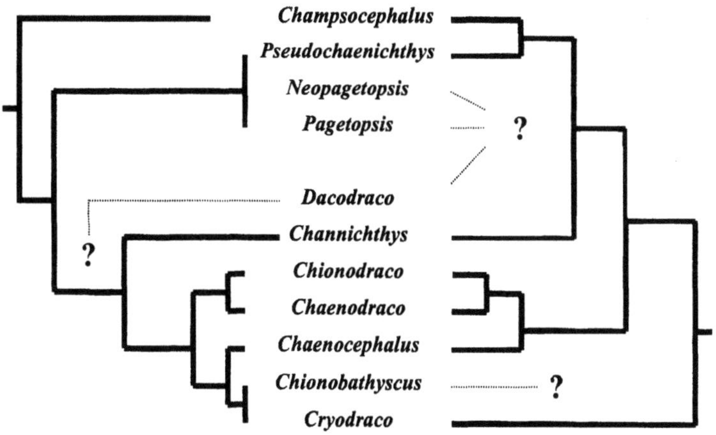

Fig. 2. Left, the cladogram proposed by Iwami [2] on the basis of 17 morphological characters. Right, the phenogram proposed by Voskoboinikova [6] on the basis of measurements of rates of osteological development. This tree was artefactually rooted on *Cryodraco*

on his preconceived view of channichthyid phylogeny. However, his intuitions are strikingly confirmed by the present study, even for the phylogenetic relationships of *Dacodraco* that he hesitated to branch as the sister-group of the clade D (Figs. 1 and 2), a position that is unambiguously supported by the two present mitochondrial genes. The only differing point concerns the relationships of *Channichthys*. Constraining the "molecular tree" to Iwami's morphological data set requires no extra step. Indeed, the molecular position of *Channichthys* is already present in one of the four equi-parsimonious trees given by PAUP from Iwami's data set. Therefore, there is no contradiction between Iwami's data set and ours. For the morphological data of Iwami [2], the molecular position of *Channichthys* implies two reversals in the branch of *Channichthys*, respectively for the characters 2 (sphenotic-pterosphenoid connection from the state "separated" to the state "bordered") and 15 (ventral fin from the state "fan or can-like shape" to the state "normal").

Voskoboinikova's phenogram [6] (Fig. 2, right) based on measurements of the rate of osteological development was initially rooted on *Cryodraco*.

If this tree is rerooted on *Champsocephalus*, it becomes similar to ours (Fig. 3) except for the position of *Channichthys*. This shows that her topology was the same as ours but there was just a rooting artefact in the phenetic method used. It can be suspected that, in case of extreme divergence of the rate of bone development in *Cryodraco* (in the present case, extremely slow) and using a phenetic algorithm that requires equal amounts of change in two sister-branches, the root will be automatically attracted toward the most divergent taxon. In the extreme case, the phenogram is directly rooted on the branch of the most divergent taxon.

Two elements confirm this interpretation. First, Voskoboinikova's tree is based on a method assuming equal amount of changes among lineages, as shown by the branches of her phenogram that end at the same level.

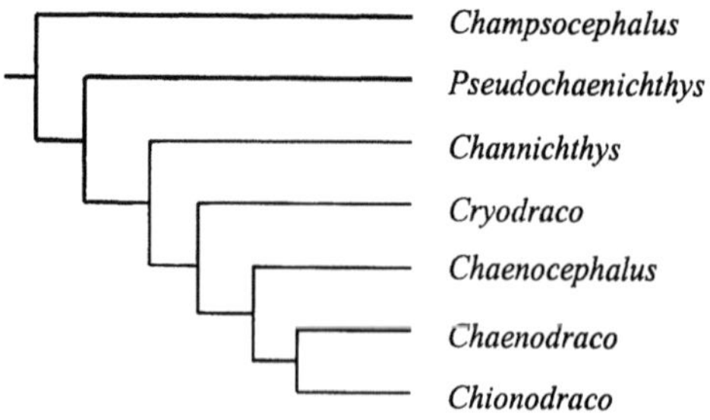

Fig. 3. Voskoboinikova's tree [6] rerooted on *Champsocephalus*

Secondly, looking at the distances used and her conclusions, *Cryodraco* is actually the most extreme taxon, i.e. it has the slowest rate of osteological development, as concluded in her abstract. However, the fact that the rerooted tree of Voskoboinikova is the same as ours is an interesting point.

This shows that the slowdown of the rate of the osteological development regularly follows the successive cladogeneses within the family without detectable homoplastic alterations of this rate. In other words, the most derived channichthyids are the most pedomorphic.

Cladogenetic Features

The examination of lifestyle, bathymetric distribution and geographic distribution of each species and mapping these three parameters on our tree (Fig. 1) provide an interesting outcome. We have first to remember some methodological aspects. A cladogram is neither an evolutionary story, nor a scenario. It is just a distribution of characters among taxa. Therefore, a node is not the direct reflection of a speciation leading to two species. The dichotomic nature of the node is simply due to the fact that the cladogram has reached its maximum explanatory power, in other words corresponds to the maximum resolution in the tree. To a single cladogram correspond many possible scenarios where nodes are speciation events [18-20]. Most of the time, a cladogram does not allow us to exhaust all the descriptive and/or causal elements that can be assumed to describe or explain speciations. For this reason, mapping the three above features has poor chance to explain something. However, in the present case, the striking conclusion of such a mapping is that each cladogenesis of Fig. 1 corresponds to the differentiation of one of these three parameters (Fig. 4), what was initially rather unexpected. We would rather expect that *at least* some nodes would not differentiate anything at all. Figure 4 shows that each node (particularly the most terminal ones E, F, H, I, J, G, K, L) can be correlated to a difference in distribution, depth or lifestyle, in such a way that each node can be correlated to a hypothetical sympatric speciation (when the lifestyles of two sister-groups differ, see nodes K, G, Fig. 4) or hypothetical allopatric speciations (when the geographic and/or bathymetric distributions of two sister-groups differ, see nodes E, F, J, H, I, J, L, Fig. 4). One must keep in mind that these correlations have no power to establish causalities; they are just elements for further research. Indeed, more complex events than described here might have occurred during the evolution of the family.

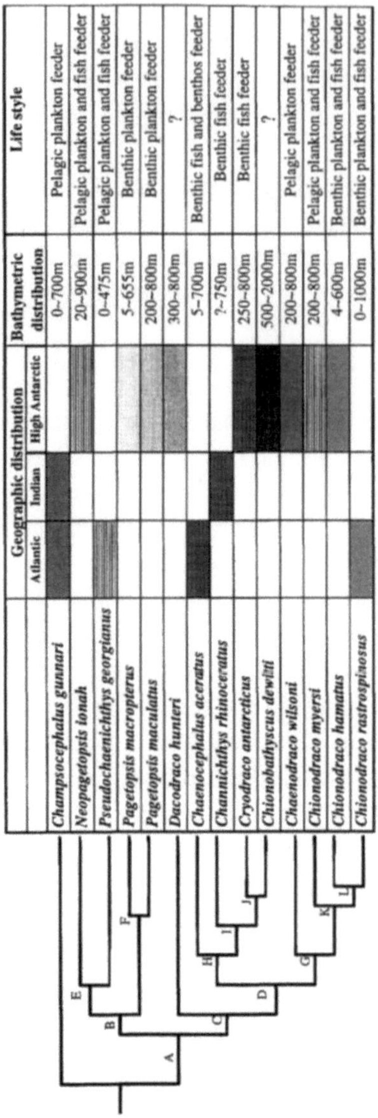

Fig. 4. Distribution of lifestyles, biogeographic and bathymetric data of each species through the channichthyid phylogeny. Data from [21-23]. The geographic distribution of channichthyids contains two major zones which differ by several hydrological parameters. The first is the Seasonal Packed-Ice Zone and its northern islands, involving two sectors: the Atlantic sector (the former of West Antarctic and South Georgia Provinces) and the Indian Sector (Kerguelen subregion). The second is the High Antarctic Zone which is confined to the shelf and upper slope areas of the Antarctic continent. Same pattern inside different squares refers to having ecological equivalence, for example: *C. hamatus* and *C. rastropinosus* have ecological equivalence. Differences in the coloring pattern indicate ecological nonequivalence

Divergence Times

Our cytochrome b sequences do not exhibit mutational saturation, neither within the Channichthyidae nor between the perciform outgroup *Trachurus* and the channichthyids. There are two techniques to infer divergence times from nonsaturated sequences. Both postulate the molecular clock. First, one can use published mitochondrial rates and apply it to one's pairwise distance matrix to infer the divergence time of a particular node. The second technique consists in calculating its own rate using a time calibration given by one species for which paleontological data provide a divergence time. *Gymnodraco* is the sister-group of channichthyids [4,5].

Let us call the node X the divergence of *Gymnodraco* from the channichthyid ancestor and node Y the divergence of *Champsocephalus* from the rest of the channichthyids. According to the first technique, assuming a rate of mitochondrial genes between 0.3% /Ma and 0.7%/Ma [4, 24-27] provides a divergence time of 35 to 15 Ma for the node X and 27 to 12 Ma for Y. Assuming a transversional rate of 0.14%/Ma [4], this divergence time is 23 Ma for X and 15 Ma for Y. According to the second technique, we have taken the perciform diversification dated back to 55 Ma [28,29] as a calibrating point. Whatever the nonnotothenioid outgroup chosen to represent a perciform, this rate, so calculated, is 0.47%/Ma. This is due to the fact that perciforms probably underwent a radiation [30], so that taking one or another perciform outgroup did not change the pairwise distances used to calculate the rate. The divergence time obtained for the node X is 23 Ma and 17.5 Ma for the node Y.

This period of time when the first channichthyid ancestor must have lived can be correlated to the development of unrestricted Antarctic circumpolar current and the initial development of the Antarctic polar front [1]. This second technique of calculation seems to be better than the former because it does not impose any extrinsic rate to our cytochrome b sequences, although it requires a correct paleontological time calibration. It is well known that 16S mtDNA sequences evolve slower than cytochrome b ones (for example see Meyer [30]). Therefore, it is questionable to apply the same transversional rate of 0.14% to both genes. This is why we prefer the second technique, which follows the cytochrome b intrinsic rate. This may also be the reason why Bargelloni and Lecointre (this Vol.) also found from 12S and 16S mtDNA a divergence time of 23 Ma for the emergence of *Eleginops*, a very early notothenioid lineage. Applying an extrinsic rate to 12S and 16S sequences may have underestimated the real divergence time of *Eleginops*. A paleontological time is needed to confirm this point of view. The other hypothesis is that the nonbovichtid notothenioid radiation may have occurred in a very short time span.

Conclusion

Our nucleotide sequence data from two mitochondrial genes of 14 species of Channichthyidae allowed reconstruction of a complete and reliable phylogeny of the family except for the relationships of *Chaenocephalus*. This study offered an example of obtaining a better resolution when the total evidence strategy is chosen. Only one node over 13 has a bootstrap proportion below 50% and 9 above 70%. Our molecular tree supports most of the clades proposed by Iwami [2] on the basis of morphological characters. For the differing points, the principle of parsimony shows that there is no real contradiction between morphology and molecules. Indeed, the molecular position of *Channichthys* requires no extra step in the morphological matrix of Iwami. A reinterpretation of Voskoboinikova's developmental data shows that the most advanced channichthyids are the most pedomorphic. Assuming a molecular clock and a calibration based on the perciform fossil record, the channichthyid family dates back to the time of the polar front formation.

Acknowledgments

This work is a part of the EPOS ESF program and was also coordinated within the European Science Foundation Network "Fishes of the Antarctic Ocean." The CNRS GDR 1005 provided financial support. We are especially grateful to Drs. C. Ozouf-Costaz, E. Pisano, C. Zimmermann, and R. Williams for having provided samples. We especially thank L. Bargelloni for having provided D-loop primer sequences and D-loop unpublished sequences of *Chionodraco hamatus* from the Ross Sea. We thank P. Ritchie for early attempts in amplifying channichthyid cytochrome b.

References

1. Eastman JT (1993) Antarctic fish biology: evolution in a unique environment. Academic Press, San Diego
2. Iwami T (1985) Osteology and relationships of the family Channichthyidae. Mem Natl Inst Polar Res, Ser E 36:1-69, Tokyo
3. Hastings PA (1993) Relationships of the fishes of the perciform suborder Notothenioidei. In: Miller RG (ed) A history and atlas of the fishes of the Antarctic Ocean. Foresta Institute for Ocean and Mountain Studies, Carson City, Nevada, CA, pp 99-107
4. Bargelloni L, Ritchie PA, Patarnello T, Battaglia B, Lambert DM, Meyer A (1994) Molecular evolution at subzero temperatures: mitochondrial and nuclear phylogenies of fishes from Antarctica (suborder Notothenioidei), and the evolution of antifreeze glycopetide. Mol Biol Evol 11:854-86

5. Ritchie PA, Lavoué S, Lecointre G (1997) Molecular phylogenetics and the evolution of Antarctic notothenioid fishes. Comp Biochem Physiol 118A:1009-1026
6. Voskoboinikova OS (1997) Osteological development of the Channichthyidae (Teleostei: Notothenioidei). Cybium 21(4):369-380
7. Chen WJ (1997) Phylogénie moléculaire des Channichthyidae: famille de Téléostéens antarctiques. Mémoire de DEA, Université Paris VI
8. Lecointre G, Bonillo C, Ozouf-Costaz C, Hureau J-C (1997) Molecular phylogeny of the Antarctic fishes: paraphyly of the Bovichtidae and no indication for the monophyly of the Notothenioidei (Teleostei). Polar Biol 18(3):193-208
9. Mullis KB, Faloona FA (1987) Specific synthesis of DNA in vitro via a polymerase catalyzed chain reaction. Methods Enzymol 155:335-350
10. Saiki RK, Gelford DH, Stoffel S, Scharf SJ, Higuchi RG, Horn GT, Mullis KB, Enlich HA (1988) Primer-directed enzymatic amplification of DNA with a thermostable DNA polymerase. Science 239:487-491
11. Philippe H (1993) MUST: a computer package of Management Utilities for Sequences and Trees. Nucl Acids Res 21:5264-5272
12. Saitou N, Nei M (1987) The neighbor-joining method: a new method for reconstructing phylogenetic trees. Mol Biol Evol 4:406-425
13. Swofford DL (1993) PAUP-phylogenetic analysis using parsimony, Version 3.1.1. Illinois Nat Hist Surv, Champaign, Illinois
14. Felsenstein J (1985) Confidence limits on phylogenies: an approach using the bootstrap. Evolution 39:783-791
15. Maddison WP, Maddison DR (1992) MacClade: analysis of phylogeny and character evolution, Version 3.02. Sinauer Associates, Sunderland, MA
16. Barrett M, Donoghue MJ, Sober E (1991) Against consensus. Syst Zool 40(4):486-493.
17. Iwami T (1994) Heterochrony in the family Channichthyidae (Perciformes, Notothenioidei) and its phylogenetic implications. Antarctic communities: species, structure and survival. Abstract:134
18. Eldredge NJ, Cracraft (1980) Phylogenetic patterns and the evolutionary process. Method and theory in comparative biology. Columbia University Press, New York
19. Nelson GJ, Platnick N (1981) Systematics and biogeography: cladistics and vicariance. Columbia University Press, New York
20. Janvier P (1988) Cladistique: théorie, objectifs et implications évolutives. In: Biosystema 2, systématique cladistique, publication de la Société Française de Systématique, Paris, p 137
21. Gon O, Heemstra PC (eds) (1990) Fishes of the Southern Ocean. JLB Smith Institute of Ichthyology, Grahamstown, South Africa
22. Kock KH (1992) Antarctic fish and fisheries. Cambridge University Press, Cambridge
23. Miller RG (1993) A history and atlas of the fishes of the Antarctic Ocean. Foresta Institute, Carson City, Nevada

24. Rand DM (1994) Thermal habit, metabolic rate and the evolution of mtDNA. Tree 9(4):125-131
25. Martin AP, Naylor GJP, Palumbi SR (1992) Rates of mitochondrial DNA evolution in sharks are slow compared with mammals. Nature 357:153-155
26. Martin AP, Palumbi SR (1993) Body size, metabolic rate, generation time, and the molecular clock. Proc Natl Acad Sci USA 90:4087-4091
27. Miya M, Nishida M (1996) Molecular phylogenetic perspective on the evolution of the deep-sea fish genus *Cyclothone* (Stomiiformes: Gonostomatidae). Ichthyol Res 43(4):375-398
28. Benton MJ (ed.) (1993) The fossil record II. Chapman & Hall, London
29. Lê HLV, Lecointre G, Perasso R (1993) A 28S rRNA based phylogeny of the gnathostomes: first steps in the analysis of conflict and congruence with morphologically based cladograms. Mol Phyl Evol 2(1):31-51
30. Meyer A (1993) Evolution of mitochondrial DNA in fishes. In: Hochachka PW, Mommsen TP (eds) Molecular biology frontiers. Biochemistry and molecular biology of fishes, vol 2. Elsevier, Amsterdam London New York Tokyo, pp 1-38

Allozymic Polymorphism and Phylogeny of the Family Channichthyidae

Olivier Clément[1,2], Catherine Ozouf-Costaz[2], Guillaume Lecointre[2] and Patrick Berrebi[1]

[1]Laboratoire Génome et Populations, UPR 9060 du CNRS, Université Montpellier II, cc 63, place E. Bataillon, 35095 Montpellier Cedex 05, France.
[2]Laboratoire d'Ichtyologie, CNRS GDR 1005, Muséum National d'Histoire Naturelle, 43 rue Cuvier, 75231 Paris Cedex O5, France.

Introduction

The suborder Notothenioidei represents a moderate radiation of teleosts into various niches, mostly at less than 500 m depths, and adaptation to subzero temperatures [1,2]. The decrease of temperature of the Antarctic continent 38 million years ago (Ma) provoked important species extinction and the dominance of Notothenoidei.

Among the Notothenioidei, all members of the icefish family Channichthyidae (11 genera and 15 species) are totally endemic to the Antarctic and subAntarctic regions and lack hemoglobin. The phylogeny established for this family, based on morphological characters [3], is not completely in agreement with the phenogram established by Voskoboinikova [4] on the basis of developmental features.

A previous genetic study [5] of the family Nototheniidae used allozymes for studying relationships within species of the *Trematomus/Pagothenia* complex. In contrast, Duhamel et al. [6], with the primary objective of assessing genetic variation and divergence between populations of exploited species in the subAntarctic zone, have shown that *Notothenia rossii* displays a rather low allozymic polymorphism (H = 0.03) and that *Champsocephalus gunnari* is completely monomorphic.

In this study, we envisaged the use of allozymes to investigate species, interrelationships within the family Channichthyidae. This is the primary objective of this paper. Since it was possible to detect genetic divergence between samples collected in different geographic sectors within two species, *Neopagetopsis ionah* and *Chionodraco myersi,* that kind of

G. di Prisco, E. Pisano, A. Clarke (Eds)
Fishes of Antarctica. A biological overview
© Springer-Verlag Italia 1998

differentiation has been also precisely analyzed, mainly by the F statistics of Wright [7], which represent the second objective of this paper.

Several species of this family are difficult to catch or rare. Nevertheless, the collection of the Muséum National d'Histoire Naturelle of Paris, constituted during several years of Antarctic expeditions, was the most complete known for biological analyses and specially for allozymes. Analysing that precious collection before allozyme inactivation was a strong stimulation.

Material and Methods

Sampling

Samples of liver and muscle of 14 species of Channichthyidae have been collected during several expeditions: EPOS leg 3 cruise on board the oceanographic ship "Polarstern" around the South Shetland Islands and the Weddell Sea (January-February 1989); AAMBER II cruise in the Prydz Bay (February 1991); THIRST cruise of the "Aurora Australis" oceanographic ship around Heard Island (September 1993) and ICOTA campaign in Adelie Land (January- February 1996).

Each species is represented by 1 to 65 individuals. Moreover, several reasons explain the various enzymatic activities observed: depending on the cruise, the tissues were either rapidly frozen in liquid nitrogen or deep-frozen between -20 °C and -80 °C on board; the storing conditions between the Antarctic and Paris were diverse (from -195 °C to -20 °C). Only muscle samples were collected around Heard Island because the livers were too parasitized to be used. Finally, the time between sampling and analyzing is variable (from 1 to 8 years) for the respective samples.

A nototheniid fish, *Trematomus lepidorhinus* (Pappenheim, 1912) from the Weddell Sea (11 specimens), has been included in the sample as an outgroup for the phylogenetic analysis.

The following list give the names of species, the sample sizes and the localities: *Chaenocephalus aceratus* Lönnberg, 1906, Shetland Islands (4 specimens); *Chaenodraco wilsoni* Regan, 1914, Weddell Sea (WS) (10 sp) + Prydz Bay (6 sp); *Champsocephalus gunnari* Lönnberg, 1905, Heard Island (10 sp); *Channichthys rhinoceratus* Richardson, 1844, Heard Island (10 sp); *Chionobathyscus dewitti* Andriashev and Neelov, 1978, WS (1 sp); *Chionodraco hamatus* Lönnberg 1905, WS (18 sp) + Adelie Land (4 sp) + Prydz Bay (6 sp); *Chionodraco myersi* DeWitt and Tyler 1960, WS (59 sp) + Prydz Bay (6 sp); *Chionodraco rastrospinosus* DeWitt and Hureau 1979, WS (4 sp); *Cryodraco antarcticus* Dollo 1900,

WS (11 sp) + Prydz Bay (6 sp); *Dacodraco hunteri* Waite 1916, WS (1 sp); *Neopagetopsis ionah* Nybelin 1947, WS (28 sp) + Prydz Bay (7 sp); *Pagetopsis macropterus* Boulenger 1907, WS (1 sp) + Prydz Bay (3 sp); *Pagetopsis maculatus* Barsukov and Permitin 1958, WS (3 sp); *Pseudochaenichthys georgianus* Norman 1937, WS (1 sp).

Allozymic Electrophoresis

The tissues were homogenized with an equal volume of Tris-HCl-EDTA-NADP pH 6.8 buffer. Some drops of CCl_4 helped the elimination of lipids from the enzymatic extract after a 30 min centrifugation at 15000 g. The extract was then frozen at –80 °C, before being analyzed.

Enzyme extraction, horizontal 12% starch gel electrophoresis and the staining solutions were carried out according to the protocols of Pasteur et al. [8]. Shaklee et al. [9] nomenclature was used for the loci.

Data Analyses

After interpretation of patterns on the gels, a genotype matrix was established. The Genetix software [10] has been used to calculate the allelic frequencies, the heterozygosity, the Nei [11] genetic distances and the estimators of the F statistics of Wright [7].

The Fis can detect the deviations from the Hardy Weinberg genotype proportions. Actually, it is the estimator f of Weir and Cockerham [11] which is calculated. The software Genetix permits estimation of the significance of the deviation of the estimator from zero by permutation method generating 1000 artificial matrices. This parameter permits detection of panmixia disequilibrium or Walhund effect in each sample.

The Fst detects the differences in heterozygote proportions between samples (species or populations). In fact, the estimator θ of Weir and Cockerham [12] is calculated. The significance of the value is also obtained by permutations.

Phylogenetic Reconstruction

With the Nei [11] genetic distances, a phylogenetic tree is built (Fig. 1) using the neighbor joining method [13].

A second phylogenetic tree is established (Fig. 2) using the Dollo parsimony method (the method allows, for each allele, only one appearance but several losses in the tree). To limit the disadvantage of small sample, a threshold of 5% is applied, considering an allele as present when its frequency is over 5%. This matrix [12 taxa, 51 characters, 612 character states among which are 30 question marks

(4.9%)] is available upon request. The PAUP software [14] has been used. The maximum parsimony trees were obtained using the Branch and Bound algorithm. Bootstrap proportions were obtained from 1000 replicates.

Results

Enzymatic variability

The allelic frequencies and the values of observed and expected heterozygosity are given in Table I. Since many labile enzymes tended to denature due to variable quality of the samples, there are some missing data (corresponding to 4.9% of question marks in the NEXUS data matrix of PAUP).

By using the intraspecific and intrasubsample Fis, plus intersubsample Fst, a clear heterogeneity appears at the intraspecific level for *Chionodraco myersi* and *Neopagetopsis ionah*. Values of estimated Fis and Fst are available upon request to the corresponding author.

1) In *Chionodraco myersi*, four populations are concerned: pop. 6: Weddell Sea - Cape Norvegia; 71°06,2'S; 012°53,8'W, 771 m depth, 10 ind. + pop. 7: Weddell Sea - Cape Norvegia, 71°12,0'S; 013°14,0'W; 402 m; 13 ind. + pop. 8: Weddell Sea - Halley Bay, 75°09,1'S; 027°34,7'W; 407 m; 30 ind. + pop. 9: Prydz Bay, 301 m; 6 ind.

2) In *Neopagetopsis ionah*, there is a significant differentiation between two populations: pop. 3: Weddell Sea, 74°36,5'S; 029°41,2'W; 799 m; 24 ind. + pop. 4: Prydz Bay, 300 m; 7 ind.

Phylogenetic reconstructions

The species *Chionobathyscus dewitti*, *Dacodraco hunteri* and *Pseudochaenichthys georgianus* were removed from the analyses because they were represented by only one individual, the reliability of the data being not established.

The neighbor - joining tree is given in Fig. 1. Note that the clade (*Channichthys* (*Chaenocephalus*, *Cryodraco*)) is well supported by a long internal branch. Also are the successive basal positions of *Champsocephalus* and *Neopagetopsis*.

The Branch and Bound search yielded two equally parsimonious trees of 66 steps, differing by the position of *Chionodraco rastrospinosus*. The tree (Fig. 2) is the strict consensus of these two Dollo parsimony trees, with the bootstrap proportions above 50%. The consistency index (CI) is 0,591 and the retention index (RI) is 0.757.

Table 1. Allelic frequencies and heterozygosity calculated for each species studied for polymorphic loci only. Monomorphic loci *AK**, *LDH-M** and *GLO** are not presented here. Empty spaces = no result. Heterozygosity values between brackets are only indicative, since species sampling was low (<5 individuals). See text for species names in full

Locus		*C.acer*	*C.ant*	*C.dew*	*C.gun*	*C.ham*	*C.myer*	*C.rastr*	*C.rhino*
CK (N)*		*3*	*15*	*1*	*10*	*14*	*61*	*2*	*9*
	100	1	1	1	0	1	1	1	1
	120	0	0	0	1	0	0	0	0
	130	0	0	0	0	0	0	0	0
IDH-M (N)*		*3*	*16*		*8*	*15*	*62*	*4*	*10*
	100	1	1		0	1	1	1	1
	112	0	0		0,25	0	0	0	0
	114	0	0		0,75	0	0	0	0
MDH-1 (N)*		*3*	*16*		*10*	*14*	*62*	*4*	*10*
	100	1	1		1	0	0,06	0	1
	250	0	0		0	1	0,94	1	0
MDh-2 (N)*		*3*	*16*		*10*	*14*	*60*	*3*	*10*
	70	0	0		0	0,14	0,42	0	0
	75	0	0		0	0	0	0	1
	100	1	1		0	0	0	0	0
	108	0	0		1	0	0	0	0
	110	0	0		0	0,86	0,58	1	0
	120	0	0		0	0	0	0	0
PGM (N)*		*3*	*16*		*9*	*15*	*62*	*4*	*9*
	90	0	0		1	0	0,02	0	0
	100	1	1		0	1	0,98	1	1
	110	0	0		0	0	0	0	0
GPI-1 (N)*		*2*	*14*		*9*	*13*	*61*	*3*	*8*
	80	0	0		0	0	0	0	0
	85	0	0		1	0	0	0	0
	100	0,25	0,32		0	0,77	0,70	0,83	0
	110	0	0		0	0	0,02	0	0
	115	0,75	0,68		0	0,23	0,26	0,17	0,31
	130	0	0		0	0	0,02	0	0,69
	135	0	0		0	0	0	0	0
ME-1 (N)*		*3*	*15*		*9*	*14*	*61*	*4*	*9*
	100	1	1		0	1	1	1	1
	108	0	0		0	0	0	0	0
	113	0	0		1	0	0	0	0
G3PDH (N)*		*1*	*16*		*10*	*13*	*60*	*2*	*9*
	80	0	0		0	0	0	0	0
	90	0	0		1	0	0	0	0
	100	1	1		0	1	1	1	1
	110	0	0		0	0	0	0	0
MDH-1L (N)*		*2*	*16*	*1*		*13*	*33*	*4*	
	100	1	0,94	1		0	0,08	0	
	250	0	0,06	0		1	0,86	1	
	252	0	0	0		0	0,06	0	
SOD (N)*		*2*	*15*	*1*		*1*	*33*	*4*	
	100	0	0,93	1		0	1	0	
	140	0	0	0		0	0	0	
	180	0	0	0		0	0	0	
	200	1	0,07	0		1	0	1	
He obs.		(0,0333)	0,0134	-	0	0,0349	0,0887	(0,0222)	0,0568
He n.b.		(0,0333)	0,0635	-	0,0364	0,0532	0,1062	(0,0222)	0,0417

Locus	C.wilso	D.hunt	N.ionah	P.geor	P. macr	P. macu	Outgroup
CK (N)*	*16*		*34*	*1*	*4*	*2*	*10*
100	1		0	1	1	1	0,1
120	0		1	0	0	0	0
130	0		0	0	0	0	0,9
IDH-M (N)*	*16*	*1*	*33*	*1*	*3*	*2*	*11*
100	0,84	0,5	0	0	1	1	1
112	0,16	0,5	1	1	0	0	0
114	0	0	0	0	0	0	0
MDH-1 (N)*	*16*		*34*	*1*	*4*	*2*	*11*
100	0		0	0	0	0	1
250	1		1	1	1	1	0
MDh-2 (N)*	*16*		*34*	*1*	*4*	*2*	*11*
70	0		0	0	0,75	0,5	0
75	0		0	0	0	0	0
100	0		0	0	0	0	0
108	0		0	0	0	0	0
110	0,97		1	1	0,25	0,5	1
120	0,03		0	0	0	0	0
PGM (N)*	*16*		*34*	*1*	*4*	*2*	*11*
90	0,03		0	0	0	0	1
100	0,88		0	1	0,75	1	0
110	0,09		1	0	0,25	0	0
GPI-1 (N)*	*15*		*34*		*4*	*3*	*11*
80	0,03		0		0	0	0
85	0		0		0	0	0
100	0,83		0		0	0	0
110	0		0		1	1	1
115	0,13		0		0	0	0
130	0		0		0	0	0
135	0		1		0	0	0
ME-1 (N)*	*16*		*34*		*3*	*2*	*11*
100	1		1		1	1	0
108	0		0		0	0	1
113	0		0		0	0	0
G3PDH (N)*	*16*		*28*		*3*	*2*	*11*
80	0		0		0	0	1
90	0,03		0		0	0	0
100	0,94		1		1	1	0
110	0,03		0		0	0	0
MDH-1L (N)*	*14*	*1*	*34*	*1*	*4*	*3*	
100	0	0	0	0	0	0	
250	1	1	1	1	1	1	
252	0	0	0	0	0	0	
SOD (N)*	*14*		*34*				*11*
100	0		0				0
140	0		0,03				0
180	0		0,97				0
200	1		0				1
He obs.	0,0437	-	0,0130	-	(0,0333)	(0,0667	0,0208
He n.b.	0,0616	-	0,0219	-	(0,0571)	(0,0444	0,0200

N: Number of interpretable results
He obs.: Observed heterozygisity
He n.b.: Expected nonbiased heterozygosity

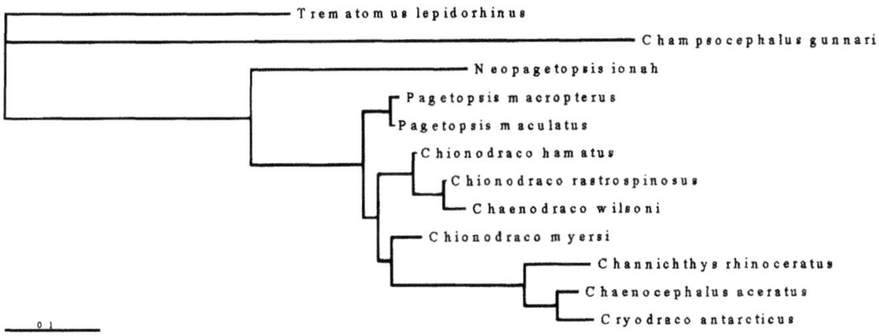

Fig. 1. The neighbor-joining tree obtained from the Nei [11] genetic distance matrix

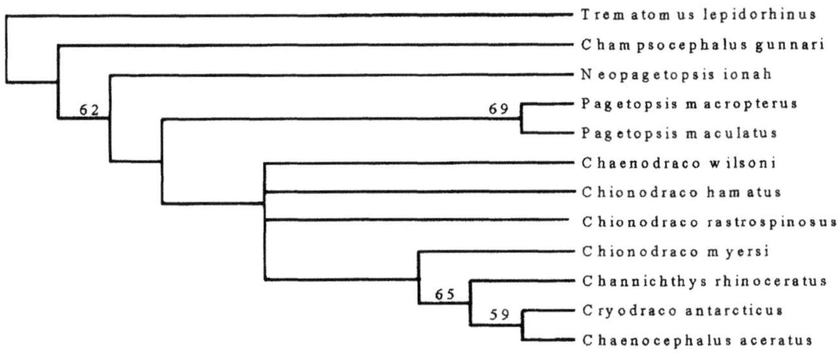

Fig. 2. The strict consensus of the two maximum parsimony trees obtained from a Branch and Bound search using PAUP 3.1.1. Tree length is 66; CI = 0.591; RI = 0.757. Bootstrap proportions above 50% are indicated (calculated using 1000 replicates)

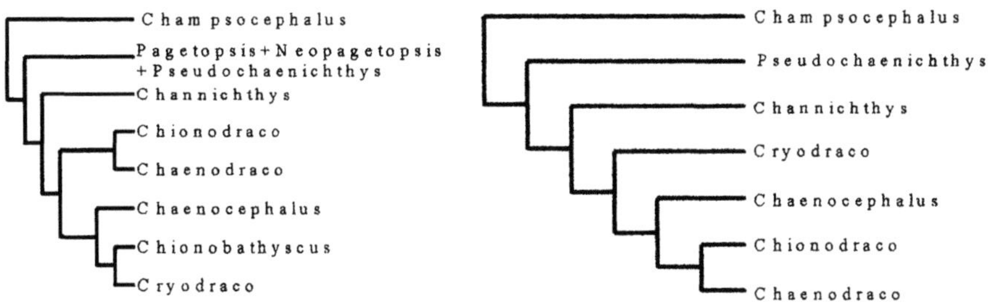

Fig. 3. *Left*: Iwami's (1985) cladogram based on morphological characters. *Dacodraco* branching point in the original tree was undetermined. *Right*: Voskoboinikova's (in press) phenogram based on the global similarity of developmental features. The original tree was rooted on *Cryodraco*; it was rerooted here on *Champsocephalus* for comparison. Original branch lengths are not presented here

The neighbour - joining tree and the Dollo parsimony trees are congruent, though the maximum parsimony trees do not solve the relations between the taxa *Chionodraco hamatus*, *Chionodraco rastrospinosus* and *Chaenodraco wilsoni*.

The most important results, in agreement in the two kinds of trees, are (1) the basal topology (*Champsocephalus* (*Neopagetopsis* (*Pagetopsis*, the rest))) found by the two methods; (2) the paraphyly of the genus *Chionodraco* and (3) the terminal clade (*Channichthys* (*Chaenocephalus*, *Cryodraco*)).

Discussion and Conclusions

Long Distance Differentiation

Based on the significant Fst, two species present a pattern of isolation by distance. *Chionodraco myersi* and *Neopagetopsis ionah* are the two first Antarctic species showing genetic divergence between two geographic areas, the Weddell Sea and Prydz Bay. Even if the observation is limited to two sites, it is new. Mitochondrial DNA analyses [15] and allozymic surveys [6] did not give such results.

One can notice that the significant long distance divergence was observed only in those species with good sample size. It is probable that if all species analyzed here had been represented by larger samples, more than two cases of differentiation would be described. On the other hand, *Champsocephalus gunnari* was not structured with sufficiently large samples in the three populations analyzed [6]. It is possible that Antarctic species are more structured due to local isolation than subAntarctic species as observed from chromosomal data (unpublished).

Structure of *Chionodraco myersi* populations in the Weddell Sea

Among the four populations considered, three are located in Weddell Sea. Among them, two (no. 6 and 8) are precisely located near Norvegia Cape, and the other in Halley Bay. The population 7 remains different from the two others. The physicochemical characteristics of that station [16] cannot explain the structure. Nevertheless, the Norvegia Cape, at the level of the slope break in the continental shelf, presents a rapid geotrophic stream running to the South (20 cm sec^{-1}). That peculiarity could provoke the isolation from the population 7. Thus there could not be any larval mixture due to the pelagic fry often permitting gene flow.

Heterozygosity in the Polar Environment

In marine fish, the mean heterozygosity is about 5-6% [17]. In our sample, except for *Chionodraco myersi*, whose heterozygosity is 10%, most species present a low polymorphism, but greater than that of the two species analyzed by the same markers by Duhamel et al. [6]. McDonald et al. [5], with seven species belonging to the *Trematomus/Pagothenia* complex, indicated that multilocus heterozygosity varied from 1.8 to 7.0% across taxa.

However, the data given here should be interpreted with caution: *Chionodraco myersi,* with a 65 individuals sample, is the best sampled species and the most polymorphic. The estimation of polymorphism level for species represented by small samples is thus doubtful.

Phylogeny of the Channichthyidae

The Dollo parsimony tree (Fig. 2) can be considered as reliable since its RI is relatively high (0.757). The most parsimonious tree maximizes synapomorphy hypotheses and minimizes the ad hoc hypotheses of homoplasy [18,19]. This rather high RI tends to show a relatively good level of structured variability in these allozymic data, an observation that is to be balanced with relatively low bootstrap proportions (below 70%).

Allozymic and morphological data provide congruent points and incongruent ones. Comparing Iwami's [3] tree (Fig. 3) and our Dollo-parsimony tree (Fig. 2), the congruent points are the basal position of *Champsocephalus*; then, the emergence of *Neopagetopsis* and *Pagetopsis* and the terminal position of the clade (*Chaenocephalus*, *Cryodraco*).

The two approaches disagree on the position of *Channichthys*, and the relative positions of *Chionodraco* and *Chaenodraco* (monophyly of the two in Iwami's [3] tree and paraphyly in our trees). Furthermore, the tree of Iwami [3] is not completely congruent with the phenogram of Voskoboinikova [4]; (see the tree Fig. 3, rerooted on *Champsocephalus* for comparison). Voskoboinikova's tree differs on the positions of *Chaenocephalus*, without being more congruent with ours.

However, Voskoboinikova's tree (originally rooted on *Cryodraco* without justification) is not a phylogenetic tree *sensu stricto* but an analysis of global similarity in developmental features. The comparison is therefore difficult to make.

Our hypothesis concerning the paraphyly of *Chionodraco* does not appear elsewhere. That paraphyly, whatever its robustness, may be due to different sample sizes in *Chionodraco* and other genera.

In conclusion, in spite of poor bootstrap proportion, these allozymic data appear informative, providing two parsimonious trees with an RI of 0.757. It seems reasonably established that *Champsocephalus* is the most basal channichthyid. The terminal clade (*Channichthys* (*Cryodraco*, *Chaenocephalus*)) is one of the best established nodes (bootstrap proportion of 65%), but not congruent with morphological data analysed so far with respect to the relationships of *Channichthys*. On one hand, one must keep in mind that *Channichthys* has 15 characters of 51 which are missing (question marks) in the NEXUS matrix. On the other hand, the morphological characters of Iwami [3] advocating a more basal position of *Channichthys* could also be reconsidered. Allozymic data do not clarify either relationships of *Chionodraco* and *Chaenodraco*, or the monophyly of *Chionodraco*.

Acknowledgments

This work is a part of the EPOS ESF program and was also coordinated within the European Science Foundation Network " Fishes of the Antarctic Ocean." The CNRS (GDR 1005, UMS 826 and UPR 9060) provided financial supports. We are specially grateful to Drs. E. Pisano and R. Williams for having provided the samples from Heard Island and Prydz Bay.

References

1. Eastman JT (1993) Antarctic fish biology, evolution in a unique environment. Academic Press, London New York
2. DeVries Al (1971) Freezing resistance in fishes. In: Hoar WS, Randall DJ (eds) Fish physiology, Vol 6, pp 157-190
3. Iwami T (1985) Osteology and relationships of the family Channichthyidae. Mem Natl Inst Polar Res ser E 36:1-69
4. Voskoboinikova OS (1997) Osteological development of the Channichthyidae. Cybium 21(4):369-380
5. McDonald MA, Smith MH, Smith MW, Novak JM, Johns PE, DeVries AL (1992) Biochemical systematics of notothenioid fishes from Antarctica. Biochem Syst Ecol 20 (3):233-241
6. Duhamel G, Ozouf-Costaz C, Cattaneo-Berrebi G, Berrebi P (1995) Interpopulation relationship in two species of Antarctic fish *Notothenia rossii* and *Champsocephalus gunnari* from the Kerguelen Islands: an allozyme study. Antarct Sci 7(4):351-356
7. Wright S (1951) The genetical structure of populations. Ann Eugenics 15:323-354
8. Pasteur N, Pasteur G, Bonhomme F, Catala J, Britton-Davidian J (1987) Manuel de génétique par électrophorèse des protéines. Lavoisier, Paris

9. Shaklee JB, Allendorf FW, Morizot DC, Whitt G (1990) Gene nomenclature for protein-coding loci in fish. Trans Am Fish Soc 119:2-15
10. Belkhir K, Borsa P, Goudet J, Chikhi L, Bonhomme F (1996) Genetix v.3.0, logiciel sous Windows TM pour la génétique des populations. Montpellier II, France
11. Nei M (1978) Estimation of average heterozygosity and genetic distance from a small number of individuals. Genetics 89:583-590
12. Weir BS, Cockerham CC (1984) Estimating F-statistics for the analysis of population structure. Evolution 38:1358-1370
13. Saitou N, Nei M (1987) The neighbour-joining method: a new method for reconstructing phylogenetic trees. Mol Biol Evol 4(4):406-425
14. Swofford DL (1993) PAUP: phylogenetic analysis using parsimony, version 3.1.1. Computer program distributed by the Illinois Natural History Survey, Champaign, Illinois
15. Williams R, Smolenski AJ, White RWG (1994) Mitochondrial DNA variation of *Champsocephalus gunnari* Lönnberg (Pisces: Channichthyidae) stocks on the Kerguelen Plateau, southern Indian Ocean. Antarct Sci 6 (3):347-352
16. Arntz W, Ernst W, Hempel I (eds) (1990) The expedition ANTARKTIS VII/4 (EPOS leg 3) and VII/5 of RV "Polarstern" in 1989. Ber. Polarforsch. 68. A. Wegener Institute for Polar and Marine Research, Bremerhaven
17. Ward RD, Woodwark M, Skibinski DO (1994) A comparison of genetic diversity levels in marine, freshwater, and anadromous fishes. J Fish Biol 44: 213-232
18. Farris JS (1979) The informative content of the phylogenetic system. Syst Zool 28: 483-519
19. Farris JS (1983) The logical basis of phylogenetic analysis. In: Platnick N, Funk VA (eds). Advances in cladistics, Vol 2. Columbia University Press, New York, pp 1-36

Origin and Mechanism of Evolution of Antifreeze Glycoproteins in Polar Fishes

Chi-Hing C. Cheng

Department of Molecular and Integrative Physiology, University of Illinois, Urbana, IL 61801, USA

Introduction

The frigid waters of the polar oceans delimit the cold extreme for marine life. This is particularly true in the case of the thermally isolated Antarctic Ocean which is perpetually near or at freezing (-1.9 °C) due to the thermal barrier imposed by the Antarctic Circumpolar Current [1]. The most fundamental survival challenge faced by teleost fishes in these waters is a physical one - the threat of being frozen. The body fluids of marine teleosts including polar species are hyposmotic to sea water, 300-600 mOsM [2,3] versus 1000 mOsM, and thus have a higher colligative freezing point than the latter, -0.56 °C to -1.1 °C versus -1.86 °C. By these simple physical considerations alone, freezing death would be unavoidable especially in the presence of ice. Unlike some reptiles and amphibians, fish cannot survive even partial freezing of their body fluids. A number of polar and subpolar fishes had overcome this environmental challenge with a biological solution - they evolved ice-binding antifreeze proteins which enabled them to successfully colonize icy habitats that were otherwise out of their reach. The impact of the evolution of these unique antifreezing proteins on organismal and ecological success is manifested most strikingly in the case of the Antarctic notothenioid fishes - a single teleost suborder (Notothenioidei) that has come to dominate today's Antarctic fish fauna in terms of species number (~50%) and biomass (≥90%) [2,4-6].

Although other forms of adaptations - biochemical, physiological, anatomical, and behavioral (other papers, this Vol.) - have also evolved and must have contributed in concert to the ecological success of the notothenioids in their frigid environment, it is fair to say that since a frozen fish could not evolve, the emergence of the antifreeze protein function to assure survival in the first place was pivotal in allowing other adaptations and organismal diversification to occur.

It is intuitively obvious that the driving force for the evolution of antifreeze proteins was the cooling of the polar oceans to freezing temperatures over geologic time. What is not obvious is where these novel

G. di Prisco, E. Pisano, A. Clarke (Eds)
Fishes of Antarctica. A biological overview
© Springer-Verlag Italia 1998

proteins came from. What are their genetic origins and the molecular mechanisms that created them? When did they appear? The availability of modern molecular techniques has allowed us to tackle these longstanding questions and answer them in some cases. This chapter reviews the recent findings on the evolution of the antifreeze glycoproteins of the Antarctic notothenioids, and the near-identical antifreeze glycoproteins of the unrelated northern cods.

Diversity of Fish Antifreeze Proteins

Four structurally different antifreeze proteins - antifreeze glycoprotein (AFGP), and type I, II, and III AFPs (antifreeze peptides) from different cold-water fishes - have been well characterized. These proteins are found in abundance in the circulation of these fishes, 10-35 mg/ml depending on the species and the severity of their cold environment. Despite their diverse structures, they all perform the same function. The generally accepted mechanism of antifreeze action is adsorption-inhibition. It is a non-colligative mechanism involving binding of antifreeze molecules to specific faces of the ice crystals that occasionally enter the fish, and inhibit ice growth at the temperatures they live in, through the so-called Kelvin effect whereby the thermodynamic driving force for ice expansion was essentially removed. Thus the fish's body fluids are preserved in a liquid state and the fish avoids freezing ([7,8] for reviews).

The first antifreeze protein discovered was the AFGP from Antarctic notothenioid fishes by DeVries [9]. AFGPs are present in the notothenioids as a family of size isoforms, all composed of repeating units of the simple glycotripetide monomer:

$(\text{-Thr-Ala-Ala-})_n$
/
N-acetylgalactosamine
/
Galactose

Eight different sizes of AFGPs were initially described; the largest was designated as AFGP1 (molecular weight 34 kDa, n=55), and the smallest as AFGP8 (molecular weight 2.6 kDa, n=4) [10,11]. This nomenclature has persisted although better protein resolution techniques have now revealed the presence of many other intermediate sizes, totaling at least 18 in the McMurdo species *Dissostichus mawsoni* and *Pagothenia borchgrevinki* [12]. In addition to size variations, minor compositional variations are found in the small AFGPs 6-8 in which the first Ala in the repeats are sometimes replaced by Pro [12,13]. Thus the notothenioid AFGPs, although simple in primary structure, are complex in protein heterogeneity (size and composition).

AFGPs were subsequently found in several Arctic and north Atlantic cods which are unrelated to notothenioids [14,15]. Cod AFGPs are identical to notothenioid AFGPs in every way, structure and protein heterogeneity. The only difference is that Thr in the tripeptide repeats is occasionally replaced by an Arg [14,15].

The three types of AFPs are small peptides or proteins without attached sugars. They were first classified based on distinctive differences in their protein sequences. Type I AFPs are Ala-rich (>50%), α-helical peptides of 3.3 - 4.5 kDa, most of which have an 11-residue repeat ending with a Thr. They are found in the pleuronectids (flounder and plaice, order Perciformes) [16,17] and the unrelated cottid sculpins (order Scorpaeniformes) [18]. Type II AFPs are 11 - 24 kDa, cystine-rich, β-structured proteins found in a highly divergent group of fishes - smelt (order Salmoniformes), herring (order Clupeiformes) [19] and sea raven (order Scorpaeniformes) [20]. Type III AFPs are 7 kDa and 14 kDa globular proteins with unbiased amino acid composition found in the closely related eel pouts (zoarcids) [21-23] and wolffish (anarhichadid) [24].

Two recurring themes are evident in the studies of the various fish antifreezes - disparate protein structures but identical protein function, and inconsistent correlation between the phyletic relatedness of the various fishes and the type of antifreeze protein they synthesize. Both indicate that fish antifreeze proteins arose independently (at least four times) and achieved remarkable functional convergence. As a monofunctional group of macromolecules, the diverse antifreeze proteins epitomize the creativity of molecular evolution. Where did they come from?

Evolutionary Enigma of Near-Identical AFGPs in Unrelated Antarctic Notothenioids and Northern Cods

For over two decades since the AFGPs were discovered, the puzzle has persisted as to how two such geographically and phylogenetically distant fishes, the modern Antarctic notothenioids (superorder Acanthopterygii, order Perciformes) and the more basal northern cods (superorder Paracanthopterygii, order Gadiformes), could have arrived at essentially the same AFGPs. An obvious possibility which had been suggested is that the two taxa inherited their AFGPs from a common ancestor [25], but this is counter to their known phylogeny [26], their distinct geographic origins [27,28], and the recent but separate glacial histories of the polar regions [1], which collectively indicate that their ancestor is ancient and would not need or have an antifreeze. The alternate possibility is that these two groups evolved their AFGPs separately, but it is equally difficult to conceive how two near-identical sets of complex proteins could arise independently by chance. As gene sequences are records of evolutionary histories, through detailed comparative analyses of the AFGP genes of Antarctic notothenioids and northern cods, we have recently unveiled part of the evolutionary mystery of the AFGPs [29-31].

AFGP Genes of Antarctic Notothenioids

Notothenioid AFGPs are encoded by a family of many member genes [12]. The first AFGP gene reported was from *Notothenia coriiceps* [32]; the sequence was later determined to be missing the signal peptide coding sequence [29]. We have now reported the complete sequences of two AFGP genes from a related nototheniid, the giant Antarctic toothfish *D. mawsoni* [29,30]. A diagrammatic representation of a *Dm (D. mawsoni*)

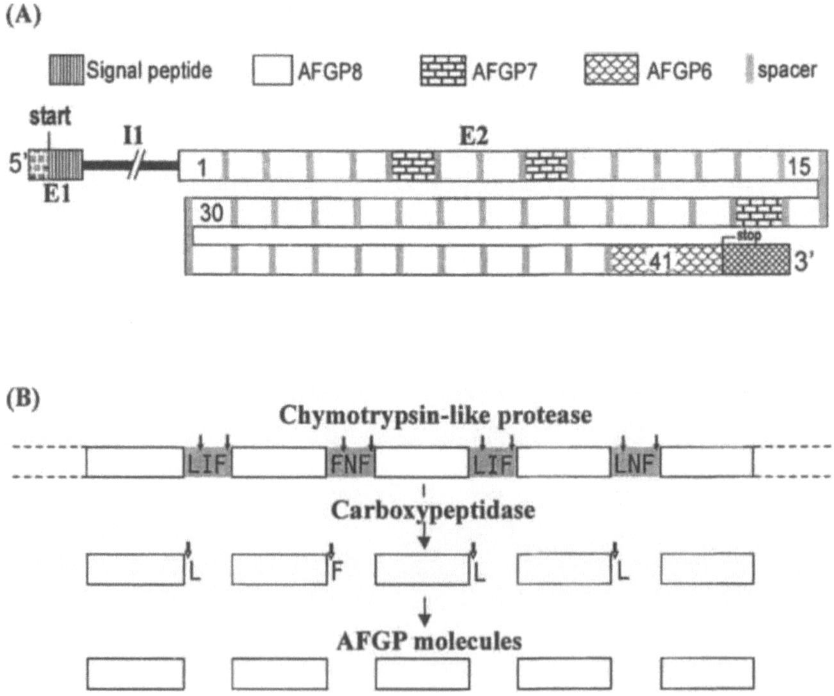

Fig. 1. A Structure of an AFGP polyprotein gene from *D. mawsoni* (clone *Dm*1A, [29]) that encodes 41 molecules of AFGP. **B** Processing of the AFGP polyprotein precursor. A chymotrypsin-like protease cleaves at Leu (L) and Phe (F), and a carboxypeptidase removes the C-terminus Leu or Phe, producing the individual mature AFGPs

AFGP gene is shown in Fig. 1A. It consists of two exons interrupted by a single large intron (I1, about 1.9 kbp). The small exon 1 (E1) encodes the 5'UTR (untranslated region) and the signal peptide, and the large exon 2 (E2) encodes the AFGPs and 3'UTR. Other notothenioid AFGP genes have the same structures (unpublished data).

Notothenioid AFGP genes have an unusual polyprotein structure in the AFGP coding region (E2) (Fig. 1A). While most protein genes encode one protein molecule per gene, each notothenioid AFGP gene encodes a polyprotein precursor containing many AFGP molecules, as many as 46 in the *Nc* (*Notothenia coriiceps*) gene [32]. These molecules are linked in tandem by 3-residue spacers with the highly conserved sequence of L/F-I/N-F (Leu/Phe-Ile/Asn-Phe) (Fig. 1B) [12,29,30,32]. Since the spacer residues are not found in the mature AFGPs, and Leu and Phe are known substrates of chymotrypsin, it is very likely that the AFGP gene is transcribed and translated into a polyprotein which is then cleaved post-translationally at the spacers by a chymotrypsin-like protease to yield the individual AFGP molecules (Fig. 1B).

The one *Nc* and two *Dm* AFGP genes that we have reported all encode the small AFGP isoforms, AFGPs 6, 7 and 8, encompassing their various Pro (substituting Ala) variants [29,30,32]. We have now also obtained another gene from *Dm* that encodes two large size isoforms, in the range of AFGP4 and AFGP1 respectively, among five other small ones (manuscript in preparation). These gene sequence data provided two pieces of important information regarding AFGP heterogeneity and abundance. First, each and every size and compositional AFGP isoform is distinctly encoded as individual copies within polyprotein genes. Thus the various lengths are not due to protein processing through splicing small AFGPs or cleaving large AFGPs into small ones. Second, a large family of polyprotein genes [12] can rapidly escalate the gene dosage that must contribute to the circulatory abundance (35 mg/ml) of the AFGP in these fishes.

Notothenioid AFGP Gene Evolved from a Pancreatic Trypsinogen Gene

The first clue to the origin of notothenioid AFGP gene was obtained from a data base search for similar sequences, using the three AFGP genes that encode small AFGPs as queries. The 3' flanking sequences of these AFGP genes were found to share over 70% identity with the 3' end of a trypsinogen cDNA from a pleuronectiform teleost, Atlantic plaice, *Pleuronectes platessa* [33]. This high degree of nucleotide sequence identity clearly signals that notothenioid AFGP and trypsinogen genes are somehow evolutionarily related. It was not obvious however how an AFGP gene could evolve from a trypsinogen gene as the amino acid sequences of the two proteins share no similarity whatsoever.

To decipher this relationship, we isolated and sequenced trypsinogen gene and cDNA from the same notothenioid, *D. mawsoni*, and compared them to the *Dm* AFGP gene and cDNA sequences. The results and comparative analyses that lead to our discovery of the molecular mechanism involved in the transformation of a trypsinogen gene to an AFGP gene have been published in detail recently [29]. The essential aspects of the evolutionary process are shown in Fig. 2, and described below.

The front and tail segments of *Dm* trypsinogen and AFGP genes where high degree of sequence identity (93-96%) resides clearly were recruited and incorporated to form the new gene. The front segment of trypsinogen gene, namely E1 and I1 inclusive of a run of gt dinucleotides, was recruited and used as is in the AFGP gene. Exon 1 which contains the coding sequence for the signal peptide would provide the secretory signal for the emerging new antifreeze protein. A large insertion (~1.7 kbp) was subsequently incorporated into I1. The tail end of trypsinogen gene, namely the last exon (exon 6) which encodes the C-terminus 50 residues of trypsin and the 3'UTR, was recruited and converted into the 3' sequence of AFGP gene starting from the penultimate codon in the AFGP coding region through the downstream 3' flanking sequence (same as 3'UTR). This region corresponds to the segment of plaice trypsinogen cDNA that shares >70% identity with AFGP 3' flanking sequence in the initial database search.

The putative ancestral sequence that would give rise to the AFGP coding sequence comprising E2 in the AFGP gene required more intensive searches. At the very end of the front segment of sequence similarity, a 9-nucleotide ThrAlaAla coding element, ac*ag*cggca (splice sequence in italics) was found straddling the junction of I1 and E2 of trypsinogen gene (Fig. 2). This 9-nt sequence corresponds very closely to the coding sequences of the ThrAlaAla repeats in the AFGP polyproteins (Fig. 2), indicating strongly that it is the extant version of the ancestral sequence that gave rise to the AFGP tripeptide repeats through iterative duplications.

It is very likely that the ancestral 9-nt element was ac*ag*c**t**gca (g for **t** substitution in extant version), reflected in gct being the majority (61%, versus 36% of gcg) of the codons for the first Ala in the AFGP tripeptide repeats (Fig. 2). A single nucleotide substitution of this codon will provide the same amino acid Ala (gct to gcg), or the periodic Pro (gct to cct) seen in this position in the tripeptide repeats. The almost exclusive bias for cct as the Pro codon (96%) (Fig. 2) in the AFGP polyprotein genes lends strong support to this reasoning.

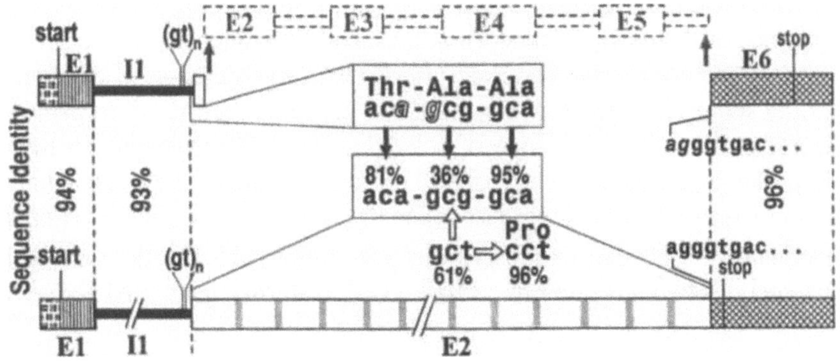

Fig. 2. Proposed evolutionary process of an AFGP gene from its trypsinogen progenitor in notothenioid fish. Two gene segments from trypsinogen gene, E1, I1 and several nucleotides of E2 at the front end, and E6 at the tail end, were recruited. E1 provided an identical signal peptide for, and E6 was converted into 3' sequence of, the new AFGP gene. Sequence identities were indicated numerically (%) and by same shading pattern in these regions. The 9-nt ThrAlaAla coding element from trypsinogen that straddles the E1/I1 junction gave rise to a new AFGP coding region through repeated duplications, as reflected in the close correspondence between its sequence and those of the tripeptide repeats (given in %) in extant AFGP genes. The codon for the first Ala was likely gct, as it would provide for the same residue Ala, or Pro replacement seen at this position through a 1-nt substitution. During this evolutionary process, the bulk of trypsinogen gene (remainder of E2 through E5, dashed bars and boxes) was removed, likely after some amount of tripeptide duplications had taken place (see text), and a stop codon (tga) for the coding region of AFGPs was established from a 1-nt shift in reading frame at the 5' end of the recruited E6. The intron splice sequences are given in italics. The origin of the spacer sequence in the polyprotein is unknown at this time

Thus, an ancestral trypsinogen gene spawned the primordial structure for an AFGP gene by bequeathing and linking together two of its segments (front and tail end) and shedding the bulk of the trypsin coding sequence in the middle (most of E2 through I5). The partly non-sense (intron) 9-nt element that translates into ThrAlaAla, positioned thus in the middle of the evolving structure, underwent iterative duplications to create an entirely new protein coding region for the tripeptide repeats of AFGP. The first duplication of the single ancestral 9-nt element probably occurred due to slippage at the repetitive gt dinucleotides immediately ahead of it during DNA replication, as repetitive sequences are prone to this process [34-36].

Figure 3 proposes a possible scheme for the first duplications. In support of the scheme, the translated sequence at the 5' end of exon 2 after the proposed first slippage replication, AlaValThrAlaAla-, is found in almost all the AFGP genes we have sequenced ([29,32], and unpublished data).

Subsequent duplications could arise from slippage at either the $(gt)_n$ sequence or the newly minted ThrAlaAla coding sequences (Fig. 3). These duplications could have occurred prior to the large deletion of the trypsin coding sequence in the evolving structure producing an AFGP/trypsinogen hybrid protein. We have indeed isolated a *Dm* AFGP/trypsinogen hybrid gene, that is, an AFGP gene that still retains the bulk of the trypsin coding region (from E2 through at least I4) as its downstream sequence without a terminator in between (manuscript in preparation). This is the same gene that encodes two large and five small AFGP isoforms mentioned above.

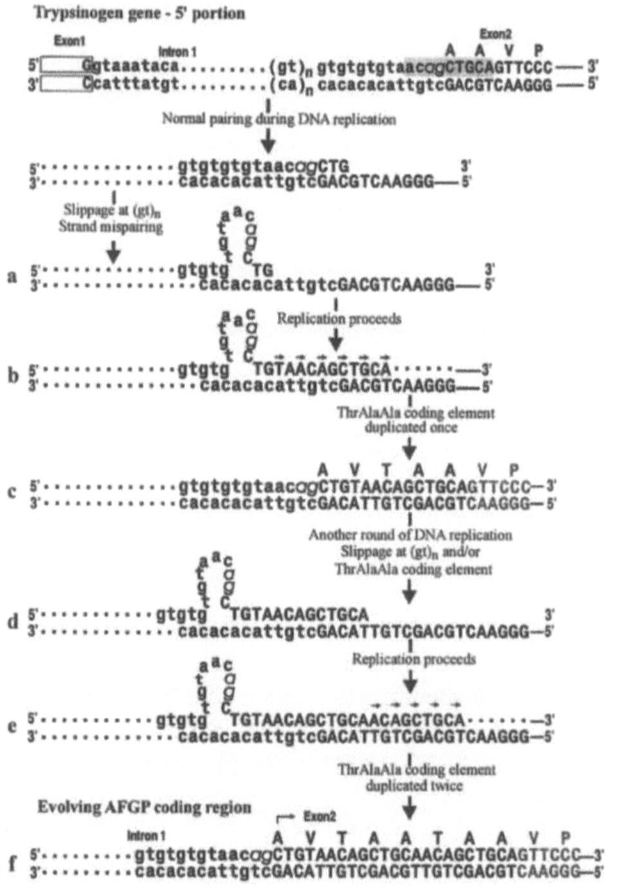

Fig. 3. A likely scheme for the first duplications of the 9-nt ancestral trypsinogen ThrAlaAla coding element, ac*ag*ctgca (highlighted; splice sequence in italics). Slippage replication and mispairing at the $(gt)_n$ immediately upstream of the coding element could lead to its first duplication (a to c). A second round of slippage replication would lead to 2 tripeptide repeats (d to f). A third round would lead to 4 tripeptide repeats, resulting in AFGP8, the smallest functional AFGP

This hybrid gene indicates that tripeptide duplication first and bulk deletion of trypsinogen later was likely the order of events in the conversion of trypsinogen to AFGP gene. The origin of the spacer sequence is presently unknown. Apparently it must have been acquired early as it was iteratively duplicated along with the ThrAlaAla coding sequence to produce the polyprotein structure. Further investigations may solve this remaining mystery in the trypsinogen to AFGP conversion.

How Fast Could the First Notothenioid AFGP Gene Emerge?

The threat of freezing is a matter of life and death, and therefore would be expected to exert tremendous selection pressure for the emergence of a freeze-avoidance function, otherwise a lineage would rapidly terminate. In the notothenioids, the first ThrAlaAla duplication in fact could have occurred before the selection pressure was in full force, as an accidental result of replication slippage at the minisatellite sequence $(gt)_n$ in intron 1 immediately ahead of the putative ThrAlaAla coding element in the ancestral trypsinogen gene (Figs. 2,3). A single or two ThrAlaAla repeats at the N-terminus of a trypsin molecule might not affect trypsin activity but could endow the hybrid protein with some measure of ice-binding activity, which would be a selective advantage that became retained as the seawater approached and hovered around freezing. The destination of trypsin secretion, the intestinal fluid, where ice crystals could enter through seawater and food ingestion, provided the logical testing site for refinement of the emerging antifreeze molecule. The smallest fully functional AFGP, AFGP8, has only 4 tripeptide repeats, which could be produced in only two additional rounds of slippage replication after the first ThrAlaAla duplication depicted in Fig. 3 (a to c). Thus the "experimental process" whereby the first functional antifreeze gene (and gene product), likely an AFGP8/trypsinogen hybrid, was generated could theoretically be very short, a matter of several rounds of DNA replications. As the environmental pressure increased and persisted, the incipient AFGP/trypsinogen hybrid gene could rapidly undergo further AFGP tripeptide expansion and whole gene duplications to escalate the gene dosage required for large AFGP output to meet the environmental exigency.

When Did Notothenioid AFGP Gene Arise?

The answer to this question will also answer the question of when the Antarctic Ocean reached freezing as the two are presumably temporally linked. Since the partly non-sense 9-nt ThrAlaAla coding element in the ancestral trypsinogen gene by itself had no functional consequence, it is reasonable to suggest that its de novo amplification must be driven by certain selection pressure to provide a new function required by the organism, in this case, the novel ice-binding function to survive the increasingly frigid, icy Antarctic water. By the use of a molecular clock rate (sequence divergence rate), 0.5-0.9% per million years estimated for the mitochondrial DNA of another teleost (salmon) [37], and the amount of sequence divergence between *Dm* trypsinogen and AFGP genes in their homologous segments (4-7%) (Fig. 2), we deduced that the conversion of the ancestral trypsinogen gene to the first AFGP gene occurred about 5-14 million years ago (mya) [29], a relatively recent event in evolutionary time.

The time frame during which the Antarctic Ocean reached freezing temperatures has been examined by physical methods (oxygen isotope ratios of planktonic sediments) extensively by paleo-oceanographers, but the estimates have not been unequivocal. The generally accepted time frame in which the Antarctic water approached present-day conditions is mid-Miocene or about 10-14 mya, subsequent to the establishment of the Antarctic Circumpolar Current at about 22 mya which lead to the thermal isolation of Antarctica [1]. This estimate agrees remarkably well with the estimated time of appearance of notothenioid AFGP gene from our molecular data. There is some evidence suggesting production of Antarctic sea ice at the high latitudes as far back as mid-Oligocene (38 - 30 mya) when glaciation occurred over land, although bottom water temperature was still above freezing [1]. Could the first emergence of notothenioid AFGPs have occurred in that time frame [38]? The estimate from molecular data indicates not so, that it is in fact a much more recent, mid-Miocene event.

Another piece of evidence that supports the mid-Miocene time frame of AFGP evolution comes from recent molecular phylogenetic analyses of the notothenioid fishes based on mitochondrial rRNA gene sequences, which showed that the extant notothenioid families emerged during a burst of rapid phyletic diversification around 7-15 mya [39]. The ancestral notothenioid stock is believed to be a bottom dweller [3,28], while the extant families occupy all niches throughout the water column. Thus it is reasonable to suggest that the ability to radiate into the cryopelagic habitats must coincide with or closely follow the emergence of the antifreeze protective function.

Three independent areas of studies, paleoclimatic history of Antarctica, evolution of notothenioid AFGPs, and molecular phylogeny of notothenioid fishes, each with its own methodological uncertainties and limitations, appear to have converged on the same mid-Miocene time frame for three temporally related events - freezing of the Antarctic water, evolution of notothenioid AFGP gene, and speciation and radiation of notothenioid fishes. It is difficult to dismiss such remarkable agreement as mere coincidence. Indeed this is a rare case in which protein evolution, organismal adaptation, and environment conditions can be linked directly through corroborating biological/molecular and physical evidence.

Innovativeness of the Evolutionary Process of Notothenioid AFGP Gene

The evolutionary process by which notothenioid AFGP genes arose is novel and creative. Most new genes evolve from existing genes or gene segments, either through duplication and divergence [40], recruitment [41] or exon shuffling [42]. Type II AFPs of sea raven, smelt and herring, the only other antifreeze protein with a known evolutionary origin - the carbohydrate binding domain of C-type lectins [19] - apparently arose through straight forward duplication and divergence. The formation of the notothenioid AFGP gene indeed utilized some aspects of these known mechanisms. What sets it apart is that an entire new protein coding region was built up from a rudimentary foundation - a minimal partly noncoding

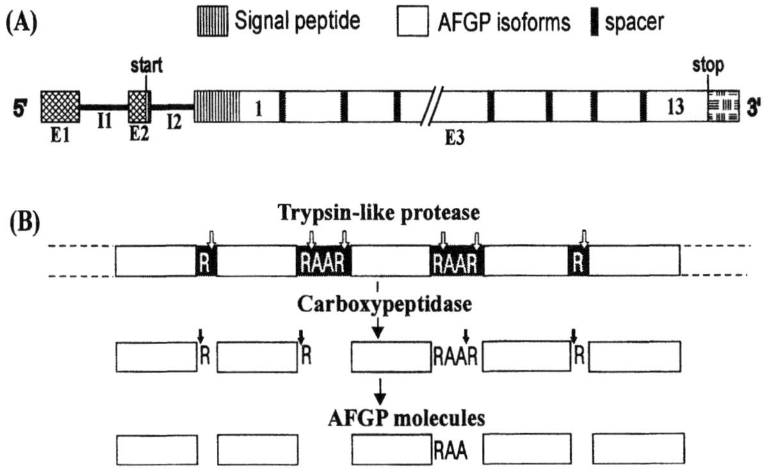

Fig. 4. A Structure of an AFGP polyprotein gene from the Arctic cod *Boreogadus saida* (clone *Bs*3L, [30]) that will produce 13 molecules of AFGP if all the spacers (Arg residues) are cleaved. **B** Processing of the AFGP polyprotein precursor. A trypsin-like protease cleaves at the carboxyl side of Arg (R), and a carboxypeptidase removes the Arg, producing the individual mature AFGPs. Not all spacers are removed as a small percentage of Arg is retained in the mature AFGPs

sequence - through repeated duplications to create a unique protein with the novel ice-binding function, thereby creating sense from partly non-sense.

Antifreeze Glycoprotein Genes of Arctic Cod

Through the use of molecular analyses we have also addressed the lingering mystery of how the northern cods arrived at essentially the same AFGPs as the unrelated Antarctic notothenioids [30]. We recently reported the first AFGP gene from a northern gadid, the Arctic cod, *Boreogadus saida* [30]; its schematic structure is shown in Fig. 4A. Other *Bs* (*B. saida*) AFGP genes have the same structure (unpublished data).

The *Bs* AFGP genes consist of three exons and two introns. Exons 1 and 2 and the two introns are small. E1 and E2 encode the 5'UTR and the translation start (Met). The large exon 3 encodes the signal peptide, an apparent AFGP precursor, and the 3'UTR. The encoded AFGP precursor consists of a long string of Thr-Ala/Pro-Ala repeats, punctuated at various intervals by an Arg (R) or ArgAlaAlaArg (RAAR), with the Arg residue occupying where a Thr is normally found. The percentage of Arg residues in the precursor encoded in these genes is always much higher (≥7%) than that in mature AFGPs (0.6-1.2%) [30]. Thus, similar to the notothenioids, Arctic cod AFGP genes appear to also encode AFGPs as a polyprotein precursor, but with some of the Arg residues serving as cleavable spacers and involving a trypsin-like protease instead (Fig. 4B).

Arctic Cod AFGP Genes Evolved Independently

The polyprotein gene structure of the AFGP coding region of the Arctic cod and notothenioid AFGP genes is where their similarity ends.

Molecular evidence from detailed comparative analyses argues strongly for independent evolution of the cod AFGP genes, and thus the near-identical AFGPs of these two unrelated fish are a rare example of protein sequence convergence [30]. The evidence is summarized below

Arctic Cod AFGP Gene Did Not Evolve From Trypsinogen Gene

The Arctic cod AFGP genes have none of the inherited features that bespeak the trypsinogen origin of the notothenioid AFGP genes, i.e. the 5' and 3' regions of high sequence identity which include identical signal peptides. Thus, assuming an endogenous genetic origin, the cod AFGP genes must have evolved from a different genomic locus which is decidedly not trypsinogen.

Nonhomologous Gene Structure

Homologous genes, i.e. genes that share a common ancestor, are highly conserved in terms of the number of exons and introns and their positions.

This is clearly not the case between the AFGP genes of the two fishes. Arctic cod AFGP genes have three exons and two introns, and the introns intervene in the 5'UTR and between translation start and signal peptide (Fig. 4A). By contrast, notothenioid AFGP genes have two exons and a single large intron, and the intron intervenes between signal peptide and the AFGP polyprotein (Fig. 1A). The AFGP genes of the two fishes are thus not homologous.

Distinct AFGP Tripeptide Coding Sequences

The codons for the tripeptide repeats in Arctic cod AFGP genes have a very different bias from that of the notothenioids. The sequence composition of one of the *Bs* genes is as follows:

Thr -	Ala -	Ala
48%act	54%gca	54%gca
47%aca	33%gcg	38%gcc

Each of the three amino acids is encoded primarily by two codons, one of which (highlighted) is seldom or never used in notothenioid AFGP genes (Fig. 2). The Pro that replace the first Ala are encoded by either cca (50%) or ccg (48%) and very likely resulted from a 1-nt substitution of gca or gcg. By contrast, the Pro in this position in notothenioid AFGPs are encoded exclusively by cct, a result of a 1-nt substitution from gct (not used in Arctic cod) to cct (Fig. 2). These differences indicate that the Arctic cod AFGP repeats are derived from duplications of an unrelated ancestral ThrAlaAla coding element.

Unrelated Spacer Sequence

The spacers of the two groups of genes share no sequence similarity at both the nucleotide and protein level and are clearly unrelated. The distinct spacers in fact invoke the use of a different protease in the processing of the polyprotein precursor. While the origin of the notothenioid spacers is unknown, the spacers in Arctic cod clearly arose from a 1-nt transversion, aca to aga, converting a Thr to an Arg, since the Arg residues are exclusively encoded by aga, and always replace a Thr in the repeats [30].

Convergent Evolution of Notothenioid and Cod AFGPs

Although the origin of the Arctic cod AFGP genes remains to be determined, the above molecular evidence is consistent with and strongly supports a separate ancestry. Thus the very similar AFGPs of the orthern cods and Antarctic notothenioids are the products of convergent evolution.

Protein sequence convergence is very rare, and difficult to establish unequivocally [43]. A priori, the probability for two very similar complex proteins to arise separately is very low if not nil. What made it possible in the case of the AFGPs is very likely the simple primary structure, repeats of just three amino acids or nine nucleotides. The apparently complex AFGP polyprotein gene structure as well as the protein heterogeneity encoded in these genes in both fish, seemingly improbable to have risen separately by chance, could in fact arise quite readily from the tendency for repetitive sequences to undergo expansion through slippage replication and unequal crossing over [34-36]. A prime example of the occurrence of such a process is the variable lengths of short tandem repeats, or length polymorphism, in mini- and microsatellite DNA prevalent in eukaryote genomes [44-46], which forms the basis for population studies [46,47].

Indeed the 9-nt AFGP tandem repeats in both fish could be regarded as minisatellite DNA-like, which could have and had undergone similar expansion process as minisatellite DNA. The immense selection pressure from freezing water temperatures drove intragene expansion of the repeats by slippage replication or unequal crossing over, as well as whole gene duplications, resulting in the two similar and apparently complex AFGP families in these two unrelated fish.

Future Studies

Our recent work on the AFGP genes demonstrates the power of molecular analyses in answering several long-standing biological questions in the field of fish antifreeze research - where did the Antarctic notothenioid AFGPs come from (trypsinogen), how did they arise (partial recruitment and de novo amplification), when did they appear (mid-Miocene), and how did the unrelated northern cods arrive at the same AFGPs (through convergent evolution).

The use of molecular techniques will undoubtedly lead to the answers for the many other remaining or new questions regarding antifreeze protein. Besides the northern cods, the origin of type I AFP of flat fishes and sculpins, and of type III AFP of eel pouts and wolffish, has yet to be determined. New antifreezes are being discovered in addition to these well known types. A recent report describes the presence of an AFP in long-horn sculpin that differs in primary sequence from all the known antifreezes [48], although it is present in extremely low circulatory concentrations and thus its physiological role as an antifreeze is dubious.

In any event, like type II AFP, this may represent another case in which the progenitor molecule could serve as an antifreeze prototype, and thus holds interest from the standpoint of structure-function relationships. We have also recently discovered an abundant AFP in an Arctic liparid fish that superficially resembles type I AFP in having high Ala content but differs in all other aspects (no sequence repeats and over two times in size) (unpublished results), and thus likely represents a new antifreeze type with a distinct genetic origin. The evolution of the diverse fish antifreeze proteins to achieve the same end function undoubtedly qualifies as one of the most interesting model systems for studies of macro-adaptation and molecular evolution. As more advances are made in deciphering the origins and evolution of these proteins, a wealth of information on adaptive molecular evolution will certainly be forthcoming.

Acknowledgments

I would like to thank the organizers of the Fishes of the Antarctic Ocean Network and especially Dr. Guido di Prisco for the invitation to give this presentation at the final conference of the Network. I greatly appreciated the travel information provided by Dr. Eva Pisano. The European Science Foundation funded my travel to the conference. I thank Dr. Joe Eastman for all his helpful discussions. This work was funded by NSF grants OPP 96-15023 to C.-H.C. Cheng and A.L. DeVries and OPP 93-17629 to A.L. DeVries.

References

1. Kennett JP (1982) Marine geology. Prentice-Hall, New Jersey
2. Prosser CL (1973) Water: osmotic balance; hormonal regulation. In: Prosser CL (ed) Comparative animal physiology. Saunders, Philadelphia, pp 1-78
3. Eastman JT (1993) Antarctic fish biology. Academic Press, California
4. Hubold G (1991) Ecology of notothenioid fish in the Weddell Sea. In: di Prisco G, Maresca M, Tota (eds) Biology of Antarctic fish. Springer-Verlag, Berlin, pp 3-22
5. Ekau W (1990) Demersal fish fauna of the Weddell Sea. Antarct Sci. 2:129-137
6. Dewitt HH (1971) Coastal and deep-water benthic fishes of the Antarctic. In: Bushnell VC (ed) Antarctic map folio series folio 15. American Geographical Society, New York, pp 1-10
7. Cheng C-HC, DeVries AL (1991) The role of antifreeze glycopeptides and peptides in the freezing avoidance of cold-water fish. In: di Prisco G (ed) Life under extreme conditions. Springer-Verlag, Berlin-Heidelberg, pp 1-14

8. Hew CL, Yang DSC (1992) Protein interaction with ice. Eur J Biochem 203:33-42
9. DeVries AL (1971) Glycoproteins as biological antifreeze agents in Antarctic fishes. Science 172:1152-1155
10. DeVries AL, Vandenheede J, Feeney RE (1971) Primary structure of freezing point-depressing glycoproteins. J Biol Chem 246:305-308
11. DeVries AL (1982) Biological antifreeze agents in coldwater fishes. Comp Biochem Physiol A73:627-640
12. Cheng C-HC (1996) Genomic basis for antifreeze glycopeptide heterogeneity and abundance in Antarctic fishes. In: Ennion S, Goldspink G (eds) Gene expression and manipulation in aquatic organisms. Cambridge, United Kingdom, pp 1-20
13. Morris HR, Thompson MR, Osuga DT, Ahmed AT, Chan SM, Vandenheede JR, Feeney RF (1978) Antifreeze glycoproteins from the blood of an Antarctic fish. J Biol Chem. 253:5155-5162
14. O'Grady SM, Schrag JD, Raymond JA, DeVries AL (1982) Comparison of antifreeze glycopeptides from Arctic and Antarctic fishes. J Exp Zool 224:177-185
15. Fletcher GL, Hew CL, Joshi SB (1982) Isolation and characterization of antifreeze glycoproteins from the frostfish, *Microgadus tomcod*. Can J Zool 60:348-355
16. Duman JG, DeVries AL (1976) Isolation, characterization and physical properties of protein antifreeze from the winter flounder, *Pseudopleuronectes americanus*. Comp Biochem Physiol B54:375-380
17. Scott GK, Davies PL, Shears MA, Fletcher GL (1987) Structural variations in the alanine-rich antifreeze proteins of the Pleuronectinae. Eur J Biochem 168:629-633
18. Hew CL, Joshi S, Wang N-C, Kao M-H, Ananthanarayanan VS (1985) Structures of shorthorn sculpin antifreeze polypeptides. Eur J Biochem 151:167-172
19. Ewart KV, Fletcher GL (1993) Herring antifreeze protein primary structure and evidence for a C-type lectin evolutionary origin. Mol Mar Biol Biotech 2:20-27
20. Ng NF, Trinh YK, Hew CL (1986) Structure of an antifreeze polypeptide precursor from the sea raven, *Hemitripterus americanus*. J Biol Chem 261:15690-15696
21. Cheng C-HC, DeVries AL (1989) Structures of antifreeze peptides from the Antarctic fish *Austrolycicthys brachycephalus*. Biochim Biophy Acta 997:55-64
22. Hew CL, Wang N-C, Joshi S, Fletcher GL, Scott GK, Hayes PH, Buettner B (1988) Multiple genes provide the basis for antifreeze protein diversity and dosage in the ocean pout, *Macrozoarces americanus*. J Biol Chem 263:12049-55
23. Wang X, DeVries AL, Cheng C-HC (1996) Antifreeze peptide heterogeneity in an Antarctic eel pout includes an unusually large major variant composed of two 7 kDa type III AFPs linked in tandem. Biochim Biophy Acta 1247:163-172

24. Scott GK, Hayes PH, Fletcher GL, Davies PL (1988) Wolffish antifreeze protein genes are primarily organized as tandem repeats that each contain two genes in inverted orientation. Mol Cell Biol 8:3670-3675
25. Scott GK, Fletcher GL, Davies PL (1986) Fish antifreeze proteins: recent gene evolution. Can J Fish Aquat Sci 43:1028-1034
26. Nelson JS (1994) Fishes of the world. Wiley, New York
27. Svetovidov AN (1948) Gadiformes. Israel program for scientific translation, Jerusalem
28. Eastman JT (1991) Evolution and diversification of Antarctic notothenioid fishes. Am Zool 31:93-109
29. Chen L, DeVries AL, Cheng C-HC (1997) Evolution of antifreeze glycoprotein gene from a trypsinogen gene in Antarctic notothenioid fish. Proc Natl Acad Sci USA 94:3811-3816
30. Chen L, DeVries AL, Cheng C-HC (1997) Convergent evolution of antifreeze glycoproteins in Antarctic notothenioid fish and Arctic cod. Proc Natl Acad Sci USA 94:3817-3822
31. Logsdon MJ, Doolittle WF (1977) Origin of antifreeze protein genes: a cool tale in molecular evolution. Proc Natl Acad Sci USA 94:3485-3487
32. Hsiao KC, Cheng C-HC, Fernandes IE, Detrich HW, DeVries AL (1990) An antifreeze glycopeptide gene from the Antarctic cod *Notothenia coriiceps neglecta* encodes a polyprotein of high peptide copy number. Proc Natl Acad Sci USA 87: 9265-9269
33. Leaver MJ, George SG (1996) unpublished. Genbank accession number X56744
34. Levinson G, Gutman GA (1987) Slipped-strand mispairing: a major mechanism for DNA sequence evolution. Mol Biol Evol 4:203-221
35. Maeda N, Smithies O (1986) The evolution of multigene families: human haptoglobin genes. Annu Rev Genet 20:81-108
36. Lewin B (1994) Genes V. Cell Press, Massachusetts
37. Martin AP, Palumbi SR (1993) Body size, metabolic rate, generation time, and the molecular clock. Proc Natl Acad Sci USA 90:4087-4091
38. Clarke A, Johnston IA (1996) Evolution and adaptive radiation of Antarctic fishes. Trends Ecol Evol 11:187-228
39. Bargelloni L, Ritchie PA, Patarnello T, Battaglia B, Lambert DM, Meyer A (1994) Molecular evolution at subzero temperatures: mitochondrial and nuclear phylogenies of fishes from Antarctica (suborder Notothenioidei) and the evolution of antifreeze glycopeptides. Mol Biol Evol 11:854-863
40. Ohta T (1989) Role of gene duplication in evolution. Genome 31:304-310
41. Piatigorsky J, Wistow G (1991) The recruitment of crystallins: new functions precede gene duplication. Science 252:1078-1079
42. Patthy L (1996) Exon shuffling and other ways of module exchange. Matrix Biol 15:301-310
43. Doolittle RF (1994) Convergent evolution: the need to be explicit. TIBS 19:15-19
44. Hamada H, Petrino MG, Kakaunaga T (1982) A novel repeated element with Z-DNA forming potential is widely found in evolutionary diverse eukaryotic genomes. Proc Natl Acad Sci USA 79:6465-6469

45. Pardue ML, Lowenhaupt K, Rich A, Nordheim A (1987) $(dC\text{-}dA)_n \cdot (dG\text{-}dT)_n$ sequences have evolutionarily conserved chromosomal locations in *Drosophila* with implications for roles in chromosome structure and fuctions, EMBO J 6:1781-1789
46. Estoup A, Presa P, Krieg F, Vaiman D, Guyomard R (1993) (CT)n and (GT)n microsatellites: a new class of genetic markers for *Salmo trutta* L. (brown trout). Heredity 71:488-496
47. Harris AS, Wright JM (1995) Nucleotide sequence and genomic organization of cichlid fish minisatellites. Genome 38:177-184
48. Deng G, Andrews DW, Laursen RA (1997) Amino acid sequence of a new type of antifreeze protein, from the longhorn sculpin, *Myoxocephalus octodecimspinosus*. FEBS Lett 402:17-20

Evolution of Sensory Systems: A Comparison of Antarctic and Deep-Sea Ichthyofauna

John C. Montgomery and John A. Macdonald

Experimental Biology Research Group, School of Biological Sciences, University of Auckland, Private Bag 92019 Auckland, New Zealand

Introduction

There are many similarities between Antarctic seas and the deep ocean the principal ones being that both are cold and dark. Antarctic high latitude basins are effectively dark for the duration of the Antarctic winter. Even during summer, ice cover extensively reduces light levels, and the continental shelf is deeper (~600 m) than for other continents (~200 m).

With respect to the fish faunas, the major difference is that the Antarctic fauna is dominated by members of the suborder Notothenioidei which belong to the most numerous order of teleost fishes, the Perciformes, whereas the deep-sea fauna is overwhelmingly composed of nonperciform fishes. The evidence suggests that Antarctic fishes have evolved under cold and dark conditions for at most 20 million years (Ma), whereas some of the nonperciform deep-sea fishes have been evolving in the cold and the dark for a much greater period, up to 220 Ma. Similar sensory environments, but different evolutionary time scales, prompt a comparison of the sensory biology of these two groups of fishes. Interestingly there are many parallels, but in every case, the sensory adaptations of the deep-sea fishes are much more extreme than those of their Antarctic counterparts. In a number of instances attempted comparisons are thwarted by a lack of data on the appropriate characteristics of Antarctic fishes, so this exercise provides a fruitful source of questions for further study. The approach of this paper is to present the broad view of the evolution of the Antarctic and deep-sea fauna, and then to compare the sensory adaptations of both faunas concentrating on the most extreme examples. Further information on the sensory biology of deep-sea fish can be found in Montgomery and Pankhurst [1].

Evolution of Antarctic and Deep-Sea Faunas

The fossil record shows that 40 Ma ago there was a diverse coastal fish fauna in Antarctica [2]. However, with the cooling of Antarctica, this early

G. di Prisco, E. Pisano, A. Clarke (Eds)
Fishes of Antarctica. A biological overview
© Springer-Verlag Italia 1998

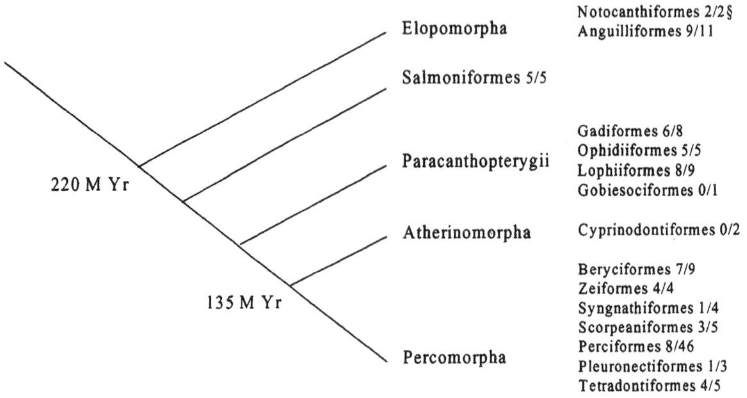

Fig. 1. Evolution of the deep-sea fauna. Timing of the branch points in the cladogram from fossil evidence [5]. §, proportion of marine families from that order recorded as having a depth range exceeding 500 m (data taken from New Zealand fish fauna [6])

fauna was replaced by a radiation of perciform fishes of the suborder Notothenioidei. Only two other groups are represented by significant species numbers: the relatively deep-water zoarcids and liparids. This paper will concentrate on the notothenioid fishes as the dominant and uniquely Antarctic element of this ichthyofauna. Notothenioids have been evolving under conditions of cold and dark for up to 20 Ma. This is based both on estimates of the degree of genetic separation of the various notothenioid lines [3] and on the geological evidence of the establishment of extensive ice-cover in Antarctica at about that time [4]. Deep-sea fishes by comparison are spread widely across all of the teleost fish groups, but particularly the nonperciform fishes (Fig. 1). The fossil evidence places the separation of the Elopomorpha from the rest of the radiation at about 220 Ma, and the percomorphs first became distinct from other teleosts at about 135 Ma [5]. Within the Percomorpha is the order Perciformes and within that again lies the suborder Notothenioidei. This phylogenetic lineage alone makes it obvious that much of the radiation of deep-sea fishes vastly predates the adaptive radiation of Antarctic notothenioids.

Olfaction/Chemosense

Olfactory receptor cells are located in the epithelium lining the floor of the nasal capsule. The olfactory epithelium is typically folded into lamellae [7], the geometry and number of which vary widely among different fish species [8] and can, within limits, provide information on the relative importance of olfaction. In deep-sea species sexual dimorphism is common, and is presumably linked with pheromonal detection of mates.

For example, in male halosaurs at maturity the anterior naris becomes large and tubular and the olfactory lamellae become enlarged and lobulated so that, as Marshall [9] describes it, the entire rosette comes to look like a sprig of broccoli. Quantitative comparison of the size and numbers of olfactory lamellae in the deep-sea hatchet fish [10] clearly shows the extent of the difference between the sexes. To date there has been no systematic investigation of the degree of development of olfactory organs in notothenioid fishes. However, from the descriptions available we can say with some confidence that there is no extreme elaboration of olfactory organs, and that any sexual dimorphism that does exist is not overtly apparent.

Vision

In response to the perpetual dim light of the deep ocean it is not surprising to see a range of visual adaptations. In the mesopelagic realm, most of the light available for vision is faint down-welling light. Correspondingly many mesopelagic species have a dorsal visual axis, which is taken to the extreme in fish like the telescope fish *Opisthoproctus*. The closest Antarctic comparison is the slight shift in visual axis reported for *Trematomus hansoni* [11] and also apparent in *T. nicolai* (Fig. 2A).

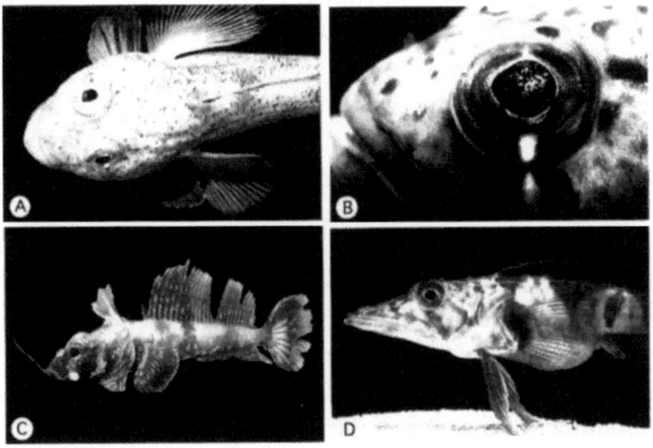

Fig. 2. Antarctic notothenioids. **A**: Dorsal view of *Trematomus nicolai.* Note that the visual axis in this species has a vertical component. **B**: Lateral view of the head and eye of *Trematomus bernacchii.* The circular lens is visible as is the gap between the lens and the anterior edge of the iris. This gap allows accommodative movement of the lens. **C**: The harpagiferid *Histiodraco velifer* with a prominent mental barbel. **D**: Lateral view of the icefish *Chionodraco hamatus* showing a characteristic posture supported on the tips of its pelvic fins. Photographs A and B by I. MacDonald, C by J. Macdonald, and D by J. Montgomery

Both the gross anatomy of the eye and the fine structure and biochemistry of the retina may be altered to enhance visual sensitivity.

Mesopelagic fishes have relatively large eyes, although relative eye size decreases for fish inhabiting depths below the level to which daylight penetrates (> 1000 m). By comparison, Antarctic fishes have a relative eyesize (eye diameter as a proportion of head length) roughly equivalent to that found in shallow water teleosts (Fig. 3). A number of deep-sea fishes also have a tubular eye, which essentially allows a large eye in a small head. Also common in deep-sea fishes is a distinct gap between the lens and the iris, the so-called aphakic space. The aphakic space allows light from obliquely placed point sources to be picked up by the whole of the lens thereby increasing the intensity of the retinal image (without an aphakic space the iris shields part of the lens). An aphakic space improves the detection of weak punctate bioluminescent sources, but can impair vision in the presence of a diffuse background illumination. A comparable aphakic space has not been reported in Antarctic fishes. The relatively small space seen on the anterior margin of the lens provides for lens movement during accommodation (Fig. 2B).

Pure rod retinas are not uncommon in deep-sea fish, but are unrecorded for Antarctic fishes. In his comprehensive study of the retinae of notothenioids, Eastman [13] showed that only *Dissostichus mawsoni* had a rod-dominated retina. In *D. mawsoni*, the ratio of rods to cones is 57:1, whereas other notothenioids had much lower ratios (e.g. *T. bernacchii* 2:1, *T. hansoni* 7:1). Rod outer segments in *Pleuragramma antarcticum* are the longest recorded [13] at about 44 μm, which is an order of magnitude less than the exceptionally long rods (400-600 μm) described in the retina of *Diretmus* [14]. Multiple rod banks, not seen in the notothenioids, are another obvious low light adaptation in deep-sea fishes, with for example up to 28 layers of rods in the fovea of *Bajacalifornia drakei* [15]. The convergence ratio of receptor cells to ganglion cells also correlates with light sensitivity. Of the Antarctic fish, *D. mawsoni* again shows the highest convergence ratio (58:1), much less than the convergence ratios of about 600:1 reported for a number of deep-sea species (orange roughy 615:1, black oreo dory 597:1, hoki 700:1, hake 623:1) [16]. Finally at the level of the fine structure of the retina, Antarctic fishes also lack the tapetum lucidum typical of deep-sea fishes [13].

In the deep-sea, the absorption spectrum of the visual pigments matches the spectrum of available light. The only information available in Antarctic fishes is an electroretinogram study [11] which shows the fish to have an action spectrum with λ_{max} of 482-528 nm which brackets the wavelength of maximum light penetration (515 nm) during the Antarctic spring.

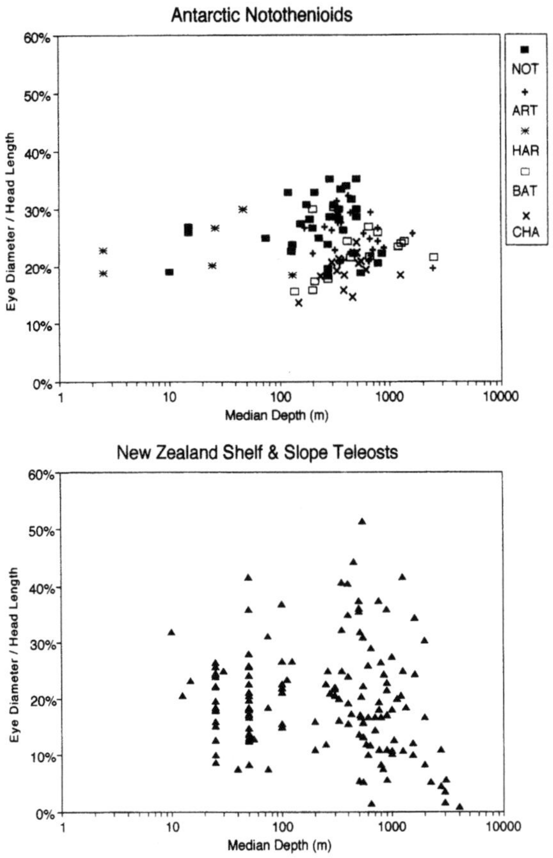

Fig. 3. Relative eye diameter (eye diameter/head length) for Antarctic notothenioids and New Zealand shelf and slope teleosts plotted as a function of the median reported depth of that species or family. Antarctic data from Gon and Heemstra [12]. NOT: Nototheniidae; ART: Artedidraconidae; HAR: Harpagiferidae; BAT: Bathydraconidae; CHA: Channichthyidae

Interestingly, later in the season the formation of ice algae significantly changes the absorption of light passing through the ice, such that the available light in the water column shifts in spectral composition to longer wavelengths (about 590 nm). Assuming the fish do not seasonally change their visual pigments, this means that later in the summer, there will be a significant mismatch between the wavelengths of available light and the λ_{max} of the fishes' visual pigment.

The final point relates as much to visual camouflage as it does to vision per se. The production of biological light is a common and important component of visual interactions in the deep-sea. Mesopelagic fishes use an intricate light production mechanism to camouflage themselves. Light produced on their ventral surface matches the color, intensity, and angular distribution of down-welling light replacing the shadow that they would otherwise cast when viewed from below. In the deeper oceans, below about 1000 m, the only visually useful light is biologically produced light. Light production by fish is used in species recognition and communication, and flashlight fishes use red-light producing photophores to detect their prey. As far as is known Antarctic fish have no photophores and do not produce light.

Octavolateralis System

The octavolateralis system comprises the hair-cell-based senses of the inner ear and the mechanosensory lateral line. As conditions for vision deteriorate, the mechanosensory lateral line becomes relatively more important. The functional unit of the lateral line system is the neuromast: a patch of hair cells with an overlying gelatinous cupula. Typically there are two main submodalities of the lateral line: superficial neuromasts on the surface of the skin, and canal neuromasts located within subdermal canals.

The primary stimulus for any neuromast is displacement of the cupula by movement of the surrounding water. Like almost all teleosts, Antarctic fishes have both superficial and canal neuromast systems (Fig. 4).

It has been recently shown that the superficial neuromast system mediates rheotactic orientation at low current flows [18], whereas the canal system is thought to be responsible for prey detection [19]. Deep-sea fishes show elaborations of the lateral line well beyond the range of

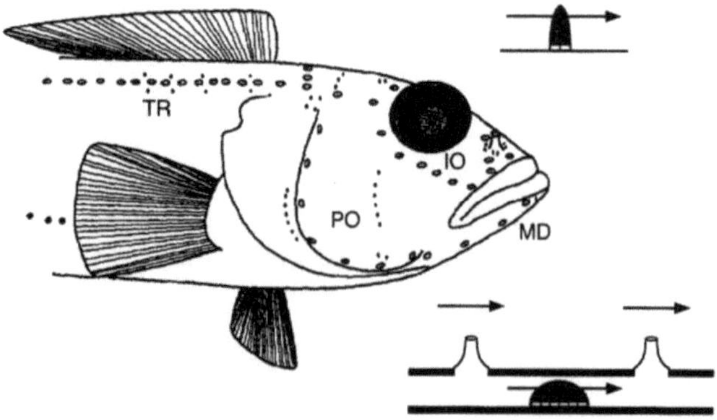

Fig. 4. Lateral line systems in the nototheniid *Trematomus bernacchii* (after Coombs and Montgomery [17]). Open circles represent pore openings to the canal lateral line system (a section of which is illustrated in the bottom right corner of the figure). Dots represent superficial neuromasts (section top right). Arrows represent the water flows which stimulate the system

morphological variation seen in the Antarctic fishes. At one extreme the superficial neuromasts of deep-sea angler fishes are elevated on papillae. At the other extreme the entire head surface of some deep-sea fish is covered with wide lateral line canals with membranous coverings. Less elaborate semimembranous canals have been reported in some icefish.

A final note on mechanosensory camouflage is that the repeated development of a tapered profile at the posterior end of deep-sea fish prompts the suggestion that this body form and swimming mode is a mechanosensory camouflage [1]. It is likely that the wake generated by this tail form will be much less than the vortex trail generated by a standard homocercal tail. Notothenioid fishes typically swim using the labriform mode. There is a possibility that this generates less of a wake than standard subcarangiform swimming, but this adaptive hypothesis has to be considered against the alternative that labriform swimming is simply a retained ancient phylogenetic trait in notothenioid fishes.

Touch

Little is known of touch or somatosensory systems in Antarctic or deep-sea fish. In general terms it would be surprising if, as the possibilities for vision declined, touch did not become relatively more important. In the deep-sea fauna, hints of this are seen in the elaborate extended fin rays of many species, such as tripod fishes, and in the common occurrence of mental (attached to the chin) barbels. Somatosensory mediated behaviors do not necessarily require obvious morphological specializations. For example, several Antarctic benthic feeders show stereotypical responses to prey touching their pelvic or anal fins [20]. In response to a touch, the fish repositioned the head to above where the prey collided with the fin. This would bring the lateral line sense organs into a position where they could detect hydromechanical stimuli from the prey. Mental barbels occur in all Antarctic plunderfish (Artedidraconidae) (Fig. 2C) and Janssen et al. [21] have shown that the barbel is used as a lure and that touching the lure initiates a predatory strike.

Although not strictly touch-related, there is a striking parallel between the icefish *Chionodraco hamatus* and the tripod fishes with respect to elongated fin rays. Tripod fishes (*Bathypterois* spp.) sit motionless, supported by three stiff elongate fin rays [9], two modified pelvic fin rays, and one elongate ray from the ventral caudal fin. They face upstream taking zooplankton brought to them by the current. *Chionodraco,* which is piscivorous, could be described as a bipod fish, since it has been observed sitting on its elongate pelvic fins (Fig. 2D). Marshall [9 p. 305] drew attention to the similarities between tripod fish and the icefish *Pagetopsis* that was described by Robilliard and Dayton [22] as perching on a sponge.

The stance of *Chionodraco* is an even more striking parallel (Fig. 2D). A number of other channichthyids also have elongate pelvic fins probably used in a similar manner. In *Cryodraco antarcticus* these may be as much as 63% of body length and bear flattened tips suitable for surface contact. Icefishes and tripod fishes seem to have converged on a similar strategy to sit motionless above the substrate with the attendant benefits that motionlessness brings to nonvisual, particularly mechanosensory, function.

Discussion

The comparison of sensory adaptations in deep-sea fishes and Antarctic fishes is summarized in Table 1. The Antarctic notothenioid fishes show a trend towards some of the same sensory adaptations that characterize fishes of the deep-sea. The dorsal visual axis of a few Antarctic species, mental barbels and elongate fin rays are the most striking examples. These

Table 1. Summary of the comparison of the sensory adaptations of deep-sea and Antarctic fishes

	Deep-sea fishes	Antarctic fishes
Olfaction/ chemosense	Sexual dimorphism	Not so far as we know
	Macrosmatism	
Vision	Dorsal visual axis	*T. hansoni and T. nicolai*
	Sensitivity adaptations	
	Large eyes	No
	Tubular eyes	No
	Aphakic space	No
	Pure rod retina	No
	Long rod outer segments	*P. antarcticum*
	400-600 μm	44 μm
	High convergence ratios	*D. mawsoni*
	600:1	58:1
	Multiple rod banks	No
	Tapeta lucida	No
	Matched visual pigment	Yes in spring
	Photophores	No
Octavolateralis system	Hearing	No information
	Gravitation receptors	No information
	Large semicircular canals	No
	Lateral line	
	Papillate neuromasts	No
	Superficial neuromasts	Yes for *Rheotaxis*
	Canal neuromasts	Standard
	Membranous canals	Partial
	Mechanosensory camouflage	Labriform probably not
Touch	Fin rays	Yes
	Mental barbels	Yes

similarities are probably an adaptive evolutionary response to the similar conditions that obtain in both environments. Despite the parallels, however, in every case, the sensory adaptations of the deep-sea fishes are much more extreme than those of their Antarctic counterparts. This apparent partial adaptation may be related to a number of factors: (1) the time scale for evolutionary changes is shorter in the Antarctic fishes by an order of magnitude. (2) The notothenioids have remained primarily inhabitants of the continental shelf and continental slope, and the majority are not found in depths greater than 1000 m; thus the conditions faced by Antarctic fishes are less extreme than in the deep-sea. (3) There are gaps in our knowledge of the sensory biology of Antarctic fishes, particularly in relation to the deeper-dwelling species, which probably lead to an underestimation of the degree of sensory specialization exhibited by the Antarctic notothenioids.

References

1. Montgomery JC, Pankhurst NW (1997) Sensory biology. In: Randall DJ, Farrell AP (eds) Deep sea fish. Fish Physiology Series, Vol. 16. Academic Press, San Diego, pp. 325-349
2. Eastman JT, Grande L (1989) Evolution of the Antarctic fish fauna with emphasis on recent notothenioids. In: Crame JA (ed) Origins and evolution of the Antarctic biota. US Geol Soc Spec Pub No. 47: 241-252
3. Bargelloni L, Ritchie PA, Patarnello T, Battaglia B, Lambert DM, Meyer A (1994) Molecular evolution at subzero temperatures: mitochondrial and nuclear phylogenies of fishes from Antarctica (suborder Notothenioidei), and the evolution of antifreeze glycopeptides. Mol Biol Evol 11:854- 863
4. Kennett P (1977) Cenozoic evolution of Antarctic glaciation, the circum-Antarctic ocean and their impact on global paleoceanography. J Geophysical Research 82: 3843-3876
5. Long JA (1995) The rise of fishes. University of New South Wales Press, Sydney
6. Paulin C, Stewart A, Roberts C, McMillan P (1989) New Zealand fish: a complete guide. National Museum of NZ Misc, Series 19: pp 279
7. Caprio J (1988). Peripheral filters and chemoreceptor cells in fishes. In: Atema J, Fay RR, Popper AN and Tavolga WN (eds) Sensory biology of aquatic animals. Springer-Verlag. New York. pp 313-338
8. Yamamoto M (1982) Comparative morphology of the peripheral olfactory organ in teleosts. In: Hara TJ (ed.) Chemoreception in fishes. Elsevier, New York, pp 39-59
9. Marshall NB (1979) Developments in deep-sea biology. Blandford Press, Poole
10. Baird RC, Jumper GY (1993) Olfactory organs in the deep sea hatchetfish *Sternoptyx diaphana* (Stomiiformes, Sternoptychidae). Bull Mar Sci 53: 1163-1167

11. Pankhurst NW, Montgomery JC (1989) Visual function in four Antarctic nototheniid fishes. J Exp Biol 142: 311-324
12. Gon O, Heemstra PC (1990) Fishes of the Southern Ocean. JLB Smith Institute of Ichthyology, Grahamstown
13. Eastman JT (1988) Ocular morphology in Antarctic notothenioid fishes. J Morphol 196: 283-306
14. Denton EJ, Locket NA (1989) Possible wavelength discrimination by multibank retinae in deep-sea fishes. J Mar Biol Assoc UK 69: 409-435
15. Locket NA (1985) The multiple bank fovea of *Bajacalifornia drakei*, an alepocephalid deep-sea teleost. Proc Roy Soc B 224: 7-22
16. Pankhurst NW (1987) Intra- and interspecific changes in retinal morphology among mesopelagic and demersal teleosts from the slope waters of New Zealand. Environ Biol Fish 19: 269-280
17. Coombs S, Montgomery JC (1994) Functional consequences of structural diversity in the lateral line system of Antarctic fish. Sensory Systems 8: 150-156 (translated from Russian)
18. Montgomery JC, Baker C, Carton AG (1997) The lateral line can mediate rheotaxis in fish. Nature (in press)
19. Montgomery JC, Macdonald JA (1987) Sensory tuning of lateral line receptors in Antarctic fish to the movements of planktonic prey. Science 235: 195-196
20. Janssen J (1996) Use of the lateral line and tactile senses in feeding in four Antarctic nototheniid fishes. Envir Biol Fish 47: 51-64
21. Janssen J, Slattery M, Jones W (1993) Feeding responses to mechanical stimulation of the barbel in *Histidraco velifer* (Artedidracondidae). Copeia 1993: 885-889
22. Robilliard GA, Dayton PK (1969) Notes on the biology of the chaenichthyid fish *Pagetopsis macropterus* from McMurdo Sound, Antarctica. Antarct J US 4: 304-306

Molecular Adaptations of Antarctic Fish Hemoglobins

Guido di Prisco

Institute of Protein Biochemistry and Enzymology, CNR, Via Marconi 12, 80125 Naples, Italy

Introduction

Following the inception in 1985 of the Italian National Programme for Antarctic Research and the consequent development of our studies on the structure and function of hemoglobin (Hb) [1-5], we have now reached a stage in which we are finally able to begin identifying firm guidelines for understanding the interplay among biochemical/physiological processes of oxygen transport, ecology and adaptive evolution.

For the study of temperature adaptations, Antarctica - more than any other habitat on earth - is indeed a unique natural laboratory. Oxygen carriers are one of the most interesting systems for studying the interrelationships between environmental conditions and molecular evolution. Hb, being a direct link between the exterior and body requirements and fulfilling its primary function under extremely variable conditions, has experienced a major evolutionary pressure to adapt and modify its functional features. In view of the role of temperature in modifying the oxygenation-deoxygenation cycle in respiring tissues, thermodynamic analysis deserves special attention.

Fishes of the perciform suborder Notothenioidei, mostly confined within Antarctic and subAntarctic waters, are the dominant component of the Southern Ocean fauna. They comprise 95 of the 174 benthic species of the continental shelf and upper continental slope, and 120 (including non-Antarctic notothenioids) of the 274 Southern Ocean species described to date [6]. Notothenioids probably appeared in the early Tertiary [7]; they began to diversify while gradually adapting to progressive cooling and isolation.

Bovichtidae, Nototheniidae (with the highest number of species, 34 Antarctic and 15 nonAntarctic), Harpagiferidae, Artedidraconidae, Bathydraconidae and Channichthyidae are the families of the suborder. Notothenioids are red-blooded, except Channichthyidae, the only known vertebrates whose blood - in the adult stage - is devoid of Hb [8].

This report summarizes the advances in the study of molecular adaptations in Antarctic fish hemoglobins. Thermodynamic analysis, X-ray

G. di Prisco, E. Pisano, A. Clarke (Eds)
Fishes of Antarctica. A biological overview
© Springer-Verlag Italia 1998

crystallography and molecular modeling have allowed us to try correlating amino acid sequence, multiplicity and oxygen binding features of hemoglobin with ecological constraints. In collaboration with W.T. Stam and J.J. Beintema, amino acid sequences are used as a source of phylogenetic information on evolution. The oxygen-transport system of nonAntarctic Bovichtidae (the most primitive notothenioid family) has been characterized for the first time, providing further indications on evolution. In collaboration with H.W. Detrich, we are also investigating the status of globin genes in Hb-less icefishes and evaluating evolutionary mechanisms leading to the Hb-less phenotype.

Antarctic Notothenioidei

Cold adaptation was developed during the last 20-30 million years of increasing isolation south of the Polar Front, a barrier to migration in both directions and thus a key factor for fish evolution. Antarctic fish are now characterized by physiological and biochemical (often unique) specializations.

The adaptive reduction in erythrocyte count and Hb content/multiplicity in the blood of Antarctic fish counterbalances the increase in blood viscosity produced by the subzero seawater temperature [9] with potentially negative physiological effects (i.e. higher demand of energy needed for circulation).

On the other hand, the coexistence of Hb-less and naturally cytopenic red-blooded species suggests that the need for an oxygen carrier in a stable, cold environment is reduced also in red-blooded fish. In fact, functional incapacitation of Hb (by means of exposure to carbon monoxide) and reduction of the hematocrit from 8-15% to <1% in cannulated specimens of *Trematomus bernacchii* [9,10] caused no discernible harm in the absence of metabolic challenges. Similar to channichthyids, red-blooded fish can carry routinely needed oxygen dissolved in the plasma. Our findings on globin genes in Hb-less species (see below), suggesting that the loss of gene expression is a primitive character (established in the ancestral channichthyid about 25 million years ago, prior to diversification within the clade), open promising pathways for evolutionary studies.

As clearly shown by the contributions in this volume, Notothenioidei are by far the most thoroughly characterized group of fish in the world. For example, our studies on their oxygen-transport system have so far been addressed to 38 out of a total of 80 red-blooded Antarctic species. This highly representative number encompasses all families and also two species of nonAntarctic notothenioids.

In the first stages of our research, the Antarctic fish fauna seemed a rather uniform and simplified community - even somewhat dull. The fish [1-5] had a single major Hb (Hb 1) and often a minor one (Hb 2, generally having the β chain in common) both displaying - with some exceptions - strong Bohr [11] and Root [12] effects (Table 1). All these species are bottom dwellers. *This was a first important correlation.*

A general trend was found in the amino acid sequences of Antarctic Hbs.

The sequences of major and minor Hbs cluster in two groups; in each group, identity is high (73-99% and 84-100%, respectively). However, the identity between major and minor Hbs is lower, ranging between 61% and 73%. The analysis of sequences, together with the similar functional features of major and minor Hbs in a given species, led us to conclude that minor Hbs are vestigial (perhaps larval) remnants, devoid of physiological significance [1].

The most surprising findings came when we examined the Hb system of three nototheniids whose life style differs from that of the sluggish benthic species: *Trematomus newnesi*, *Pagothenia borchgrevinki* (two active cryopelagic species) and *Pleuragramma antarcticum* (a pelagic, sluggish but migratory fish). Each species turned out to have a unique oxygen-transport system, and each system appeared adjusted to the fish's specific mode of life.

Table 1. The oxygen-transport system of Notothenioidei [1-5]

Family	No. of species examined	Major Hb components	Bohr and Root effects
Bovichtidae	1	1 (95%)	strong [a]
Nototheniidae	16	1 (95-99%)	strong [b]
Bathydraconidae	8	1 (90-99%)	strong [c]
Artedidraconidae	9	1 (99%)	weak
Harpagiferidae	1	1 (99%)	(not investigated)

[a] Bohr and Root effects were measured according to [13,14]. Bohr coefficients ($\Delta \log P_{50}/\Delta pH$) higher than -0.6 and lower than -0.3 denote *strong* and *weak* Bohr effects, respectively. *Strong* and *weak* Root effects correspond to Hb oxygenation at atmospheric pressure, pH 6.0, lower than 40% and higher than 80%, respectively.

[b] Except in *Aethotaxis mitopteryx* [15]

[c] Except in *Gymnodraco acuticeps* [16]

Trematomus newnesi

The Hb system of *T. newnesi* [17] is made of Hb C, Hb 1 (with the α chain in common) and Hb 2 (having the β chain in common with Hb 1). This notothenioid is the only species having two major Hbs, only one of which (Hb C, intriguingly but a trace component in all other notothenioids examined) displays pH and organophosphate regulation (Table 2). The Hill coefficient indicates cooperativity of oxygen binding in Hb 1 and Hb 2 in the whole physiological pH range, and loss of cooperativity in Hb C at lower pH (a typical feature of all Root-effect Hbs). Changes in temperature bring about a small effect on the oxygen affinity of the most abundant component Hb 1, as shown by the small overall ΔH of oxygen binding, which keeps constant in the pH range 8.0-6.5 [17]. Thus, this Hb does not require significant amounts of energy during the oxygenation-deoxygenation cycle, which may well be an example of molecular adaptation to the extreme conditions in the Antarctic.

This Hb system can ensure oxygen binding at the gills (via Hb 1) and controlled delivery to tissues (via Hb C) *also when active behavior produces acidosis*. High levels of Hb C, conceivably redundant in other notothenioids (which count on Root- and Bohr-effect Hb), compensate for lack of proton/effector regulation of Hb 1 and Hb 2.

Table 2. The oxygen-transport system of *Trematomus newnesi* [17]

Hb component	Percent of total	Bohr and Root effects	
		no ATP	+ 3 mM ATP
Hb C	20-25	strong	enhanced
Hb 1	70-75	weak or absent	not enhanced
Hb 2	3-5	weak or absent	not enhanced

Pleuragramma antarcticum

This pelagic migratory species [18] is the most abundant nototheniid, and has great biological importance in the circumAntarctic pelagic system. *P. antarcticum* has three major Hbs (Hb 1, Hb 2 and Hb 3), the highest multiplicity within notothenioids (Table 3). Hb 1 has the α chain in common with Hb 2 and the β with Hb 3. The amino acid sequences of the four globins [19-21] show high identity between Hb 1 of *P. antarcticum* and of other species, and between the chains of Hb 2 and Hb 3 not shared

Table 3. The oxygen-transport system of *Pleuragramma antarcticum* [19-21]

Hb component	Percent of total	Bohr and Root effects	
		no ATP	+3 mM ATP
Hb 1	25-30	strong	enhanced
Hb 2	20-25	strong	enhanced
Hb 3	45-50	strong	enhanced

by Hb 1 and those of minor Hbs. The identity with globins of non-Antarctic fish Hbs is lower, similar to other Antarctic fish [1]. The three Hbs display strong, effector-enhanced Root and Bohr effects; for example, the Bohr coefficient ($\varphi = \Delta \log P_{50}/\Delta pH$) ranges from -0.9 to -1.23 at 20 °C.

Although the oxygen-binding features are similar, these Hbs show important differences in thermodynamic behavior (Table 4). Hb 1 and Hb 3 show a very strong enthalpy change at pH 8.0, enhanced by chloride and organophosphates in the former, but drastically decreased in the latter; the heat of oxygenation of Hb 2, in the presence and absence of effectors, is much lower. A dramatic decrease is observed at lower pH in Hb 3 and Hb 2 (ΔH approaches zero); in contrast, Hb 1 retains high oxygenation enthalpy, especially when effectors are absent.

These observations clearly indicate a stronger Bohr effect at physiological temperatures in Hb 1 (in the presence of effectors) and Hb 3

Table 4. Heat of oxygenation of *Pleuragramma antarcticum* Hbs [19-21]

Hb component	ΔH (kcal/mol oxygen) [a]			
	no effector		+ 100 mM NaCl, 3 mM ATP	
	pH 7.0	pH 8.0	pH 7.0	pH 8.0
Hb 1	-12.8	-15.3	-8.6	-17.4
Hb 2	-3.6	-6.4	-1.8	-8.1
Hb 3	-0.1	-16.5	-4.1	-7.6

[a] The overall oxygenation enthalpy change ΔH (kcal/mol oxigen; 1 kcal = 4.184 kJ), corrected for the heat of oxygen solubilization (-3 kcal/mol oxigen), was calculated by the integrated van't Hoff equation:

$\Delta H = - 4.574\ [(T_1 \cdot T_2)/T_1 - T_2)]\ \log P_{50}/1000$

(P_{50} is the partial pressure of oxygen required to achieve Hb half saturation) (in their absence). In addition, the moderate effect of temperature on Hb 2 in the pH range 7.0-8.0 and on Hb 3 at pH 7.0 is indicative of energy-saving mechanisms of oxygen loading and unloading.

From a thermodynamic standpoint, the oxygen-transport system of *P. antarcticum* is one of the most specialized ever found in fish. It appears designed to fit its unusual pelagic life style through refined molecular adaptation in the thermodynamic features of each Hb. Rather than the response to face acidosis, this Hb system may have developed, as the main adaptive feature, the response to the need to optimize oxygen loading and unloading during seasonal migrations through water masses which can have different and fluctuating temperatures.

Thus *P. antarcticum* relies on three Hbs, which differ functionally mainly *in thermodynamic behavior, rather than pH and organophosphate regulation*.

Pagothenia borchgrevinki

This cryopelagic nototheniid has a higher Hb amount than other notothenioids [22], which could be linked to the mode of life of this species, which has a more active behavior than most notothenioids. It has five Hbs [23], among which Hb C is in trace amounts and Hb 1 reaches 80% of the total (Table 5). Hb 1, Hb 2 and Hb 3 bind oxygen with weak pH-dependence (they have weak Bohr and Root effects), not significantly influenced by organophosphates. Hb 0 has a strong Bohr effect and (neglecting Hb C) is the only component with a strong, effector-enhanced, Root effect. The heats of oxygenation are lower than those of Hbs of temperate fish; in each Hb, differences were detected in the absence and presence of the effectors. The high multiplicity of functionally distinct Hbs indicates that *also this active, cryopelagic species has a very specialized Hb system. The structure/function relationship in each Hb is currently under investigation.*

Table 5. The oxygen-transport system of *Pagothenia borchgrevinki* [23]

Component	Percent of total	Bohr and Root effects	
		no ATP	+ 3 mM ATP
Hb C	< 1	strong	enhanced
Hb 0	5-10	strong	enhanced
Hb 1	70-80	weak	not enhanced
Hb 2	5-10	weak	slightly enhanced
Hb 3	5-10	weak	slightly enhanced

Artedidraconidae

In the adult stage, these benthic fish have a single Hb. The amino acid sequences of *Artedidraco orianae* and *Pogonophryne scotti* Hbs have very high identity with the major Hbs of species of other families.

The Hbs of several artedidraconids (Tamburrini et al., unpublished; [24]) show a modest Bohr effect, and no Root effect; 3 mM ATP enhances the Bohr effect and induces a weak Root effect (Table 6).

However, ATP was never found in the erythrocytes at saturating concentration even after the fish had recovered from the stress of capture, as demonstrated - in addition to quantitative assays - by further induction of the Root effect upon addition of ATP to intact erythrocytes or unstripped hemolysate. The physiological role of organophosphates, therefore, appears critical in the pH regulation of Hbs of species of this family, and calls for further analysis.

Moreover, cooperativity of oxygen binding is virtually absent, which implies that the Hb-oxygen dissociation curve is hyperbolic rather than sigmoidal, not allowing large volumes of oxygen to be bound or released in response to small changes in the blood oxygen partial pressure. In *A. orianae* Hb, the enhancement of the Bohr effect suggests an efficient coupling of oxygen and proton transport only at very low temperatures [25]. The lack of subunit cooperativity in Hbs of several artedidraconids is a striking finding. In a Triassic reptilian relict, *Sphenodon punctatus* [26], the exceptionally low resting metabolic rate and low temperature of this primitive vertebrate is considered consistent with reduced Bohr effect, absence of cooperative oxygen binding and high oxygen affinity of its Hbs.

Table 6. The oxygen-binding system of artedidraconids [3-5, 24]

Species	Bohr and Root effects; effect of organophosphates
Artedidraco orianae	weak (Root only with ATP)
Artedidraco shackletoni	weak, only with ATP
Histiodraco velifer	weak
Pogonophryne scotti	weak (Root only with ATP)
Dolloidraco longedorsalis	weak (in hemolysate)
Pogonophryne sp. 1	weak (Root only with ATP)
Pogonophryne sp. 2	weak (Root only with ATP)
Pogonophryne sp. 3	weak (Root only with ATP)

These remarkable characteristics are shared by the Hbs of *A. orianae* and *P. scotti*. It is astonishing that *Hbs of far-from-primitive Artedidraconidae are displaying functional properties typical of primitive organisms.* If extreme conditions have driven evolution towards low metabolism and decreased dependence on Hb-mediated oxygen transport, this functional behavior may well be consistent with the less critical physiological role of Hb. The lack of cooperative oxygen binding is even more astonishing in the light of the high sequence identity of these Hbs (to be consequently regarded as "modern") with the "normal" major Hbs of the other notothenioids, and raises interesting questions on the mode of function of multisubunit molecules.

NonAntarctic Notothenioidei

Morphology [27,28] and karyology [29] indicate Bovichtidae as the most primitive notothenioid family [30-32], which comprises eleven species, living in temperate waters except *Bovichtus elongatus* (Antarctic Peninsula).

A recently proposed classification [33] separates the new family Pseudaphritidae from Bovichtidae.

The bovichtid (or pseudaphritid) *Pseudaphritis urvillii*, common in estuaries and the lower portion of Australian rivers, is considered a relict species. No information was available on the Hb system of this primitive family.

Like most notothenioids, *P. urvillii* has a major (Hb 1) and a minor component (Hb 2). Hb 1 has a strong Bohr effect; the Root effect is ATP-induced [34-36].

Although the Hb multiplicity closely resembles that of sedentary Antarctic notothenioids, the oxygen affinity of Hb 1 is much higher; in fact, $\log P_{50}$ stays below zero in the pH range 8.0-7.0. This feature can be ascribed to the widely different constraints arising from the habitat of this nonAntarctic species, and is likely to be reflected in changes in the primary structure. We have indeed found two substitutions which may alter the geometry of the invariantly hydrophobic heme pocket: in α87(F) and α 92(FG3) Glu and Met replace Leu. Glu α87(F) is found only in this species; interestingly, all Antarctic Notothenioidei have Gln, whose codon differs from that of Glu by a single base change at the first position.

Although *P. urvillii* has never developed cold adaptation, the globin amino acid sequences reveal high identity with those of the other notothenioids.

Table 7. Sequence identity (%) between Hb 1 of *P. urvillii* and other notothenioids[a]

Species	*P. urv.*[b]	*N. ang.*[b]	*P. ant.*	*G. acut.*	*C. maw.*	*A. mit.*	*T. bern.*	*T. newn.*
(α chains)								
N. coriiceps	80	99	95	82	83	83	89	87
T. newnesi	78	94	89	92	90	90	97	
T. bernacchii	80	96	92	91	91	90		
A. mitopteryx	73	91	88	84	84			
C. mawsoni	78	89	87	93				
G. acuticeps	76	89	85					
P. antarcticum	81	94						
N. angustata[b]	78							

	P. urv.[b]	*N. ang.*[b]	*P. ant.*	*G. acut.*	*C. maw.*	*A. mit.*	*T. bern.*	*T. newn.*
(β chains)								
N. coriiceps	77	93	86	80	88	82	90	86
T. newnesi	77	85	84	80	84	83	93	
T. bernacchii	82	91	90	83	87	86		
A. mitopteryx	77	84	84	77	80			
C. mawsoni	75	87	82	85				
G. acuticeps	75	81	80					
P. antarcticum	77	88						
N. angustata[b]	81							

[a]The amino acid sequences are from [1, 15-17, 20] and D'Avino and di Prisco, unpublished
[b]NonAntarctic species

Most of the residues which differentiate Antarctic notothenioids from temperate fish are found in *P. urvillii* Hb. Following the general trend, the identity of Hb 1 with Antarctic notothenioids (Table 7) is higher (73-81% for the α and 75-82% for the β chains) than with any temperate fish (60-63% and 60-66%, respectively; not shown). In addition, the ß chain of Hb 2 which is not in common with Hb 1 is highly similar to those of the minor Antarctic Hbs (85-88%; not shown). It should however be noted that the identity between *P. urvillii* and Antarctic Hb 1 is close to, or lower than, the low extreme of the ensemble of values of Antarctic notothenioids (82-99% and 77-93% for the α and β chains, respectively).

These data argue in favor of a common origin of this species within notothenioids, but also suggest that the major Hb has undergone modifications (possibly adaptive) only to a limited extent. If sequence mutations in Antarctic fish are indeed related to the development of cold adaptation, this may imply *divergence during the first stages of the cooling process*, and in any case before the event which gave origin to antifreeze glycoproteins. In contrast, the sequence of Hb 1 and Hb 2 of the non-Antarctic nototheniid *Notothenia angustata* [37], as well as the finding that its genome contains antifreeze genes (Cheng, personal communication) suggest that this fish was cold adapted prior to its quite recent migration from Antarctic to temperate waters.

Globin Genes in Channichthyidae

Channichthyidae, the most phyletically derived notothenioid family, have Hb-less blood. These unique vertebrates carry oxygen in physical solution at approx. 10% of the carrying capacity of red-blooded notothenioids. They developed physiological adaptations that maintain adequate tissue oxygenation, e.g. enhanced gas exchange by highly vascularized gills and skin, and increased cardiac output, circulatory volume and heart size.

Using cDNAs that encode the α and ß globins of Hb 1 of red-blooded *Notothenia coriiceps*, we have compared the hybridization patterns of genomic DNAs from three species representing primitive and advanced genera (*Chaenocephalus aceratus*, *Champsocephalus gunnari*, and *Chionodraco rastrospinosus*) with those from four red-blooded notothenioids, the nototheniids *N. coriiceps*, *Gobionotothen gibberifrons* (Antarctic) and *N. angustata* (temperate) and the bathydraconid *Parachaenichthys charcoti*. As expected, the genomes of the four red-blooded fish yield strong hybridization signals for both α- and ß-globin cDNA probes. The channichthyid genomes show high-stringency hybridization signals when probed with α-globin cDNA, but fail to hybridize at moderate stringency to the ß-globin probe, suggesting that icefish genomes share retention of DNA sequences closely related to the α-globin gene of red-blooded notothenioids and loss of the ß-globin locus [38-40]. The assessment of steady-state globin mRNA levels in hematopoietic and nonhematopoietic tissues (including the cellular component of blood) of *C. aceratus* reveals neither α- nor ß-globin transcripts. Thus, the α-globin-related icefish sequences are not transcriptionally active.

In the light of these results, the most plausible mechanism leading to the Hb-less phenotype might be deletion of the ß-globin locus in the ancestral channichthyid; the α-globin genes, no longer under positive selection pressure, would have accumulated mutations which caused loss of gene expression without complete loss of sequence information.

The Molecular Modeling and Phylogenetic Approaches

Antarctic fish Hbs display large differences in functional characteristics despite high identity in primary structure. Thus the ensemble of Antarctic Hbs is a greatly simplified system, which allows us to apply molecular modeling as a useful approach to structure/function relationships, to tackle not only intriguing questions pertaining to the Antarctic extreme environment (e.g. the lack of subunit cooperativity found in Hbs of Artedidraconidae), but also questions of general bearing (e.g. the molecular basis of the Root effect, a very important functional feature of fish Hb).

The major Hbs of *T. bernacchii* and *T. newnesi* only differ by four residues in the α chain and ten in the β [17,41]. Despite this high identity, the functional behavior of these Hbs is very different. *T. newnesi* Hb 1 has no Root effect and a weak Bohr effect, whereas both effects are very large in *T. bernacchii* Hb 1. These Hbs are therefore an exceptional model for the study of Hb structure-function relationship and, in particular, for the study of the molecular basis of the Root effect. The differences in the sequences of the two proteins include neither the residues considered responsible for the Root effect nor those involved in the binding of organophosphates [42].

Moreover, the analysis of the two structures does not evidentiate substantial structural differences, as also established by crystallographic studies of *T. newnesi* Hb 1 in the R and T states (D'Avino et al., unpublished). Space limitations do not allow a detailed description of these results, which will be reported elsewhere.

The above considerations imply that the structure-function interplay may entail very refined and often unpredictable mechanisms. The limited number of residues involved makes these two functionally distinct Hbs the simplest system so far available. Site-directed mutagenesis, currently being initiated, is likely to become another tool of choice - in close combination with molecular modeling - to tackle the molecular basis of the Root effect, as well as to shed light on other open questions.

Notothenioid cladograms and phylogenetic trees have been built from the amino acid sequences of Antarctic and nonAntarctic Hbs (Stam et al., this Vol.; [43,44]). The trees agree with those obtained by morphological analysis and other molecular approaches (see Bargelloni and Lecointre, this Vol.); they give strong support to the monophyly of Antarctic notothenioids, with nonAntarctic *P. urvillii* as ancestral taxon. In agreement with our Hb studies, they also indicate that Bovichtidae (and/or Pseudaphritidae [33]) diverged before the formation of the Antarctic Polar Front. The grouping of nonAntarctic *N. angustata* with two Antarctic nototheniid species also supports our hypothesis of recent migration from Antarctic to temperate waters [37].

Acknowledgments

This research is in the framework of the Italian National Programme for Antarctic Research. It would not have been possible without the outstanding contributions of L. Camardella, V. Carratore, C. Caruso, M.A. Ciardiello, E. Cocca, R. D'Avino, R. De Luca, A. Fago, A. Riccio, M. Romano, M. Tamburrini and the late B. Rutigliano.

References

1. di Prisco G, D'Avino R, Caruso C, Tamburrini M, Camardella L, Rutigliano B, Carratore V, Romano M (1991) The biochemistry of oxygen transport in red-blooded Antarctic fish. In: di Prisco G, Maresca B, Tota B (eds) Biology of Antarctic fish. Springer-Verlag, Berlin Heidelberg New York, pp 263-281
2. di Prisco G, Camardella L, Caruso C, D'Avino R, Fago A, Kunzmann A, Tamburrini M (1993) Antarctic fish and cold adaptation: structure and function of haemoglobin. Abstr 18th Lorne Conference "Protein structure and function", Lorne, Australia, p 13
3. di Prisco G, Giardina B (1996) Temperature adaptation: molecular aspects. In: Johnston IA, Bennett AF (eds) Animals and temperature. Phenotypic and evolutionary adaptation. Soc Exptl Biol, Seminar Series 59. Cambridge University Press, Cambridge, pp 23-51
4. di Prisco G (1997) Physiological and biochemical adaptations in fish to a cold marine environment. In: Battaglia B, Valencia J, Walton DWH (eds) Proc SCAR 6th Biol Symp, Venice (Antarctic communities: species, structure and survival). Cambridge University Press, Cambridge, pp 251-260
5. di Prisco G, Tamburrini M, D'Avino R (1997) Oxygen-transport systems in extreme environments: multiplicity and structure/function relationship in hemoglobins of Antarctic fish. In: Pörtner HO, Playle R (eds) Cold ocean physiology. Soc Exptl Biol, Seminar Series 66. Cambridge University Press, Cambridge
6. Gon O, Heemstra PC (eds) (1990) Fishes of the Southern Ocean. JLB Smith Institute of Ichthyology, South Africa
7. Eastman JT (1993) Antarctic fish biology. Evolution in a unique environment. Academic Press, San Diego
8. Ruud JT (1954) Vertebrates without erythrocytes and blood pigment. Nature 173:848-850
9. Wells RMG, Macdonald JA, di Prisco G (1990) Thin-blooded Antarctic fishes: a rheological comparison of the haemoglobin-free icefishes *Chionodraco kathleenae* and *Cryodraco antarcticus* with a red-blooded nototheniid, *Pagothenia bernacchii*. J Fish Biol 36:595-609
10. di Prisco G, Macdonald JA, Brunori M (1992) Antarctic fish survive exposure to carbon monoxide. Experientia 48:473-475
11. Riggs AF (1988) The Bohr effect. Ann Rev Physiol 50:181-204

12. Brittain T (1987) The Root effect. Comp Biochem Physiol 86B:473-481
13. Giardina B, Amiconi G (1981) Measurement of binding of gaseous and nongaseous ligands to hemoglobins by conventional spectrophotometric procedures. Methods Enzymol 76:417-427
14. D'Avino R, di Prisco G (1989) Hemoglobin from the Antarctic fish *Notothenia coriiceps neglecta*. 1. Purification and characterisation. Eur J Biochem 179:699-705
15. D'Avino R, Fago A, Kunzmann A, di Prisco G (1992) The primary structure and oxygen-binding properties of the high-Antarctic fish *Aethotaxis mitopteryx* DeWitt. Polar Biol 12:135-140
16. Tamburrini M, Brancaccio A, Ippoliti R, di Prisco G (1992) The amino acid sequence and oxygen-binding properties of the single hemoglobin of the cold-adapted Antarctic teleost *Gymnodraco acuticeps*. Arch Biochem Biophys 292:295-302
17. D'Avino R, Caruso C, Tamburrini M, Romano M, Rutigliano B, Polverino de Laureto P, Camardella L, Carratore V, di Prisco G (1994) Molecular characterization of the functionally distinct hemoglobins of the Antarctic fish *Trematomus newnesi*. J Biol Chem 269:9675-9681
18. Hubold G (1985) On the early life history of the high-Antarctic silverfish *Pleuragramma antarcticum*. In: Siegfried WR, Condy PR, Laws RM (eds) Proc 4th SCAR Biol Symp (Antarctic nutrient cycles and food webs). Springer-Verlag, Berlin Heidelberg New York, pp 445-451
19. Tamburrini M, D'Avino R, Fago A, Carratore V, Kunzmann A, di Prisco G (1994) The unique hemoglobin system of *Pleuragramma antarcticum*, a high-Antarctic fish with holopelagic mode of life. Abstr SCAR 6th Biol Symp, Venice, p 261
20. Tamburrini M, D'Avino R, Fago A, Carratore V, Kunzmann A, di Prisco G (1996) The unique hemoglobin system of *Pleuragramma antarcticum*, an Antarctic migratory teleost. Structure and function of the three components. J Biol Chem 271:23780-23785
21. Tamburrini M, D'Avino R, Carratore V, Kunzmann A, di Prisco G (1997) The hemoglobin system of *Pleuragramma antarcticum*: correlation of hematological and biochemical adaptations with life style. Comp Biochem Physiol, 118A:1037-1044
22. Macdonald JA, Wells RMG (1991) Viscosity of body fluids from Antarctic notothenioid fish. In: di Prisco G, Maresca B, Tota B (eds) Biology of Antarctic fish. Springer-Verlag, Berlin, Heidelberg, New York, pp 163-178
23. Tamburrini M, di Prisco G (1994) The unique features of the hemoglobin system of the Antarctic teleost *Pagothenia borchgrevinki*. “Proteine '94”, Abstr A25, p 43
24. Tamburrini M, Romano M, di Prisco G (1996) Unique oxygen-binding characteristics of the hemoglobins of two Antarctic fish species of the family Artedidraconidae. 3rd Workshop (Antarctic fish: ecology, lifestyle and adaptive evolution. Comparison with Arctic fish) of the European Science Foundation Network “Fishes of the Antarctic Ocean”, Saint Rémy-lès-Chevreuse, France

25. Tamburrini M, Romano M, Coletta M, di Prisco G (1995) The hemoglobin of the Antarctic teleost *Artedidraco orianae*. Amino acid sequence and ligand binding properties. Abstr 23rd Meet Fed Eur Biochem Soc, P66.13
26. Tetens V, Brittain T, Christie DL, Robb J, Wells RMG (1984) Characterization and function of isolated hemoglobins from the tuatara, *Sphenodon punctatus* (Reptilia: O. Rhynchocephalia). Comp Biochem Physiol 79B:119-123
27. Iwami T (1985) Osteology and relationships of the family Channichthyidae. Mem Natl Inst Polar Res, Tokyo, Ser E, No 36, 1-69
28. Balushkin AV (1984) Morphological bases of the systematics and phylogeny of nototheniid fishes. Acad Sci USSR, Zool Inst Leningrad, 1-140
29. Prirodina VP (1986) Karyotypes of *Cottoperca gobio* (Bovichthyidae, Notothenioidei) as compared to karyotypes of other Notothenioidei. USSR Acad Sci Proc Zool Inst, Leningrad 153: 67-71
30. Hureau J-C (1986) Relations phylogénétiques au sein des Notothenioidei. Océanis 12(5):367-376
31. Andriashev AP (1987) A general review of the Antarctic bottom fish fauna. In: Kullander SO, Fernholm B (eds) Proc 5th Congr Eur Ichthyol, Stockholm. Swedish Museum of Natural History, Stockholm, pp 357-372
32. Bargelloni L, Ritchie PA, Patarnello T, Battaglia S, Lambert DM, Mayer A (1994) Molecular evolution at subzero temperatures: mitochondrial and nuclear phylogenesis of fishes from Antarctica (suborder Notothenioidei), and the evolution of antifreeze glycopeptides. Mol Biol Evol 11:854-863
33. Balushkin AV (1992) Classification, phylogenetic relationships, and origins of the families of the suborder Notothenioidei (Perciformes). J Ichthyol 32(7):90-110
34. D'Avino R, di Prisco G (1997) The hemoglobin system of Antarctic and non-Antarctic notothenioid fishes. Comp Biochem Physiol, 118A:1045-1049
35. D'Avino R, Fago A, di Prisco G (1994) Structural and functional characterisation of the hemoglobin of an estuarine fish. Ital J Biochem 43(5):221-222A
36. D'Avino R, Romano M, Carratore V, di Prisco G (1995) Structure and function of the main hemoglobin of *Pseudaphritis urvillii*, an estuarine temperate Bovichtid. 2nd Workshop (Relationships linking ecology, lifestyle and adaptive evolution in Antarctic fish) of the European Science Foundation Network "Fishes of the Antarctic Ocean", Liège, Belgium
37. Fago A, D'Avino R, di Prisco G (1992) The hemoglobins of *Notothenia angustata*, a temperate fish belonging to a family largely endemic to the Antarctic Ocean. Eur J Biochem 210:963-970
38. Cocca E, Ratnayake-Lecamwasam M, Parker SK, Camardella L, Ciaramella M, di Prisco G, Detrich HW III (1995) Genomic remnants of α-globin genes in the hemoglobinless Antarctic icefishes. Proc Natl Acad Sci USA 92:1817-1821
39. Detrich HW III, Cocca E, Ratnayake-Lecamwasam M, Parker SK, Camardella L, Ciaramella M, di Prisco G (1995) ß-Globin gene deletion in the hemoglobinless Antarctic icefishes. 37th Am Soc Hematol Meet, Seattle, USA

40. Ratnayake-Lecamwasam M, Parker SK, Detrich HW III, Cocca E, Camardella L, Ciaramella M, di Prisco G (1995) Do the hemoglobinless Antarctic icefishes possess globin genes? Antarctic J US 30:179-182
41. Camardella L, Caruso C, D'Avino R, di Prisco G, Rutigliano B, Tamburrini M, Fermi G, Perutz MF (1992) Haemoglobin of the Antarctic fish *Pagothenia bernacchii*. Amino acid sequence, oxygen equilibria and crystal structure of its carbomonoxy derivative. J Mol Biol 224:449-460
42. Perutz MF, Brunori M (1982) Stereochemistry of cooperative effects in fish and amphibian haemoglobins. Nature 299:421-426
43. Stam WT, Beintema JJ, D'Avino R, Tamburrini M, di Prisco G (1996) Molecular evolution of hemoglobin of Antarctic fishes (Notothenioidei). 3rd Workshop (Antarctic fish: ecology, lifestyle and adaptive evolution. Comparison with Arctic fish) of the European Science Foundation Network "Fishes of the Antarctic Ocean", Saint Rémy-lès-Chevreuse, France
44. Stam WT, Beintema JJ, D'Avino R, Tamburrini M, di Prisco G. (1997) Molecular evolution of hemoglobins of Antarctic fishes (Notothenioidei). J Mol Evol, 45:437-445

Evolutionary Studies on Teleost Hemoglobin Sequences

Wytze T. Stam[1], Jaap J. Beintema[2], Rossana D'Avino[3], Maurizio Tamburrini[3], Ennio Cocca[3], and Guido di Prisco[3]

[1]Department of Marine Biology
[2]Department of Biochemistry, University of Groningen, Kerklaan 30, 9751 NN Haren, The Netherlands
[3]Institute of Protein Biochemistry and Enzymology, CNR, Naples, Italy

Introduction

During cold adaptation, the blood of Antarctic fish has acquired some features which clearly differentiate these fish from fish of temperate and tropical climates. A much lower erythrocyte and hemoglobin content counteracts the temperature-induced viscosity increase in the blood and, consequently, cardiac work is greatly facilitated. At the extreme end of such evolution, the blood of the Notothenioid family Channichthyidae (a unique case among vertebrates) is characterized by the total lack of hemoglobin.

We have initiated investigations on the relationship between the molecular structure and the biological function in hemoglobins of Antarctic teleosts belonging to the suborder Notothenioidei. Hemolysates of these fish species generally contain a single major hemoglobin and often a second minor component (about 5% of the total). The markedly reduced hemoglobin multiplicity of Antarctic fish in comparison with teleosts living in temperate and tropical waters may be expected, since hemoglobin multiplicity is likely to be linked with the variability of the environment [1], and the Antarctic waters are a stable habitat.

In a previous paper [2] amino acid sequences of α- and β-chains of human hemoglobin, and of hemoglobins of coelacanth and 24 teleost fish species, including 13 Notothenioidei, were analyzed using maximum parsimony. Trees were derived for the α- and β-chains separately, and for tandemly arranged sequences, using the human and coelacanth sequences as outgroups in all analyses. The topologies of the trees of the α- and β-chains were highly congruent, and indicated a specific pattern of gene duplications and gene expression of teleost hemoglobins which has not yet been analyzed at the genomic level.

G. di Prisco, E. Pisano, A. Clarke (Eds)
Fishes of Antarctica. A biological overview
© Springer-Verlag Italia 1998

Tree topologies showed four sequence clusters of paralogous teleost hemoglobin sequences, which may originate from three gene duplications in ancestral teleost species. The two most recently derived clusters consisted of the sequences of the "major" and "minor" hemoglobin chains of the Notothenioids, respectively, together with those of tuna (*Thunnus thynnus*) and red gurnard (*Chelidonichthys kumu*), which is a scorpaeniform. The Notothenioids belong to the teleost order Perciformes [3], together with tuna and snapper (*Chrysophrys auratus*), but not with red gurnard. The position of sequenced hemoglobin chains of snapper in another cluster in the trees indicated a paralogous relationship with the hemoglobins of the other Perciformes. The α- and β-chains of the major components of the Notothenioids formed a monophyletic group in all investigated trees, with the nonAntarctic *Pseudaphritis* as their sister taxon. The minor chains also were a monophyletic group. The families Nototheniidae and Bathydraconidae appeared to be paraphyletic.

Here we present new analyses of teleost hemoglobins in which we have included deduced amino acid sequences of two sets of adjacent α- and β-globin genes from *Notothenia coriiceps* (unpublished) and of the almost completed sequences of one of the hemoglobins of *Cottoperca gobio* (10 and 6 residues still are unknown in the α- and β-chains, respectively; unpublished).

Methods

Sequences were analyzed using maximum parsimony under the heuristic search option (tree bisection-reconnection, local and global swapping, simple and random stepwise addition) using the PAUP computer program, version 3.1.1 [4]. Gaps (deletions) are considered as "missing data". Searches were done under the "protpars" assumption, in which the number of codon differences between replaced amino acids is taken into account [5]. Bootstrap analysis (100 replicates) using a 50% majority rule was used to provide an estimate of tree stability. In all analyses the hemoglobins of human and coelecanth were used as outgroups.

Results and Discussion

Two most parsimonious trees (MPT) requiring 685 substitutions were found for the α-chains and three, each requiring 826 substitutions, for the β-chains. As the trees were again highly congruent, we decided to analyze trees of tandemly arranged α- and β-chains as in our previous investigations [2]. However, now we arranged in tandem not only globin chains which occur in the same hemoglobin components, like trout (*Oncorhynchus mykiss*) HbI and HbIV, and which may be derived from adjacent α- and β-globin genes on the same chromosome, but also "minor" notothenioid hemoglobin chains like those of *Pleuragramma antarcticum* (Hb3 and Hb2) and *Trematomus newnesi* (Hb2 and HbC), which are at congruent positions in the separate α- and β-trees, but which do *not* occur in one hemoglobin component. On the contrary, in *P. antarcticum* three major hemoglobin components occur: Hb1 consisting of the two "major" chains, Hb2 with the "major" α- and the "minor" β-, and Hb3 with the "minor" α- and the "major" β-chain [6].

Three tandemly arranged MPT trees were obtained, each requiring 1293 substitutions. They differ in details of the topology near trout HbI and electric eel Hb. One of the three trees is shown in Fig. 1. Bootstrap values using a 50% majority rule are also indicated.

Strong support (100% bootstrap values) is again obtained for the monophyly of teleost hemoglobins, of the cluster of "major" and "minor" Notothenioid, *T. thynnus* and *C. kumu* hemoglobins and of the "major" Notothenioid hemoglobin chains (except *Pseudaphritis urvillii*).

Notothenia coriiceps Hemoglobin P1 and P2

Two sets of adjacent α- and β-hemoglobin genes have been sequenced from DNA of *N. coriiceps*. The deduced amino acid sequences differ from those expressed in the adult and may be juvenile sequences. The α-chains are identical, while the β-chains differ at only four positions, which may indicate that the two sets originate from a recent gene duplication. These deduced sequences group in the tree (Fig. 1) with the "minor" notothenioid chains. However, they diverge from them before species divergences in the notothenioids, but - in the β-tree - after divergence of sequences of *P. urvillii* and *C. gobio* (not shown).

Cottoperca gobio Hemoglobin 1

The incomplete sequences of *C. gobio* hemoglobin occupy in the trees different positions. In the α-trees *C. gobio* groups with the "major" Notothenioid chains: in one of the two most parsimonious trees as the first diverging taxon, in the other one monophyletic with *P. urvillii* (not shown). In the most parsimonious β-trees *C. gobio* groups with the "minor" Notothenioid chains (diverging before *P. urvillii*, not shown).

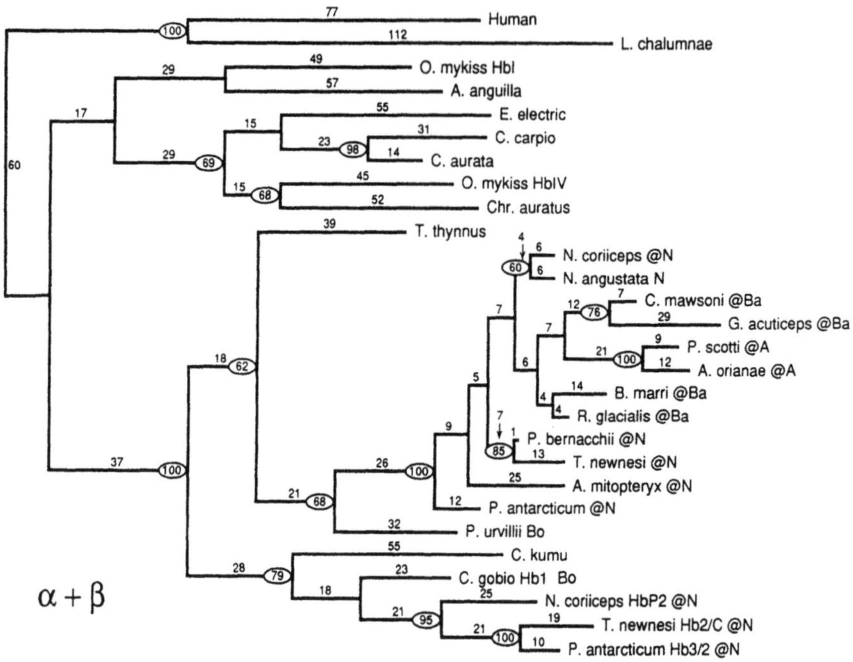

Fig. 1. One out of three most parsimonious trees for tandemlyarranged α– and β–chains of hemoglobin sequences using human and coelcanth sequences as outgroup. Numbers above the branches are steps (amino acid changes, which are obtained under the accelerated transformation; ACCTRAN), and circled numbers are bootstrap percentages (>50%). Hb = hemoglobin, @ = Antarctic species, Bo = Bovichtidae, N = Nototheniidae, A = Artedidraconidae, B = Bathydraconidae. References to amino acid sequences can be found in [2], together with more information about species, higher taxa and protein/subunit (α or β) identity

In the tree of the tandemly arranged chains, *C. gobio* still groups with the "minor" chains. In our previous paper [2] we already discussed evidence from several sources [7,8] that the Bovichtidae are not monophyletic, and that other *Bovichtus* species diverged from the other Notothenioidei before *P. urvillii* [9]. The *C. gobio* hemoglobin data also do not support monophyly of the Bovichtidae. However, discrimination between paralogous and orthologous relationships between hemoglobin chains is necessary before definite conclusions can be made.

Conclusion

The deduced amino acid hemoglobin sequences from *N. coriiceps* DNA are grouped with certainty and those from *C. gobio* perhaps with the "minor" Notothenioid chains. This may indicate that during Notothenioid evolution there has been specialization leading to dominant expression of only the "major" hemoglobin genes, with only one component found in several Bathydraconids and in the Artedidraconids, and loss of hemoglobin expression in the Channichthyids [10].

Acknowledgments

This study is in the framework of the Italian National Programme for Antarctic Research and has been supported by the European Science Foundation Network "Fishes of the Antarctic Ocean".

References

1. Riggs A (1970) Properties of fish hemoglobins. In: Hoar WS, Randall DJ (eds) Fish physiology, Vol IV, Academic Press, New York, pp 209-252
2. Stam WT, Beintema JJ, D'Avino R, Tamburrini M, di Prisco G (1997) Molecular evolution of hemoglobins of Antarctic fishes (Notothenioidae). J Mol Evol 45:437-445
3. Nelson JS (1994) Fishes of the world. 3rd edition. John Wiley, New York. 600 pp
4. Swofford DL (1993) PAUP (phylogenetic analysis using parsimony), version 3.1.1. User's manual by Swofford DL, Begle DP, Laboratory of Molecular Systematics, Smithsonian Institution, Washington DC, 257 pp
5. Felsenstein J (1989) PHYLIP - phylogeny inference package (version 3.2). Cladistics 5:164-166
6. Tamburrini M, D'Avino R, Fago A, Carratore V, Kunzmann A, di Prisco G (1996) The unique hemoglobin system of *Pleuragramma antarcticum*, an Antarctic migrating teleost. Structure and function of the three components. J Biol Chem 271:23780-23785
7. Balushkin AV (1992) Classification, phylogenetic relationships, and origins of the family suborder Notothenioidei (Perciformes). J Ichthyol 30:132-147
8. Pisano E, Ozouf-Costaz C, Hureau J-C, Williams R (1995) Chromosome differentiation in the subAntarctic Bovichtidae species *Cottoperca gobio* (Günther, 1861) and *Pseudaphritis urvillii* (Valenciennes, 1832) (Pisces, Perciformes). Antarct Sci 7:381-386
9. Ritchie PA, Lavoué S, Lecointre G (1997) Molecular phylogenetics and the evolution of Antarctic notothenioid fishes. Comp Biochem Physiol 118A:1009-1026
10. di Prisco G, Tamburrini M, D'Avino R (1997) Oxygen-transport systems in extreme environments: multiplicity and structure/function relationship in hemoglobins of Antarctic fish. In: Pörtner HO, Playe R (eds) Cold ocean physiology. Soc Exptl Biol Seminar Series 66. Cambridge University Press, Cambridge

Subject Index

MIX
Papier aus verantwortungsvollen Quellen
Paper from responsible sources
FSC® C105338

If you have any concerns about our products,
you can contact us on
ProductSafety@springernature.com

In case Publisher is established outside the EU,
the EU authorized representative is:
Springer Nature Customer Service Center GmbH
Europaplatz 3, 69115 Heidelberg, Germany

Printed by Libri Plureos GmbH
in Hamburg, Germany